Reversible and DNA Computing

Reversible and DNA Computing

Hafiz Md. Hasan Babu
Department of Computer Science and Engineering
University of Dhaka, Bangladesh

Registered Offices
John Wiley & Sons, Inc., 111 River Street, Hoboken, NJ 07030, USA
John Wiley & Sons Ltd, The Atrium, Southern Gate, Chichester, West Sussex, PO19 8SQ, UK

Editorial Office
The Atrium, Southern Gate, Chichester, West Sussex, PO19 8SQ, UK

For details of our global editorial offices, customer services, and more information about Wiley products visit us at www.wiley.com.

Wiley also publishes its books in a variety of electronic formats and by print-on-demand. Some content that appears in standard print versions of this book may not be available in other formats.

Library of Congress Cataloging-in-Publication Data

Names: Babu, Hafiz Md. H., 1966- author.
Title: Reversible and DNA computing / Hafiz Md. Hasan Babu, University of
   Dhaka, Bangladesh.
Description: First edition. | Hoboken, NJ : John Wiley & Sons, Inc., [2020]
   | Includes bibliographical references and index.
Identifiers: LCCN 2020009452 (print) | LCCN 2020009453 (ebook) | ISBN
   9781119679424 (hardback) | ISBN 9781119679363 (adobe pdf) | ISBN
   9781119679431 (epub)
Subjects: LCSH: Molecular computers.
Classification: LCC QA76.887 .B33 2020 (print) | LCC QA76.887 (ebook) |
   DDC 006.3/842–dc23
LC record available at https://lccn.loc.gov/2020009452
LC ebook record available at https://lccn.loc.gov/2020009453

Cover Design: Wiley
Cover Image: Courtesy of Hafiz Babu

Set in 9.5/12.5pt STIXTwoText by SPi Global, Chennai, India

*To my beloved parents and also to my
beloved wife, daughter, & son, who made
it possible for me to write this book.*

Contents

List of Figures   *xvii*
List of Tables   *xxix*
About the Author   *xxxi*
Preface   *xxxiii*
Acknowledgments   *xxxv*
Acronyms   *xxxvii*
Introduction   *xxxix*

Part I   Reversible Circuits   *1*
An Overview About Reversible Circuits   *1*

1   **Reversible Logic Synthesis**   *5*
1.1   Reversible Logic   *5*
1.2   Reversible Function   *5*
1.3   Reversible Logic Gate   *6*
1.4   Garbage Outputs   *6*
1.5   Constant Inputs   *7*
1.6   Quantum Cost   *7*
1.7   Delay   *8*
1.8   Power   *8*
1.9   Area   *8*
1.10   Hardware Complexity   *9*
1.11   Quantum Gate Calculation Complexity   *9*
1.12   Fan-Out   *10*
1.13   Self-Reversible   *10*
1.14   Reversible Computation   *10*
1.15   Area   *11*
1.16   Design Constraints for Reversible Logic Circuits   *11*
1.17   Quantum Analysis of Different Reversible Logic Gates   *12*
1.17.1   Reversible NOT Gate (Feynman Gate)   *12*
1.17.2   Toffoli Gate   *12*
1.17.3   Fredkin Gate   *13*

1.17.4	Peres Gate	*13*
1.18	Summary	*13*

2	**Reversible Adder and Subtractor Circuits**	*15*
2.1	Reversible Multi-Operand n-Digit Decimal Adder	*15*
2.1.1	Full Adder	*15*
2.1.2	Carry Skip Adder	*19*
2.1.2.1	Design of Carry Skip Adder	*20*
2.1.3	Carry Look-Ahead Adder	*24*
2.2	Reversible BCD Adders	*26*
2.2.1	Design Procedure of the Reversible BCD Adder	*27*
2.2.1.1	Properties of the Reversible BCD Adder	*28*
2.2.2	Design Procedure of the Reversible Carry Skip BCD Adder	*31*
2.2.2.1	Properties of the Reversible Carry Skip BCD Adder	*32*
2.3	Reversible BCD Subtractor	*34*
2.3.1	Carry Look-Ahead BCD Subtractor	*36*
2.3.2	Carry Skip BCD Subtractor	*36*
2.3.3	Design of Conventional Reversible BCD Subtractor	*37*
2.3.3.1	Reversible Nine's Complement	*37*
2.3.3.2	Reversible BCD Subtractor	*38*
2.3.3.3	Reversible Design of Carry Look-Ahead BCD Subtractor	*40*
2.3.3.4	Reversible Design of Carry Skip BCD Subtractor	*40*
2.4	Summary	*41*

3	**Reversible Multiplier Circuit**	*43*
3.1	Multiplication Using Booth's Recoding	*43*
3.2	Reversible Gates as Half Adders and Full Adders	*44*
3.3	Some Signed Reversible Multipliers	*45*
3.4	Design of Reversible Multiplier Circuit	*45*
3.4.1	Some Quantum Gates	*46*
3.4.2	Recoding Cell	*46*
3.4.3	Partial Product Generation Circuit	*49*
3.4.4	Multi-Operand Addition Circuit	*52*
3.4.5	Calculation of Area and Power of $n \times n$ Multiplier Circuit	*52*
3.5	Summary	*64*

4	**Reversible Division Circuit**	*67*
4.1	The Division Approaches	*67*
4.1.1	Restoring Division	*67*
4.1.2	Nonrestoring Division	*67*
4.2	Components of Division Circuit	*68*
4.2.1	Reversible MUX	*68*
4.2.2	Reversible Register	*68*
4.2.3	Reversible PIPO Left-Shift Register	*68*
4.2.4	Reversible Parallel Adder	*70*

4.3   The Design of Reversible Division Circuit   *71*
4.4   Summary   *74*

5   Reversible Binary Comparator   *75*
5.1   Design of Reversible *n*-Bit Comparator   *75*
5.1.1   BJS Gate   *75*
5.1.2   Reversible 1-Bit Comparator Circuit   *76*
5.1.3   Reversible MSB Comparator Circuit   *77*
5.1.4   Reversible Single-Bit Greater or Equal Comparator Cell   *78*
5.1.5   Reversible Single-Bit Less Than Comparator Cell   *79*
5.1.6   Reversible 2-Bit Comparator Circuit   *79*
5.1.7   Reversible *n*-Bit Comparator Circuit   *79*
5.2   Summary   *85*

6   Reversible Sequential Circuits   *87*
6.1   An Example of Design Methodology   *87*
6.2   The Design of Reversible Latches   *89*
6.2.1   The SR Latch   *89*
6.2.2   The D Latch   *91*
6.2.2.1 The D Latch with Outputs Q and $\overline{Q}$ *91*
6.2.2.2 The Negative Enable Reversible D Latch   *92*
6.2.3   T Latch   *93*
6.2.4   The JK Latch   *93*
6.3   The Design of Reversible Master–Slave Flip-Flops   *94*
6.4   The Design of Reversible Latch and the Master–Slave Flip-Flop with Asynchronous SET and RESET Capabilities   *95*
6.5   Summary   *97*

7   Reversible Counter, Decoder, and Encoder Circuits   *99*
7.1   Synthesis of Reversible Counter   *99*
7.1.1   Reversible T Flip-Flop   *99*
7.1.2   Reversible Clocked T Flip-Flop   *99*
7.1.3   Reversible Master–Slave T Flip-Flop   *100*
7.1.4   Reversible Asynchronous Counter   *101*
7.1.5   Reversible Synchronous Counter   *102*
7.2   Reversible Decoder   *103*
7.2.1   Reversible Encoder   *104*
7.3   Summary   *106*

8   Reversible Barrel Shifter and Shift Register   *107*
8.1   Design Procedure of Reversible Bidirectional Barrel Shifter   *107*
8.1.1   Reversible 3 × 3 Modified BJN Gate   *108*
8.1.2   Reversible 2's Complement Generator   *109*
8.1.3   Reversible Swap Condition Generator   *110*
8.1.4   Reversible Right Rotator   *111*

8.1.4.1   (4, 3) Reversible Right Rotator   *112*
8.1.4.2   Generalized Reversible Right Rotator   *112*
8.1.5   Reversible Bidirectional Barrel Shifter   *113*
8.2   Design Procedure of Reversible Shift Register   *113*
8.2.1   Reversible Flip-Flop   *113*
8.2.1.1   Reversible SISO Shift Register   *114*
8.2.1.2   Reversible SIPO Shift Register   *114*
8.2.1.3   Reversible PISO Shift Register   *115*
8.2.1.4   Reversible PIPO Shift Register   *115*
8.2.1.5   Reversible Universal Shift Register   *118*
8.3   Summary   *121*

9   **Reversible Multiplexer and Demultiplexer with Other Logical Operations**   *123*
9.1   Reversible Logic Gates   *123*
9.1.1   RG1 Gate   *123*
9.1.2   RG2 Gate   *123*
9.2   Designs of Reversible Multiplexer and Demultiplexer with Other Logical Operations   *124*
9.2.1   The R-I Gate   *124*
9.2.2   The R-II Gate   *126*
9.3   Summary   *128*

10   **Reversible Programmable Logic Devices**   *129*
10.1   Reversible FPGA   *129*
10.1.1 3×3 Reversible NH Gate   *130*
10.1.2 4×4 Reversible BSP Gate   *130*
10.1.3   4-to-1 Reversible Multiplexer   *130*
10.1.4   Reversible D Latch   *131*
10.1.5   Reversible Write-Enabled Master–Slave Flip-Flop   *132*
10.1.6   Reversible RAM   *132*
10.1.7   Design of Reversible FPGA   *132*
10.2   Reversible PLA   *134*
10.2.1   The Design Procedure   *134*
10.2.1.1   Delay Calculation of a Reversible PLA   *139*
10.2.1.2   Delay Calculation of AND Plane   *139*
10.2.1.3   Delay Calculation of Ex-OR Plane   *140*
10.2.1.4   Delay of Overall Design   *140*
10.3   Summary   *141*

11   **Reversible RAM and Programmable ROM**   *143*
11.1   Reversible RAM   *143*
11.1.1 3×3 Reversible FS Gate   *143*
11.1.2   Reversible Decoder   *144*
11.1.3   Reversible D Flip-Flop   *145*

11.1.4   Reversible Write-Enabled Master–Slave D Flip-Flop   *146*
11.1.5   Reversible Random Access Memory   *146*
11.2   Reversible PROM   *148*
11.2.1   Reversible Decoder   *149*
11.2.2   Design of Reversible PROM   *149*
11.3   Summary   *154*

12   **Reversible Arithmetic Logic Unit**   *155*
12.1   Design of ALU   *155*
12.1.1   Conventional ALU   *155*
12.1.2   The ALU Based on Reversible Logic   *155*
12.1.2.1   The Reversible Function Generator   *156*
12.1.2.2   The Reversible Control Unit   *156*
12.2   Design of Reversible ALU   *158*
12.3   Summary   *159*

13   **Reversible Control Unit**   *161*
13.1   An Example of Control Unit   *161*
13.2   Different Components of a Control Unit   *161*
13.2.1   Reversible HL Gate   *161*
13.2.2   Reversible BJ Gate   *162*
13.2.3   Reversible 2-to-4 Decoder   *163*
13.2.4   Reversible 3-to-8 Decoder   *165*
13.2.5   Reversible n-to-2^n Decoder   *165*
13.2.6   Reversible JK Flip-Flop   *168*
13.2.7   Reversible Sequence Counter   *168*
13.2.8   Reversible Instruction Register   *168*
13.2.9   Control of Registers and Memory   *169*
13.2.10   Construction Procedure and Complexities of the Control Unit   *170*
13.3   Summary   *172*

Part II   **Reversible Fault Tolerance**   *173*
            An Overview About Fault-Tolerance and Testable Circuits   *173*

14   **Reversible Fault-Tolerant Adder Circuits**   *177*
14.1   Properties of Fault Tolerance   *177*
14.1.1   Parity-Preserving Reversible Gates   *178*
14.2   Reversible Parity-Preserving Adders   *180*
14.2.1   Fault-Tolerant Full Adder   *180*
14.2.2   Fault-Tolerant Carry Skip Adder   *181*
14.2.3   Fault-Tolerant Carry Look-Ahead Adder   *183*
14.2.4   Fault-Tolerant Ripple Carry Adder   *184*
14.3   Summary   *185*

15 **Reversible Fault-Tolerant Multiplier Circuit** *187*
15.1 Reversible Fault-Tolerant Multipliers *187*
15.1.1 Reversible Fault-Tolerant $n \times n$ Multiplier *187*
15.1.2 LMH Gate *188*
15.1.3 Partial Product Generation *188*
15.1.4 Multi-Operand Addition *190*
15.2 Summary *192*

16 **Reversible Fault-Tolerant Division Circuit** *193*
16.1 Preliminaries of Division Circuits *193*
16.1.1 Division Algorithms *193*
16.2 The Division Method *194*
16.2.1 Floating-Point Data and Rounding *195*
16.2.2 Correctly Rounded Division *195*
16.2.3 Correct Rounding from One-Sided Approximations *196*
16.2.4 The Algorithm for Division Operation *196*
16.3 Components of a Division Circuit *199*
16.3.1 Reversible Fault-Tolerant MUX *200*
16.3.2 Reversible Fault-Tolerant D Latch *200*
16.4 The Design of the Division Circuit *201*
16.4.1 Reversible Fault-Tolerant PIPO Left-Shift Register *201*
16.4.2 Reversible Fault-Tolerant Register *203*
16.4.3 Reversible Fault-Tolerant Rounding Register *204*
16.4.4 Reversible Fault-Tolerant Normalization Register *204*
16.4.5 Reversible Fault-Tolerant Parallel Adder *204*
16.4.6 The Reversible Fault-Tolerant Division Circuit *205*
16.5 Summary *210*

17 **Reversible Fault-Tolerant Decoder Circuit** *211*
17.1 Transistor Realization of Some Popular Reversible Gates *211*
17.1.1 Feynman Double Gate *211*
17.1.2 Fredkin Gate *211*
17.2 Reversible Fault-Tolerant Decoder *213*
17.3 Summary *219*

18 **Reversible Fault-Tolerant Barrel Shifter** *221*
18.1 Properties of Barrel Shifters *221*
18.2 Reversible Fault-Tolerant Unidirectional Logarithmic Rotators *222*
18.3 Fault-Tolerant Unidirectional Logarithmic Logical Shifters *224*
18.4 Summary *229*

19 **Reversible Fault-Tolerant Programmable Logic Devices** *231*
19.1 Reversible Fault-Tolerant Programmable Logic Array *231*
19.1.1 The Design of RFTPLA *232*
19.2 Reversible Fault-Tolerant Programmable Array Logic *235*

19.2.1   The Design of AND Plane of RFTPAL   *236*
19.2.2   The Design of Ex-OR Plane of RFTPAL   *238*
19.3   Reversible Fault-Tolerant LUT-Based FPGA   *240*
19.3.1   Reversible Fault-Tolerant Gates   *240*
19.3.2   Proof of Fault-Tolerance Properties of the MSH and MSB Gates   *240*
19.3.3   Physical Implementation of the Gates   *241*
19.3.4   Reversible Fault-Tolerant D Latch, Master–Slave Flip-Flop and 4×1
          Multiplexer   *242*
19.3.5   Reversible Fault-Tolerant n-Input Look-Up Table   *244*
19.3.6   Reversible Fault-Tolerant CLB of FPGA   *244*
19.4   Summary   *246*

20   **Reversible Fault-Tolerant Arithmetic Logic Unit**   *249*
20.1   Design of n-bit ALU   *249*
20.1.1   A 4×4 Parity-Preserving Reversible Gate   *249*
20.1.2   1-Bit ALU   *251*
20.1.2.1 Group-1 PP Cell   *251*
20.1.2.2 Group-2 PP Cell   *252*
20.1.2.3 Group-3 PP Cell   *253*
20.1.2.4 n-bit ALU   *255*
20.2   Summary   *259*

21   **Online Testable Reversible Circuit Using NAND Blocks**   *261*
21.1   Testable Reversible Gates   *261*
21.2   Two-Pair Rail Checker   *265*
21.3   Synthesis of Reversible Logic Circuits   *266*
21.4   Summary   *268*

22   **Reversible Online Testable Circuits**   *269*
22.1   Online Testability   *269*
22.1.1   Online Testable Approach Using R1, R2, and R Gates   *269*
22.1.2   Online Testable Approach Using Testable Reversible Cells (TRCs)   *270*
22.1.3   Online Testable Circuit Using Online Testable Gate   *271*
22.1.4   Online Testing of ESOP-Based Circuits   *271*
22.1.5   Online Testing of General Toffoli Circuit   *272*
22.2   The Design Approach   *272*
22.2.1   The UFT Gate   *272*
22.2.2   Analysis of the Online Testable Approach   *276*
22.3   Summary   *278*

23   **Applications of Reversible Computing**   *279*
          Why We Need to Use Reversible Circuits   *280*
          Applications of Reversible Computing   *280*
23.1   Adiabatic Systems   *281*
23.2   Quantum Computing   *282*

23.3   Energy-Efficient Computing   *283*
23.4   Switchable Program and Feedback Circuits   *283*
23.5   Low-Power CMOS   *284*
23.6   Digital Signal Processing (DSP) and Nano-Computing   *284*

Part III   DNA Computing   *287*
An Overview About DNA Computing   *287*

24   Background Studies About Deoxyribonucleic Acid   *291*
24.1   Structure and Function of DNA   *291*
24.2   DNA Computing   *293*
24.2.1   Watson-Crick Complementary   *294*
24.2.2   Adleman's Breakthrough   *294*
24.3   Relationship of Binary Logic with DNA   *295*
24.4   Welfare of DNA Computing   *295*
24.5   Summary   *297*

25   A DNA-Based Approach to Microprocessor Design   *299*
25.1   Basics of Microprocessor Design   *299*
25.2   Characteristics and History of Microprocessors   *300*
25.3   Methodology of Microprocessor Design   *301*
25.4   Construction of Characteristic Tree   *302*
25.5   Traversal of the Tree   *302*
25.6   Encoding of the Traversed Path to the DNA Sequence   *304*
25.6.1   Gene Pool   *305*
25.6.2   Potency Factor   *305*
25.7   Combination of DNA Sequences   *305*
25.8   Decoding the Output String   *306*
25.9   Processor Evaluation   *307*
25.10   Post-Processing   *307*
25.11   Gene Pool Update   *309*
25.12   Summary   *309*

26   DNA-Based Reversible Circuits   *311*
26.1   DNA-Based Reversible Gates   *311*
26.2   DNA-Based Reversible NOT Gate   *311*
26.3   DNA-Based Reversible Ex-OR Gate   *311*
26.4   DNA-Based Reversible AND Gate   *312*
26.5   DNA-Based Reversible OR Gate   *313*
26.6   DNA-Based Reversible Toffoli Gate   *315*
26.6.1   Fan-out Technique of a DNA-Based Toffoli Gate   *316*
26.6.2   DNA-Based Reversible NOT Operation   *317*
26.6.3   DNA-Based Reversible AND Operation   *317*
26.6.4   DNA-Based Reversible OR Operation   *318*

26.6.5   DNA-Based Reversible Ex-OR Operation   *318*
26.6.6   Properties of DNA-Based Reversible Toffoli Gate   *319*
26.6.7   DNA-Based Reversible Fredkin Gates   *319*
26.7   Realization of Reversible DNA-Based Composite Logic   *321*
26.8   Summary   *322*

27   Addition, Subtraction, and Comparator Using DNA   *323*
27.1   DNA-Based Adder   *323*
27.2   DNA-Based Addition/Subtraction Operations   *325*
27.2.1   Addition and Subtraction Operations   *325*
27.2.2   Procedures of DNA-Based Reversible Addition/ Subtraction Operations   *325*
27.3   DNA-Based Comparator   *329*
27.3.1   Sequence Design   *330*
27.3.2   Estimation of Rate Constant   *331*
27.4   Summary   *331*

28   Reversible Shift and Multiplication Using DNA   *333*
28.1   DNA-Based Reversible Shifter Circuit   *333*
28.1.1   Procedures of DNA-Based Shifter Circuit   *333*
28.2   DNA-Based Reversible Multiplication Operation   *336*
28.3   Summary   *339*

29   Reversible Multiplexer and ALU Using DNA   *341*
29.1   DNA-Based Reversible Multiplexer   *341*
29.1.1   The Working Procedures of DNA-Based Multiplexer Circuit   *342*
29.2   DNA-Based Reversible Arithmetic Logic Unit   *345*
29.2.1   Procedures of DNA-Based ALU   *345*
29.2.2   Properties of the DNA-Based ALU   *347*
29.3   Summary   *349*

30   Reversible Flip-Flop Using DNA   *351*
30.1   The Design of a DNA Fredkin Gate   *351*
30.2   Simulating the Fredkin Gate by Sticking System   *351*
30.2.1   Simulating the Fredkin Gate by Enzyme System   *353*
30.3   Simulation of the Reversible D Latch Using DNA Fredkin Gate   *355*
30.3.1   Simulation of the Reversible Sequential Circuit Using DNA Fredkin Gate   *355*
30.4   DNA-Based Biochemistry Technology   *356*
30.5   Summary   *357*

31   Applications of DNA Computing   *359*
31.1   Solving the Optimization and Scheduling Problems Like the Traveling Salesman Problem   *360*
31.2   Parallel Computing   *362*
31.3   Genetic Algorithm   *363*
31.4   Neural System   *363*

31.5   Fuzzy Logic Computation and Others   *364*
31.6   Lift Management System   *364*
31.7   DNA Chips   *364*
31.8   Swarm Intelligence   *365*
31.9   DNA and Cryptography Systems   *365*
31.10   Monstrous Memory Capacity   *366*
31.11   Low-Power Dissipation   *367*
31.12   Summary   *367*

Conclusion   *369*
Copyright Permission of Third-Party Materials   *371*
Bibliography   *373*
Index   *389*

List of Figures

Figure 1.1   A $k \times k$ reversible gate   *6*

Figure 1.2   Popular reversible gates   *6*

Figure 1.3   Reversible Feynman gate   *7*

Figure 1.4   Quantum realization of reversible Fredkin (FRG) gate   *7*

Figure 1.5   The quantum representation of reversible HNG gate   *8*

Figure 1.6   Block diagram of the reversible FRG gate   *9*

Figure 1.7   Quantum representation of a reversible FRG gate   *10*

Figure 1.8   Toffoli gates as self-reversible   *10*

Figure 1.9   Quantum cost calculation of Feynman gate   *12*

Figure 1.10   Quantum circuit of Toffoli gate   *12*

Figure 1.11   Quantum circuit of Fredkin gate   *13*

Figure 1.12   Quantum circuit of a Peres gate   *13*

Figure 2.1   Multi-operand n-digit decimal adder   *16*

Figure 2.2   The architecture of reversible multi-operand n-digit decimal adder   *16*

Figure 2.3   The architecture of reversible single-digit block of m-operand n-digit adder   *16*

Figure 2.4 4×4 reversible FAG gate   *17*

Figure 2.5   Reversibility of 4×4 reversible FAG gate   *18*

Figure 2.6   Reversible FAG gate as full adder   *18*

Figure 2.7   The n-bit carry skip adder circuit   *21*

Figure 2.8   Single block carry skip adder circuit   *21*

Figure 2.9   Reversible CSA for single digit   *21*

Figure 2.10   Reversible carry skip circuit   *22*

Figure 2.11   Reversible partial full adder   *25*

Figure 2.12 n-bit CLA adder circuit   *25*

Figure 2.13   A 1-digit BCD adder's overflow detection logic   *28*

Figure 2.14   A 1-bit BCD adder correction logic circuit   *29*

Figure 2.15   A 1-digit BCD adder   *29*

Figure 2.16   A carry skip 1-digit BCD adder   *33*

Figure 2.17   Nine's complement circuit   *35*

Figure 2.18   Modified nine's complement circuit   *35*

Figure 2.19   Modified conventional BCD subtractor   *36*

Figure 2.20   CLA BCD subtractor   *37*

Figure 2.21   Carry skip BCD subtractor   *38*

Figure 2.22   Reversible nine's complement   *39*

Figure 2.23   Reversible BCD subtractor   *39*

Figure 2.24   Reversible CLA BCD subtractor   *40*

Figure 2.25   Reversible logic implementation of the carry skip BCD adder   *41*

Figure 3.1   Process of 4×4 multiplications   *44*

Figure 3.2   Block diagram of a reversible HNG gate   *45*

Figure 3.3   The quantum representation of a reversible HNG gate   *45*

Figure 3.4   Symbols of the controlled-T and controlled-T^+ gate.   *46*

Figure 3.5 5×5 BSJ gate and its corresponding input–output mapping   *48*

Figure 3.6   Quantum realization of 5×5 BSJ gate   *48*

Figure 3.7 3×3 MPG and its corresponding input–output mapping   *48*

Figure 3.8   Quantum realization of 3×3 MPG   *48*

Figure 3.9   Quantum analysis of 3×3 MPG   *49*

Figure 3.10   A compact quantum realization of 3×3 MPG   *49*

Figure 3.11   Block diagram of R cell   *49*

Figure 3.12   Construction of R cell   *49*

Figure 3.13 3×3 MTG and its corresponding input–output mapping   *50*

Figure 3.14   Quantum realization of 4×4 MTG   *50*

Figure 3.15 3×3 MFRG and its corresponding input–output mapping   *50*

Figure 3.16   Quantum realization of 4×4 MFRG   *50*

Figure 3.17   A Compact quantum realization of 4×4 MFRG   *52*

Figure 3.18 16×16 PPG array   *52*

Figure 3.19   Gate level diagram of a 4×4 PPG for reversible Booth's multiplier   *53*

Figure 3.20   Block diagram of an $n \times n$ PPG for reversible Booth's multiplier   *53*

Figure 3.21   Gate level diagram of a 4×4 MOA for reversible Booth's multiplier   *54*

Figure 3.22   Diagram of an $n \times n$ MOA for reversible Booth's Multiplier   *54*

Figure 3.23   (a) 6×6 MOA (b) 8×8 MOA   *58*

Figure 3.24   Critical path for an 8×8 PPG for reversible Booth's multiplier   *62*

Figure 3.25   Critical path for a 6×6 MOA for reversible Booth's multiplier   *62*

Figure 4.1   Two-input *n*-bit reversible MUX   *68*

Figure 4.2   A clocked D Flip-Flop   *68*

Figure 4.3   An *n*-bit reversible D flop-flop   *69*

Figure 4.4   Implementation of the characteristic function of Equation (4.2.3.1)   *69*

Figure 4.5   The structure of the basic cell for the reversible PIPO left-shift register   *70*

Figure 4.6   The block diagram of the basic cell for the reversible PIPO left-shift register   *70*

Figure 4.7   An *n*-bit reversible PIPO left-shift register   *70*

Figure 4.8   An $(n + 1)$-bit parallel adder (carry-out ignored)   *71*

Figure 4.9   Illustration of the division circuit   *71*

Figure 5.1   Reversible BJS gate   *76*

Figure 5.2   Quantum realization of the BJS gate   *76*

Figure 5.3   Reversible HLN gate   *77*

Figure 5.4   Quantum realization of the HLN gate   *77*

Figure 5.5   BJS gate works as reversible 1-bit comparator   *78*

Figure 5.6   BJS gate works as reversible MSB comparator   *78*

Figure 5.7   The GE comparator cell   *78*

Figure 5.8   Block diagram of the single-bit GE comparator cell   *79*

Figure 5.9   The LT comparator cell   *79*

Figure 5.10   Block diagram of the single-bit LT comparator cell   *79*

Figure 5.11   Reversible 2-bit comparator   *80*

Figure 5.12   Reversible *n*-bit comparator   *80*

Figure 6.1   Mapping of JK latch (Equation $Q^+ = M \cdot E + \overline{E} \cdot Q$) on the Fredkin gate   *88*

Figure 6.2   Mapping of variable M (Equation $J \cdot \overline{Q} + \overline{K} \cdot Q$) on the Fredkin gate   *88*

Figure 6.3   Reversible design of *JK* latch with minimal garbage outputs   *88*

Figure 6.4   Conventional cross-coupled SR latch   *89*

Figure 6.5   Peres gate based SR latch without enable   *89*

Figure 6.6   Reversible SR latch based on modified truth table   *90*

Figure 6.7   Reversible gated SR latch based on modified truth table   *91*

Figure 6.8   Fredkin gate-based D latch with one Feynman gate   *91*

Figure 6.9   Fredkin gate-based D latch with two Feynman gates   *92*

Figure 6.10   Fredkin gate-based negative enable reversible D latch with only output Q *92*

Figure 6.11   Fredkin gate-based negative enable reversible D latch with outputs Q and $\overline{Q}$ *92*

Figure 6.12   Peres gate-based T latch   *93*

Figure 6.13   Reversible T latch with outputs Q and $\overline{Q}$ *93*

Figure 6.14   Reversible JK latch with outputs Q and $\overline{Q}$ *94*

Figure 6.15   Reversible master–slave D flip-flop   *94*

Figure 6.16   Reversible master–slave T flip-flop   *95*

Figure 6.17   Reversible master–slave JK flip-flop   *95*

Figure 6.18   Reversible master–slave SR flip-flop   *95*

Figure 6.19   Application of the Fredkin gate to avoid the fan-out   *96*

Figure 6.20   Asynchronous reset of the Q and R outputs of the Fredkin gate   *96*

Figure 6.21   Asynchronous set of the Q and R outputs of the Fredkin gate   *96*

Figure 6.22   Fredkin gate-based asynchronous set/reset D latch   *96*

Figure 6.23   Reversible asynchronous set/reset master–slave D flip-flop   *97*

Figure 7.1   Reversible T flip-flop   *100*

Figure 7.2   Reversible clocked T flip-flop for synchronous counter   *100*

Figure 7.3   Reversible clocked T flip-flop for asynchronous counter   *100*

Figure 7.4   Block diagram of 3×3 MPG gate   *101*

Figure 7.5   Quantum representation of *3×3* MPG gate   *101*

Figure 7.6   Reversible master–slave T flip-flop   *101*

Figure 7.7   4-bit reversible asynchronous counter   *101*

Figure 7.8   4-bit reversible synchronous counter   *102*

Figure 7.9   Quantum implementation of a reversible 2–*to*–4 decoder   *103*

Figure 7.10   Measurement of the quantum delay for the reversible 2–*to*–4 decoder unit   *104*

Figure 7.11   Reversible 2–*to*–4 decoder   *105*

Figure 7.12   Reversible 4–to–2 encoder   *105*

Figure 8.1   Simple block diagram of the barrel shifter   *108*

Figure 8.2   Block diagram of the reversible MBJN gate   *108*

Figure 8.3   Quantum realization of the reversible MBJN gate   *108*

Figure 8.4   Reversible 2-bit 2's complement generator   *109*

Figure 8.5   Reversible 3-bit 2's complement generator   *110*

Figure 8.6   Reversible 3-bit swap condition generator   *110*

Figure 8.7   Reversible 4-bit swap condition generator   *111*

Figure 8.8   A (4, 3) reversible right rotator   *112*

Figure 8.9   Block diagram of (8, 7) reversible bidirectional barrel shifter   *113*

Figure 8.10   Structure of the reversible clocked D flip-flop   *114*

Figure 8.11   Block diagram of the reversible clocked D flip-flop   *114*

Figure 8.12   *n*-bit reversible SISO shift register   *114*

Figure 8.13 n-bit reversible SIPO shift register   *115*

Figure 8.14 n-bit reversible PISO shift register   *115*

Figure 8.15   Implementation of the characteristic function of Equation (8.2.1.4.1)   *116*

Figure 8.16   Basic cell for the reversible PIPO shift register   *117*

Figure 8.17   Block diagram for the reversible PIPO shift register   *117*

Figure 8.18 n-bit reversible PIPO shift register   *117*

Figure 8.19   Implementation of the characteristic function of Equation (8.2.1.5.1)   *119*

Figure 8.20   Basic cell for the reversible universal shift register   *120*

Figure 8.21   Block diagram for the reversible universal shift register   *120*

Figure 8.22 n-bit reversible universal shift register   *120*

Figure 9.1   Reversible gate 1 (RG1)   *124*

Figure 9.2   Reversible Gate 2 (RG2)   *124*

Figure 9.3   Reversible R-I gate   *124*

Figure 9.4   Transistor level realization of reversible R-I gate   *125*

Figure 9.5   Realization of 2:1 multiplexer using reversible R-I gate   *126*

Figure 9.6   Realization of 1:2 demultiplexer using reversible R-I gate   *126*

Figure 9.7   Realization of two-input XOR using reversible R-I gate   *126*

Figure 9.8   Realization of two-input AND gate using reversible R-I gate   *126*

Figure 9.9   Reversible R-II gate   *126*

Figure 9.10   Transistor level circuit for the reversible R-II gate   *127*

Figure 9.11   Realization of 2:1 multiplexer using reversible R-II gate   *128*

Figure 9.12   Realization of two-input XOR and half adder using reversible R-II gate   *128*

Figure 9.13   Realization of two-input AND gate using reversible R-II gate   *128*

Figure 10.1   Block diagram of 3×3 reversible NH gate   *130*

Figure 10.2   Quantum realization of 3×3 reversible NH gate   *130*

Figure 10.3   Block diagram of 4×4 reversible BSP gate   *130*

Figure 10.4   4-to-1 reversible MUX   *131*

Figure 10.5   Reversible D latch   *131*

Figure 10.6   Reversible Write-Enabled Master–Slave flip-flop   *132*

Figure 10.7   Block diagram of a reversible RAM   *132*

Figure 10.8   A reversible logic element of Plessey FPGA   *133*

Figure 10.9 3×3 Reversible MUX gate   *135*

Figure 10.10   Different uses of a Feynman gate   *135*

Figure 10.11   One template of toffoli gate   *135*

Figure 10.12   Two templates of MUX gate   *135*

Figure 10.13 Ex-OR plane realization for the function *F* based on the Algorithm 10.2.1.1 *136*

Figure 10.14 Design of reversible PLAs for multi-output function *F* *138*

Figure 10.15 Delay calculation of AND plane: (a-b) delay propagation path of a gate and a cross-point and (c) overall delay propagation path for AND plane *140*

Figure 10.16 Delay calculation of Ex-OR plane: (a-b) delay propagation path of a gate and a cross-point and (c) overall delay propagation path for Ex-OR plane *140*

Figure 11.1 Block diagram of 3×3 reversible FS gate *144*

Figure 11.2 Quantum realization of 3×3 reversible FS gate *144*

Figure 11.3 2×2^2 Reversible decoder *145*

Figure 11.4 3×2^3 Reversible decoder *146*

Figure 11.5 $n \times 2^n$ reversible decoder *146*

Figure 11.6 Reversible D flip-flop *147*

Figure 11.7 Reversible write-enabled master–slave D flip flop *147*

Figure 11.8 Reversible RAM *149*

Figure 11.9 Reversible ITS decoder *149*

Figure 11.10 Quantum representation of reversible ITS decoder *150*

Figure 11.11 Block diagram of reversible PROM *151*

Figure 11.12 Reversible TI gate *151*

Figure 11.13 Different uses of Feynman gate *152*

Figure 11.14 Template of Toffoli gate *152*

Figure 11.15 Two templates of TI gate *152*

Figure 11.16 The combined design of AND plane and Ex-OR plane *153*

Figure 12.1 Logic diagram of a conventional ALU *156*

Figure 12.2 The reversible function generator *157*

Figure 12.3 Block diagram of reversible function generator *157*

Figure 12.4 The reversible control unit *158*

Figure 12.5 Block diagram of the reversible control unit *158*

Figure 12.6 The design of 16-bit reversible ALU *159*

Figure 13.1 Block diagram of a 16-bit control unit *162*

Figure 13.2 Block diagram of reversible HL gate *162*

Figure 13.3 Quantum realization of reversible HL gate *162*

Figure 13.4 Block diagram of reversible BJ gate *163*

Figure 13.5 NAND implementation of reversible BJ gate *163*

Figure 13.6 2-to-4 Reversible decoder using FG and FRG gate *164*

Figure 13.7 2-to-4 Reversible decoder using HL gate *164*

Figure 13.8   Reversible 3-to-8 decoder (Approach 1)   *165*

Figure 13.9   Reversible 3-to-8 decoder (Approach 2)   *166*

Figure 13.10   Reversible n-to-2^n decoder (Approach 1)   *166*

Figure 13.11   Reversible n-to-2^n decoder (Approach 2)   *167*

Figure 13.12   Reversible JK flip flop   *168*

Figure 13.13   4-bit reversible sequence counter   *168*

Figure 13.14   16-bit reversible instruction register   *169*

Figure 13.15   Reversible control gates associated with AR   *170*

Figure 14.1   Feynman double gate   *178*

Figure 14.2   Feynman double gate preserves fault tolerance over input–output unique mapping   *178*

Figure 14.3   Fredkin gate   *178*

Figure 14.4   New fault-tolerant gate   *178*

Figure 14.5   Parity-preserving HC gate   *179*

Figure 14.6   Parity-preserving IG gate   *179*

Figure 14.7   Parity-preserving IG gate as a NOT gate   *180*

Figure 14.8   Parity-preserving IG gate as AND gate and Ex-OR gate   *180*

Figure 14.9   Parity-preserving IG gate as Ex-OR gate, Ex-NOR gate and OR gate   *180*

Figure 14.10   Quantum representation of NFT gate   *181*

Figure 14.11   Design of single NFT full adder   *181*

Figure 14.12   Design of fault-tolerant CSA   *183*

Figure 14.13   Design of 4-bit fault-tolerant CLA   *183*

Figure 14.14   FTFA circuit   *184*

Figure 14.15   Fault-tolerant ripple carry adder   *184*

Figure 15.1   Working procedure of a 4×4 multiplier circuit   *188*

Figure 15.2 4×4 Reversible LMH gate   *188*

Figure 15.3   Quantum realization of LMH gate   *188*

Figure 15.4 4×4 Partial product generator circuit   *190*

Figure 15.5   Generalized architecture of fault-tolerant PPG   *190*

Figure 15.6 4×4 multi-operand addition circuit   *191*

Figure 15.7   Generalized architecture of fault-tolerant MOA   *192*

Figure 16.1   Illustration of the decomposition of a binary number   *197*

Figure 16.2   Example of a division operation   *198*

Figure 16.3   2-input n-bit reversible fault-tolerant MUX   *199*

Figure 16.4   Block diagram of RR gate   *199*

Figure 16.5   Reversible fault-tolerant D latch using RR gate   *199*

Figure 16.6   Block diagram of F2PG gate   *201*

Figure 16.7   Reversible fault-tolerant PIPO left-shift register   *203*

Figure 16.8   Reversible fault-tolerant register   *203*

Figure 16.9   Reversible fault-tolerant rounding register   *204*

Figure 16.10   Reversible fault-tolerant normalization register   *204*

Figure 16.11   Reversible fault-tolerant NFTFAG   *205*

Figure 16.12   Quantum representation of NFTFAG   *205*

Figure 16.13   NFTFAG as a reversible fault-tolerant full adder   *205*

Figure 16.14 $(n + 1)$-bit reversible fault-tolerant parallel adder   *207*

Figure 16.15 $(n + 1)$-bit reversible fault-tolerant parallel adder   *207*

Figure 16.16   Block diagram of the 2-bit reversible fault-tolerant division circuit   *209*

Figure 17.1   Block diagram of F2G   *212*

Figure 17.2   Quantum equivalent realization of F2G   *212*

Figure 17.3   Transistor realization of F2G   *212*

Figure 17.4   Block diagram of FRG   *213*

Figure 17.5   Quantum equivalent realization of FRG   *213*

Figure 17.6   Transistor realization of FRG   *213*

Figure 17.7   1-to-2 Reversible fault-tolerant decoder   *214*

Figure 17.8   Block diagram of the 2-to-4 RFD   *214*

Figure 17.9   Block diagram of the 3-to-8 RFD   *214*

Figure 17.10   Schematic diagram of the 2-to-4 RFD   *215*

Figure 17.11   Block diagram of the n-to-2^n decoder   *216*

Figure 17.12   Combinations of the two 2×2 quantum primitive gates   *217*

Figure 18.1   Adaptive structure of (n, k) logarithmic barrel shifter   *222*

Figure 18.2 $(4, 2)$ Reversible fault-tolerant unidirectional logarithmic barrel shifter   *222*

Figure 18.3 $(8, 3)$ Reversible fault-tolerant unidirectional logarithmic barrel shifter   *223*

Figure 18.4 (n, k) Reversible fault-tolerant unidirectional logarithmic right rotator   *225*

Figure 18.5 $(4, 2)$ Reversible fault-tolerant unidirectional logarithmic logical shifter   *226*

Figure 18.6 $((8, 3)$ Reversible fault-tolerant unidirectional logarithmic logical shifter   *227*

Figure 18.7 (n,k) reversible fault-tolerant logarithmic logical shifter (circuit for right logical shift)   *228*

Figure 19.1   AND Ex-OR programmable logic array   *232*

Figure 19.2   Four different orientations   *232*

Figure 19.3   Realization of multi-output function (F) based on Algorithm 19.1.1.1   *235*

Figure 19.4   Realization of multi-output function *F* based on Algorithm 19.1.1.1 and Algorithm 19.1.1.2   *237*

Figure 19.5   Different representations of Fredkin gate   *237*

Figure 19.6   The design of AND plane of reversible fault-tolerant PAL   *237*

Figure 19.7   Feynman extension gate (FEG)   *239*

Figure 19.8   The design of the Ex-OR plane of a reversible fault-tolerant PAL using Feynman extension gates   *239*

Figure 19.9   The block diagram of reversible fault-tolerant MSH gate   *240*

Figure 19.10   The quantum realization of reversible fault-tolerant MSH gate with quantum cost = 6   *240*

Figure 19.11   The block diagram of reversible fault-tolerant MSB gate   *241*

Figure 19.12   The quantum realization of reversible fault-tolerant MSB gate with quantum cost = 12   *241*

Figure 19.13   The transistor realization of reversible fault-tolerant MSH gate   *241*

Figure 19.14   The transistor realization of reversible fault-tolerant MSB gate   *244*

Figure 19.15   The block diagram of reversible fault-tolerant D latch   *244*

Figure 19.16   The block diagram of reversible fault-tolerant master–slave flip-flop   *244*

Figure 19.17   The block diagram of a reversible fault–tolerant 4×1 multiplexer   *245*

Figure 19.18   The block diagram of reversible fault-tolerant three-input LUT   *245*

Figure 19.19   The block diagram of reversible fault-tolerant four-input LUT   *245*

Figure 19.20   The block diagram of reversible fault-tolerant CLB of FPGA   *246*

Figure 20.1 4×4 Reversible fault-tolerant UPPG gate   *250*

Figure 20.2   Quantum realization of LMH gate   *250*

Figure 20.3   Circuit structure of Group-1 PP cell   *252*

Figure 20.4   Compressed block diagram of Group-1 PP cell   *252*

Figure 20.5   Block diagram of Group-2 PP $(Ceil)_{-1}$ *253*

Figure 20.6   Block diagram of Group-2 PP cell   *254*

Figure 20.7   Block diagram of Group-3 PP cell   *255*

Figure 20.8   Reversible fault-tolerant 2-bit ALU   *255*

Figure 20.9   *n*-bit reversible fault-tolerant ALU   *256*

Figure 21.1   Reversible logic gate R1   *262*

Figure 21.2   Reversible logic gate R2   *262*

Figure 21.3   Reversible logic gate R3   *263*

Figure 21.4   Realizations of OR and Ex-OR Gates Using R1 Gate   *264*

Figure 21.5   Realizations of Ex-NOR and NAND gates using R1 gate   *264*

Figure 21.6   Realization of NOR gate using R1 gate   *264*

Figure 21.7   Realization of AND gate using R1 gate   *264*

Figure 21.8   The testable logic block using R1 and R2 gates   *265*

Figure 21.9   Two-pair rail checker   *265*

Figure 21.10   Testable block embedded with two-pair two-rail checkers   *266*

Figure 21.11   Realization of NAND gate using R1 and R2 gates   *267*

Figure 21.12   Reversible NAND block implementation for the function $ab + cd$ *267*

Figure 21.13   Implementation of signal duplication   *267*

Figure 22.1   Block diagram of R1 gate   *270*

Figure 22.2   Block diagram of R2 gate   *270*

Figure 22.3   Block diagram of R gate   *270*

Figure 22.4   Construction of a testable block (TB)   *270*

Figure 22.5   Block diagram of a testable block (TB)   *271*

Figure 22.6   Block diagram of UFT gate   *272*

Figure 22.7   Compact representation of a UFT gate   *273*

Figure 22.8   Quantum realization of a UFT circuit   *273*

Figure 22.9   AND and EX-OR operations of UFT gate   *275*

Figure 22.10   OR operation of UFT gate   *275*

Figure 22.11   NAND and NOT operations of UFT gate   *275*

Figure 22.12   EX-OR and EX-NOR operations of UFT gate   *276*

Figure 22.13   NOR operation of the UFT gate   *277*

Figure 22.14   Nontestable circuit for $f = abc \oplus c' = ab + c'$ *277*

Figure 22.15   Online testable circuit for Example 22.2.2.1   *277*

Figure 22.16   Nontestable full adder using ESOP technique   *277*

Figure 22.17   Online testable full adder circuit   *278*

Figure 23.1   Reversible computer dissipates less heat than a conventional computer   *280*

Figure 23.2   Reversible computer has the same number of outputs and inputs   *281*

Figure 23.3   Working mechanism of a reversible computer   *282*

Figure 23.4   Reversible logic gates   *282*

Figure 23.5   Back-up states of a reversible computing system   *284*

Figure 24.1   Hydrogen bonds of the interior DNA   *292*

Figure 24.2   DNA structure   *292*

Figure 24.3   Structure of DNA   *293*

Figure 24.4   Ligation process of DNA   *293*

Figure 24.5   HPP on seven vertices   *294*

Figure 24.6   DNA replication process   *296*

Figure 25.1   Overview of the methodology   *303*

Figure 25.2   Part of the tree structure   *304*

Figure 25.3   Structure of a node   *304*

Figure 25.4 DNA sequence of a node for 20 nodes tree *306*

Figure 25.5 DNA combination *306*

Figure 26.1 Operation of DNA-based reversible NOT gate (DRNG) *312*

Figure 26.2 Operation of DNA-based reversible Ex-OR gate *313*

Figure 26.3 Operation of DNA-based reversible AND gate *314*

Figure 26.4 Operation of DNA-based reversible OR gate *315*

Figure 26.5 Overall procedures of DNA hybridization for a DNA-based Toffoli gate *316*

Figure 26.6 DNA-based Toffoli gate as NOT gate *317*

Figure 26.7 DNA-based Toffoli gate as AND gate *318*

Figure 26.8 DNA-based Toffoli gate as OR gate *318*

Figure 26.9 DNA-based Toffoli gate as Ex-OR gate *319*

Figure 26.10 Procedures of DNA hybridization for DNA-based Fredkin gate *320*

Figure 26.11 DNA hybridization of selection operation between two ANDed products using DNA-based Fredkin gate *320*

Figure 26.12 Gate-level representation of reversible half-adder using Toffoli gate *321*

Figure 26.13 Reversible DNA-based half-adder *321*

Figure 27.1 Gate-level representation of a reversible full-adder circuit *324*

Figure 27.2 Operation of DNA-based reversible full-adder circuit *324*

Figure 27.3 DNA-based reversible adder/subtractor circuit *325*

Figure 27.4 DNA hybridization of logical AND operation using DNA-based Toffoli gate *326*

Figure 27.5 DNA hybridization of logical NOT operation using DNA-based Toffoli gate *326*

Figure 27.6 DNA hybridization of logical AND operation between complemented and noncomplemented literals using DNA-based Toffoli gate *327*

Figure 27.7 DNA hybridization of logical Ex-OR operation using DNA-based Toffoli gate *327*

Figure 27.8 DNA hybridization of selection operation using DNA-based Fredkin gate *328*

Figure 27.9 The working principle of DNA comparator *329*

Figure 28.1 Operation of DNA-based reversible NOT gate *335*

Figure 28.2 The four basic biochemical events of the shifter circuit *336*

Figure 28.3 The working procedures of DNA-based reversible multiplication operation *337*

Figure 28.4 An example of DNA-based reversible multiplication operation *338*

Figure 29.1 Block diagram of DNA-based reversible multiplexer circuit *342*

Figure 29.2 DNA hybridization of the first Fredkin gate for selection operation *342*

Figure 29.3   DNA hybridization of second Fredkin gate for selection operation   *343*

Figure 29.4   DNA hybridization of third Fredkin gate for selection operation   *344*

Figure 29.5   DNA hybridization of the fourth Fredkin gate for selection operation   *344*

Figure 29.6   DNA hybridization of the fifth Fredkin gate for selection operation   *345*

Figure 29.7   Diagram of DNA-based reversible logic unit   *346*

Figure 29.8   Diagram of DNA-based reversible logic unit   *347*

Figure 30.1   Fredkin gate symbol and its working procedure as a conditional switch   *352*

Figure 30.2   Forming dsDNA by hybridizing the two complementary ssDNA and forming two complementary ssDNA by melting the dsDNA   *352*

Figure 30.3   a. A one-to-one mapping between binary bit and DNA strands with fixed length 4. b. A one-to-one mapping between binary information and DNA string by concentrating the DNA strands. c. The simulating of binary information computing by DNA biochemistry operations   *352*

Figure 30.4   a. Fredkin gate with three inputs and three outputs. b. Simulation of the Fredkin gate in detail; b1. When the input is 1, the output is also 1. b2. When the input x and the corresponding biochemistry operations. b3. When the input is y and the corresponding biochemistry operations   *353*

Figure 30.5   a. The Fredkin gate with three inputs and three outputs and three outputs and the one-to-one mapping between the three inputs and the DNA strands. b. Simulation of the Fredkin gate in detail; b1. When the input is 1 using amplification, the output is still 1. b2. When the input is x adding *Smal*, the output is y. b3. When the input is y is adding ligase enzyme, the output is x. c. The progress of cutting x into two parts using *Smal*   *354*

Figure 30.6   a. The Fredkin gate with three inputs and three outputs. b. The DNA Fredkin gate based on the sticking system. c. The DNA Fredkin gate based on the enzyme system   *355*

Figure 30.7   a. The D latch based on Fredkin gate. b. The DNA D latch based on sticking system. c. The DNA D latch based on enzyme system   *355*

Figure 30.8   DNA reversible master–slave D flip-flop with the CP clock pulse, which can be simulated by DNA strands   *356*

Figure 31.1   DNA double helix   *360*

Figure 31.2   DNA computer solving a shortest path problem   *361*

Figure 31.3   Traveling salesman problem   *361*

Figure 31.4   Parallel computing   *363*

Figure 31.5   Concept of DNA chips   *365*

Figure 31.6   Swarm intelligence in nature   *366*

Figure 31.7   Swarm intelligence in nature   *366*

Figure 31.8   Huge memory capacity of DNA   *367*

Figure 31.9   Low-power DNA computers   *368*

List of Tables

Table 2.1   Reversibility of 4 × 4 Reversible FAG Gate   *18*

Table 3.1   Radix-4 Booth's Recoding   *44*

Table 3.2   Truth Table for Recoding Cell   *47*

Table 4.1   Function Table for Reversible PIPO Left-Shift Register   *69*

Table 5.1   Truth Table of the BJS Gate   *76*

Table 5.2   Truth Table of the BJS Gate   *77*

Table 5.3   Truth Table of the 1-Bit Binary Comparator   *77*

Table 6.1   Modified Truth Table of the SR Latch   *90*

Table 7.1   Truth Table for the Reversible 2–*to*–4 Decoder   *104*

Table 7.2   Truth Table for the Reversible 4–to–2 Encoder   *106*

Table 8.1   Truth Table of Reversible PIPO Shift Register   *116*

Table 8.2   Function Table for Reversible Universal Shift Register   *118*

Table 9.1   Truth Table of the B-I Gate   *125*

Table 9.2   Truth Table of the R-II Gate   *127*

Table 10.1   The Products of Functions   *136*

Table 11.1   Truth Table of 3 × 3 Reversible FS Gate   *144*

Table 11.2   Truth Table of Reversible ITS Decoder   *150*

Table 11.3   Truth Table of Reversible ITS Gate   *152*

Table 11.4   The Product of Functions   *153*

Table 13.1   Reversibility of 4 × 4 Reversible HL Gate   *163*

Table 13.2   Reversibility of 4 × 4 Reversible BJ Gate   *164*

Table 14.1   Truth Table of the Parity-Preserving IG Gate   *179*

Table 14.2   Input-Output Patterns of a Full Adder   *181*

Table 15.1   Truth Table of 4 × 4 Reversible LMH Gate   *189*

Table 16.1   Truth Table of 4 × 4 Reversible RR Gate   *200*

Table 16.2   Control Inputs of a Fault-Tolerant Reversible Left-Shift Register   *201*

Table 16.3   Truth Table of the Fault-Tolerant F2PG Gate   *202*

Table 16.4   Truth Table of the NFTFAG   *206*

Table 17.1   Truth Table of F2G and FRG Gate   *214*

Table 17.2   Truth Table of 1-to-2 Decoder with One Constant Input   *216*

Table 17.3   Truth Table of Figure 17.12 (a)   *217*

Table 17.4   Truth Table of Figure 17.12 (b)   *218*

Table 19.1   Frequency Matrix Based on Multi-Output Function, F   *233*

Table 19.2   Calculation of Number of Product of Functions   *233*

Table 19.3   Truth Table of Reversible Fault Tolerant MSH Gate   *242*

Table 19.4   Truth Table of Reversible Fault-Tolerant MSB Gate   *243*

Table 20.1   Truth Table of 4 × 4 Reversible Fault-Tolerant UPPG Gate   *250*

Table 20.2   Truth Table of Group-I PP Cell   *253*

Table 20.3   Different Functions of Fault-Tolerant ALU   *254*

Table 21.1   Truth Table of Reversible Logic Gate R1   *262*

Table 21.2   Truth Table of Reversible Logic Gate R2   *263*

Table 21.3   Truth Table of Reversible Logic Gate R3   *263*

Table 22.1   Truth Table of UFT Gate   *273*

Table 27.1   Sequence list   *330*

About the Author

Dr. Hafiz Md. Hasan Babu is currently working as the pro-vice-chancellor of National University, Bangladesh. He is now on deputation from the Department of Computer Science and Engineering, University of Dhaka, Bangladesh. He is also the former chairman of the same department. For his excellent academic and administrative capability, he also served as the professor and founder chairman of the Department of Robotics and Mechatronics Engineering, University of Dhaka, Bangladesh. He served as a World Bank senior consultant and general manager of the Information Technology & Management Information System Departments of Janata Bank Limited, Bangladesh. Dr. Hasan Babu was the World Bank resident information technology expert of the Supreme Court Project Implementation Committee, Supreme Court of Bangladesh. He was also the information technology consultant of Health Economics Unit and Ministry of Health and Family Welfare in the project "SSK (Shasthyo Shurokhsha Karmasuchi) and Social Health Protection Scheme" under the direct supervision and funding of German Financial Cooperation through KfW.

Professor Dr. Hafiz Md. Hasan Babu received his M.Sc. degree in Computer Science and Engineering from the Brno University of Technology, Czech Republic, in 1992 under the Czech Government Scholarship. He obtained the Japanese Government Scholarship to pursue his PhD from the Kyushu Institute of Technology, Japan, in 2000. He also got DAAD (Deutscher Akademischer Austauschdienst) Fellowship from the Federal Republic of Germany.

Professor Dr. Hasan is a very eminent researcher. He was awarded the best paper awards in three reputed international conferences. In recognition of his valuable contributions in the field of Computer Science and Engineering, he received the Bangladesh Academy of Sciences Dr. M O. Ghani Memorial Gold Medal Award for the year 2015, which is one of the most prestigious research awards in Bangladesh. He was also awarded the UGC (University Grants Commission of Bangladesh) Gold Medal Award-2017 for his outstanding research contributions in computer science and engineering. He has written more than 100 research articles published in reputed international journals (*IET Computers & Digital Techniques, IET Circuits and Systems, IEEE Transactions on Instrumentation and Measurement, IEEE Transactions on VLSI Systems, IEEE Transactions on Computers, Elsevier Journal of Micro-electronics, Elsevier Journal of Systems Architecture, Springer Journal of Quantum Information Processing* etc.) and joined international conferences. According to Google Scholar, Prof. Hasan has already received around 1257 citations with h-index 16 and i10-index 30. He is a regular reviewer of reputed international journals and international conferences. He presented invited talks and chaired scientific sessions or worked as a member of the organizing committee or international advisory board in many international conferences held in different countries. For his excellent research record, he has also been appointed as the associate editor of *IET Computers and Digital Techniques*, published by the Institution of Engineering and Technology of the United Kingdom.

Professor Dr. Hasan was appointed as a member of the prime minister's ICT Task Force Committee, Government of the People's Republic of Bangladesh on recognition of his national and international level contributions in Engineering Sciences. He is currently the president of Bangladesh Computer Society and also the president of Internet Society, Bangladesh chapter. He has been recently appointed as a part-time member of Bangladesh Accreditation Council to ensure the quality of higher education in Bangladesh.

Preface

Reversible computing is called a backward deterministic system such that every state of the system has at most one predecessor. Hence, there is no pair of distinct states that goes to the same state. Reversible computing intends computation using reversible operations i.e., procedures which can be easily and exactly reversed. In technical terms, a reversible computation performs a bijective transformation of its local configuration space. When reversible computing is maintained at the highest levels of computation, the computer architectures, programming languages, and algorithms provide opportunities for interesting applications such as bidirectional debuggers, rollback mechanisms for speculative executions in parallel and distributed systems, and error and intrusion detection techniques. DNA computation emerged about 25 years ago as an exciting new research field at the intersection of computer science, biology, engineering, and mathematics. Although anticipated by Feynman as long ago as the 1950s, the notion of performing computations at a molecular level was only realized in 1994, by Adleman's seminal work on computing with DNA. Since then the field has blossomed rapidly, with significant theoretical and experimental results being reported regularly.

Reversible and DNA Computing has three parts with 31 chapters, covering reversible circuits, fault-tolerant reversible circuits, and DNA computing. This book focuses on state-of-the-art research on reversible computing, reversible fault tolerance, and DNA-based reversible circuits with their intended application.

In the first part of the book reversible circuits are illustrated and explained in a way that readers will understand, from the basics of reversible circuits to their applications in different types of arithmetic and logical units. In reversible computing, fault tolerance is an important part for the robust operation. Therefore, the second part of the book is designed with the fundamental concepts of fault tolerance and its application in the reversible computing. Various arithmetic and logical circuits are designed with fault-tolerant support, and it will give the confidence of designing the new reversible circuits for quantum computing. In addition, reversible and DNA computing are the new face of research for information processing and operation. The third part of the book consists of the most recent DNA applications in the circuit level, which supports the reversible computing. As a whole, from the reversible and DNA computing book a core researcher, academician, and student will get the guidelines of reversible and DNA circuits and its applications.

Dhaka, Bangladesh
January, 2020

Acknowledgments

I would like to express my sincere gratitude and special appreciation to the researchers and my beloved students who are working in the field of reversible and DNA computing. The contents of this book have been compiled from a wide variety of research works, which are listed at the end of this book.

I am grateful to my parents and family members for their endless support. Most of all, I want to thank my wife Mrs. Sitara Roshan, daughter Ms. Fariha Tasnim, and son Md. Tahsin Hasan for their invaluable cooperation in completing this book.

Finally, I am also thankful to all of those who have provided their support and important time to finish this book.

Acronyms

ALU	arithmetic logic unit
BCD	binary coded decimal
BJS	Babu-Jamal-Saleheen
CAD	computer-aided design
CSA	carry skip adder
CLA	carry look-ahead adder
CPU	central processing unit
DAG	directed acyclic graph
DFS	depth first search
DSP	digital signal processing
DNA	deoxyribonucleic acid
DRAG	DNA-based reversible AND gate
DRNG	DNA-based reversible NOT gate
DROG	DNA-based reversible OR gate
DXRG	DNA-based reversible Ex-OR gate
DRFA	DNA-based reversible full adder
DSM	deep sub micron
ESOP	exclusive sum-of-products
FPGA	field programmable gate array
F2G	Feynman double gate
FG	Feynman gate
FRG	Fredkin gate
FPU	floating-point unit
FPGA	field programmable gate array
FTFA	fault-tolerant full adder
HLN	Hasan-Lafifa-Nazir
HPP	Hamiltonian path problem
IR	instruction register
MIG	modified Islam gate
MOA	multi-operand addition
NFT	new fault tolerant
PCR	polymerase chain reaction
PFA	partial full adder

PG	Peres gate
PAL	programmable array logic
PLA	programmable logic array
PIPO	parallel-in parallel-out
PISO	parallel-in serial-out
PPG	partial product generation
PROM	programmable read-only memory
QCA	quantum-dot cellular automata
LDPC	low-density parity-check
RAM	random access memory
RFD	reversible fault-tolerant decoder
RFTPLA	reversible fault-tolerant programmable logic array
RG	reversible gate
SOP	sum-of-products
SISO	serial-in serial-out
SIPO	serial-in parallel-out
SNFA	single NFT full adder
TB	testable block
TG	Toffoli gate
UPPG	universal parity-preserving gate

Introduction

The limit of energy dissipation during computation is fundamentally based on the apparent thermodynamic paradox of Maxwell's demon. Maxwell described the system as follows:

> For we have seen that the molecules in a vessel full of air at uniform temperature are moving with velocities by no means uniform, though the mean velocity of any great number of them, arbitrarily selected, is almost exactly uniform. Now let us suppose that such a vessel is divided into two portions, A and B, by a division in which there is a small hole, and that a being, who can see the individual molecules, opens and closes this hole, so as to allow only the swifter molecules to pass from A to B, and only the slower ones to pass from B to A. He will thus, without expenditure of work, raise the temperature of B and lower than that of A, in contradiction to the second law of thermodynamics.

The demon has been depicted in various ways. Some show the demon inside the chamber with gas, some have it outside. Any analysis must be sure to include the thermodynamic effects within the demon itself in the energy and entropy accounting. Some images give the demon a light source to aid in measurement of the particles speed, indicating the lack taken by some authors to explain the paradox of attributing the entropy increase to dissipation during measurement.

Szilard, nearly 60 years after Maxwell first postulated the demon, attempted to resolve the paradox by arguing that the process of measurement required dissipation. Although he did notice entropy generation of $k\ln 2$ when the demon was reset. But it was not until much later that researchers firmly placed the source of dissipation in the erasure of information. When the demon measures a particle, he must set a bit indicating the speed of the particle. The hole between the portions of the vessel is controlled by the state of this bit. Once a particle has been directed to the correct portion, the demon must reset the bit in preparation for the next measurement value. This resetting is the logically irreversible event that saves the second law. Measurement may be performed reversibly; information destruction, rather than information acquisition, has a thermodynamic cost. In any irreversible process, entropy must increase. The required entropy increase during irreversible bit erasure is a function of the process by which it is done, the time taken for erasure, and the temperature of the system, but the increase must be at least zero. However, the required energy dissipation must be at least $kTln2$, where k is the Boltzmann constant and T is the operating temperature.

This book is divided into three parts, namely (i) Reversible Circuits, (ii) Fault-Tolerant and Online Testable Circuits, and (iii) DNA Computing. The first part starts with some backgrounds and preliminary studies about reversible logic synthesis and some popular reversible gates. Then many of the reversible gates are included in different chapters. Some approaches of designing different reversible adders and subtractor circuits, signed multiplier, sequential division circuit, low power n-bit binary comparator, reversible latches and flip-flops, n-bit synchronous and asynchronous counter, barrel shifter, shift register, field programmable gate array (FPGA), programmable logic array (PLA), random access memory (RAM), programmable read-only memory (PROM), the arithmetic logic unit (ALU) implementation and finally, the control unit is presented.

The second part is divided into two types of contents; namely reversible fault-tolerant circuits and reversible online testable circuits. In fault-tolerant part, some backgrounds and preliminary studies about reversible fault-tolerant logic gates are included in different chapters. Some approaches of designing different reversible fault-tolerant adders, multiplier circuit, floating-point division circuit, decoder, shifter and rotator techniques, programmable logic devices (programmable logic array, programmable array logic and field programmable gate array) and finally, arithmetic logic unit (ALU) with its QCA implementation are described. In addition, online testable part describes the realization of reversible circuits using NAND blocks and designs of some reversible online testable circuits.

The third part is also divided into two parts, namely general or irreversible DNA and reversible DNA. In irreversible DNA, some backgrounds and preliminary studies about DNA and a DNA-based approach of microprocessor design automation are shown. In reversible DNA, some DNA-based reversible gates as well as reversible circuits, reversible addition and subtraction mechanism, reversible shifting and multiplication techniques and finally reversible multiplexer and ALU are presented.

Part I

Reversible Circuits

An Overview About Reversible Circuits

The number of output bits is relatively small compared to the number of input bits in most computing tasks. For example, in a decision problem, the output is only one bit (yes or no) and the input can be as large as desired. However, computational tasks in communication, computer graphics, digital signal processing, and cryptography require that all the information encoded in the input should be preserved in the output. One might expect to get further speed-ups by adding instructions to allow computation of an arbitrary reversible function. The problem of chaining such instructions together provides one motivation for studying reversible computation and reversible logic circuits, that is, logic circuits comprising of gates computing reversible functions.

Reversible logic is an emerging research area. Interest in reversible logic is sparked by its applications in several technologies, such as quantum, CMOS, optical and nanotechnology. Reversible implementations are also found in thermodynamics and adiabatic CMOS. Power dissipation in modern technologies is an important issue, and overheating is a serious concern for both manufacturer (impossibility of introducing new, smaller scale technologies, limited temperature range for operating the product) and customer (power supply, which is especially important for mobile systems). One of the main benefits that reversible logic is theoretically zero power dissipation in the sense that, independently of underlying technology, irreversibility means heat generation.

Reversible circuits are also interesting because the loss of information associated with irreversibility implies energy loss. Some reversible circuits can be made asymptotically energy-lossless as their delay is allowed to grow arbitrarily large. Currently, energy losses due to irreversibility are dwarfed by the overall power dissipation, but this may change if power dissipation improves. In particular, reversibility is important for nanotechnologies where switching devices with gain are difficult to build.

Reversible and DNA Computing, First Edition. Hafiz Md. Hasan Babu.
© 2021 John Wiley & Sons Ltd. Published 2021 by John Wiley & Sons Ltd.

The advancement in higher-level integration and fabrication process has emerged in better logic circuits and energy loss has also been dramatically reduced over the last decades. This trend of reduction of heat in computation also has its physical limit. It is well understandable that in logic computation every bit of information loss generates $kTln2$ joules of heat energy where k is Boltzmann's constant of $1.38 \times 10^{-23} J/K$ and T is the absolute temperature of the environment. At room temperature, the dissipating heat is around $2.9 \times 10^{-21} J$. Energy loss limit is also important as it is likely that the growth of heat generation causing information loss will be noticeable in future.

Reversible circuits are fundamentally different from traditional irreversible ones. In reversible logic, no information is lost, i.e., the circuit that does not lose information is reversible. Zero energy dissipation would be possible if the network consists of reversible gates only. Thus, reversibility will be an essential property for the future circuit design. Quantum computation is also gaining popularity as some exponentially hard problems can be solved in polynomial time. It is known that quantum computation is reversible. Thus, research in reversible logic is helpful for the development of future technologies; it has the potential to methods of quantum circuit construction resulting in more powerful computers. Quantum technology is not the only one where reversibility is used.

Reversible logic has also found its applications in several other disciplines such as nanotechnology, DNA technology, and optical computing. In computers, numbers are stored in straight binary format. Due to inherent characteristics of floating-point numbers and limitations on storing formats, not all floating-point numbers can be represented with desired precision. So, the computing in decimal format is gaining popularity because precision can be avoided in this format.

This part starts with some backgrounds and preliminary studies about reversible logic synthesis and some popular reversible gates that are given in Chapter 1. Many of the reversible gates are included in different chapters. Some approaches of designing different reversible adders (full adder, carry skip adder, carry look-ahead adder, and ripple carry adder) and subtractor circuits are given in Chapter 2. Chapter 3 presents the design of a reversible signed multiplier which is based on Booth's recoding. A design of sequential division circuit using reversible logic is shown in Chapter 4. The hardware has its application in the design of a reversible arithmetic logic unit. The design of a reversible low power n-bit binary comparator is discussed in Chapter 5. Chapter 6 shows the designs of reversible latches and flip-flops are presented, which are being optimized in terms of quantum cost, delay, and garbage outputs. The contents of this chapter have optimized the reversible sequential circuit designs in terms of reversible gates and garbage outputs. The designs of reversible D latch and D flip-flop are also discussed with asynchronous set/reset capability. A reversible design for both n-bit synchronous and asynchronous counter is described in Chapter 7. This chapter also describes a synthesized design of the reversible counter that is optimized in terms of quantum cost, delay, and garbage outputs.

Chapter 8 is divided into two parts, namely, reversible barrel shifter and reversible shift register. In the first part of this chapter, the design methodology of a reversible barrel shifter is presented. In the second part of this chapter, reversible logic synthesis is carried out for SISO shift register.The key contribution of this part is the reversible realization of SIPO, PISO, PIPO and universal shift registers. Two reversible gates have been introduced in Chapter 9, named as R-I gate and R-II gate, for realizing reversible combinational logic circuits. These two gates can be used for realization of basic logical functions such as AND, XOR, and MUX. Chapter 10 is also divided into two parts, namely, reversible field programmable gate array (FPGA) and reversible programmable logic array (PLA). In the first part of this chapter, a design of reversible architecture of the logic element of Plessey FPGA is described, which results in significant power savings. In the second part of this chapter, a design of reversible PLA is described, which is able to realize multi-output ESOP (exclusive-OR sum-of-product) functions by using a 3×3 reversible gate, called MG (MUX gate).

Chapter 11 is also divided into two parts namely reversible random access memory (RAM) and reversible programmable read-only memory (PROM). In the first part of this chapter, the reversible logic synthesis of RAM is described with a 3×3 reversible gate named as FS. In the way of designing a reversible RAM, an $n \times 2n$ reversible decoder, reversible D flip-flop, and write-enabled master–slave D flip-flops are also designed. In the second part of this chapter, a reversible PROM design is described. A reversible decoder named ITS and another reversible gate TI are also introduced. In addition, for designing the reversible PROM, an AND-plane and an Ex-OR plane are also described. The design of programmable reversible logic gate structures is presented in Chapter 12, which implements ALU and presents its different uses. Finally, a reversible control unit is presented in Chapter 13. Two 4×4 reversible gates, namely HL gate and BJ gate, are introduced to design reversible decoder and JK flip-flop.

1

Reversible Logic Synthesis

Reversible logic plays a vital role at present time and it has different areas for its applications, namely low-power CMOS, quantum computing, nanotechnology, cryptography, optical computing, DNA computing, digital signal processing (DSP), quantum-dot cellular automata (QCA), digital communications, and computer graphics. It is not possible to realize quantum computing without implementation of reversible logic. The main purposes of designing reversible logic circuits are to decrease quantum cost, depth of the circuits, and the number of garbage outputs. This chapter explains the basic reversible logic gates for more complex system, which may have reversible circuits as a primitive component and can execute complicated operations using quantum computers. The reversible circuits form the basic building block of quantum computers, as all quantum operations are reversible. This chapter presents the information related to the primitive reversible gates and helps researchers in designing higher complex computing circuits using reversible gates.

1.1   Reversible Logic

In this section, basic definitions and ideas related to reversible logic are presented. Formal definitions of reversible gate, garbage output, and the popular reversible gates, along with their input–output vectors, are presented here. Illustrative figures and examples are also included in respective discussions.

1.2   Reversible Function

The multiple-output Boolean function $F(x_1, x_2, ..., x_n)$ of n Boolean variables is called reversible if:

1. The number of outputs is equal to the number of inputs.
2. Any output pattern has a unique pre-image.

In other words, the functions that perform permutations of the set of input vectors are referred to as reversible functions.

Reversible and DNA Computing, First Edition. Hafiz Md. Hasan Babu.
© 2021 John Wiley & Sons Ltd. Published 2021 by John Wiley & Sons Ltd.

1.3 Reversible Logic Gate

Reversible logic has unique mapping between input and output bit pattern. A unit logic entity is represented as a gate. The gates or circuits that do not lose information are called reversible gates or circuits.

Property 1.3.1 A reversible circuit is a circuit in which the number of input and the number of output is equal and there is one-to-one mapping between input and output vectors.

Let us consider the gate shown in Figure 1.1. According to the definition, the gate is a reversible gate, because it has k number of inputs and k number of outputs and the gate is known as $k \times k$ reversible gate. Without the NOT gate, classical logic gates are called irreversible, since they cannot determine the input vector states from the output vector states uniquely.

Example 1.1 There can be any number of dimensions for a reversible gate, but lower dimension is always preferable for designing efficient circuits. Popular reversible gates, Feynman gate (FG), Toffoli gate (TG), Peres gate (PG), Fredkin gate (FRG), Feynman double gate (F2G), and new fault-tolerant gate (NFTG), are shown in Figure 1.2.

1.4 Garbage Outputs

The output (outputs) of a reversible gate that is (are) not used as input to other gate or the output (outputs) that is (are) not treated as a primary output is (are) called garbage output

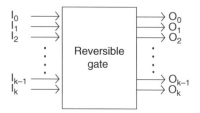

Figure 1.1 A $k \times k$ reversible gate.

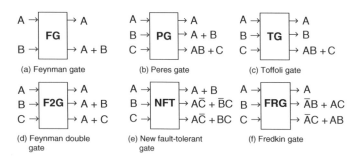

(a) Feynman gate

(b) Peres gate

(c) Toffoli gate

(d) Feynman double gate

(e) New fault-tolerant gate

(f) Fredkin gate

Figure 1.2 Popular reversible gates.

Figure 1.3 Reversible Feynman gate.

(outputs). The unutilized outputs from a gate are called garbage outputs. A heavy price is paid for every garbage output. So, for any circuit design, the fewer the garbage outputs, the better.

Example 1.2 When a Feynman gate (FG) is used for Ex-OR (exclusive-OR, $\oplus$) operation of two inputs, an extra output is generated at the output part of the FG in addition to the Ex-OR output. This additional output is known as garbage output. In Figure 1.3, the garbage output of a gate is shown. Here, A is the garbage output.

1.5 Constant Inputs

Constant inputs are the inputs of a reversible gate (or circuit) that are either set to 0 or 1.

Example 1.3 If the complement of the input A from Figure 1.3 is needed, then B is set to 1 and $Q = A$.

1.6 Quantum Cost

The quantum cost of a circuit is the total number of 2×2 quantum primitives that are used to realize corresponding quantum circuit. Basically, the quantum primitives are matrix operations, which are applied on qubits state.

Example 1.4 The quantum realization of reversible Fredkin (FRG) gate is shown in Figure 1.4. Each quantum Ex-OR gate and quantum V or V^+ gate requires 1 (one) quantum cost. The reversible FRG gate has four quantum Ex-OR gates, two quantum V gates, and one quantum V^+ gate. So, the quantum cost of reversible FRG gate seems 7 (seven). But, we know if a quantum Ex-OR gate and a quantum V or V^+ gate exist angularly (denoted by angular box), then the quantum cost is treated as 1. From the figure, we see that there exists two angular boxes, and each angular box is treated as 1 quantum cost. As a result, the total quantum cost of reversible FRG gate is 5 (five).

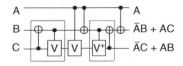

Figure 1.4 Quantum realization of reversible Fredkin (FRG) gate.

Example 1.5   The cost of all 2×2 gates is the same, and it is 1. For 1×1 gate, the cost is 0. Every circuit can be constructed from those 1×1 and 2×2 quantum primitives, and the cost of circuit is the total sum of required 2×2 gates.

1.7 Delay

The delay of a logic circuit is the maximum number of gates in a path from any input line to any output line. The definition is based on two assumptions: (i) Each gate performs computation in one unit time and (ii) all inputs to the circuit are available before the computation begins.

Example 1.6   The delay of each 1×1 and 2×2 reversible gate is taken as unit delay 1. Any 3×3 reversible gate can be designed from 1×1 reversible gates and 2×2 reversible gates, such as CNOT gate, controlled-V, and controlled-V^+ gates (V is a square root of NOT gate and V^+ is its hermitian). Thus, the delay of a 3×3 reversible gate can be computed by calculating its logical depth when it is designed from smaller 1×1 and 2×2 reversible gates.

1.8 Power

Power of a gate is defined by the energy. Energy of a basic quantum gate is 142.3 meV. Quantum circuits can be implemented with the basic quantum gates and the number of quantum gates depends on the number of basic quantum gates needed to realize it. That means the total number of required quantum gates in the quantum representation of a reversible quantum circuit or gate. So, the power of a reversible gate can be defined as follows: *Power = Number of quantum gates × Energy of a basic quantum gate*

Example 1.7   Figure 1.5 shows the quantum realization of the reversible HNG gate. From this figures, it is seen that the quantum realization of reversible HNG gate requires total six quantum gates. So, the power of the reversible HNG gate is (6×142.3) meV $= 853.8$ meV, where the number of quantum gates of HNG circuit is 6.

1.9 Area

The area of a reversible gate is defined by the feature size. This size varies according to the number of quantum gates. The size of the basic quantum gates ranges from 50–300 Å. The Angstrom (Å) is a unit equal to 10^{-10} m (one ten-billionth of a meter), or 0.1 nm. Its symbol is the Swedish letter Å. So, the area of a reversible gate can be defined as follows:
    Area = Number of quantum gates × Size of a basic quantum gate

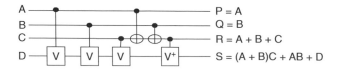

Figure 1.5   The quantum representation of reversible HNG gate.

Example 1.8   Figure 1.5 shows the quantum realization of the reversible HNG gate. From this figures, it is seen that the quantum realization of reversible HNG gate requires total six quantum gates. So, the area of the reversible HNG gate is $((50 \times 6)\text{ Å} - (300 \times 6)\text{Å}) = (300 \text{ Å} - 1800 \text{ Å})$, where the number of quantum gates of HNG circuit is 6.

1.10   Hardware Complexity

The hardware complexity of a reversible logic circuit specifies the total number of Ex-OR operations, NOT operations, and AND operations used in the circuit. Consequently, the hardware complexity can be determined using the following equation:

$$T = \alpha + \beta + \delta \tag{1.10.1}$$

where

T = Hardware complexity (total logical operations)
α = A two input EX-OR gate logical operation
β = A two input AND gate logical operation
δ = A NOT gate logical operation

Example 1.9   Figure 1.6 shows the block diagram of a reversible Fredkin (FRG) gate. The figure describes that there is only one NOT operation, two EX-OR operations, and four AND operations. So, the hardware complexity of the reversible FRG gate is $T = 2\alpha + 4\beta + 1\delta$.

1.11   Quantum Gate Calculation Complexity

The quantum gate calculation complexity of the quantum representation of a reversible circuit specifies the total number of quantum gates (NOT gates, CNOT gates, and controlled-V (controlled-V^+) gates) used in the quantum representation of a reversible circuit. Consequently, the quantum gate calculation complexity can be determined using the following equation:

$$Q = \rho + \sigma + \Omega \tag{1.11.1}$$

where

Q = Quantum gate calculation complexity
ρ = A quantum NOT gate
σ = A quantum CNOT gate
Ω = A quantum controlled-V (controlled-V^+) gate

$$
\begin{aligned}
A &\rightarrow && \rightarrow P = A \\
B &\rightarrow \boxed{\textbf{FRG}} && \rightarrow Q = \bar{A}B + AC \\
C &\rightarrow && \rightarrow R = \bar{A}C + AB
\end{aligned}
$$

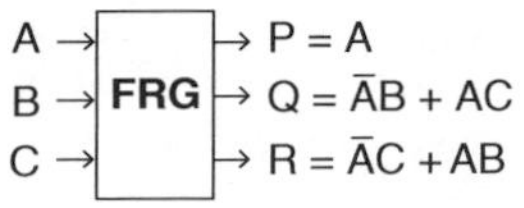

Figure 1.6   Block diagram of the reversible FRG gate.

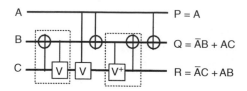

Figure 1.7 Quantum representation of a reversible FRG gate.

Example 1.10 Figure 1.7 shows the quantum representation of a reversible Fredkin (FRG) gate. The figure describes that there is only one NOT operation, four quantum CNOT operations, and three quantum controlled-V (controlled-V^+) operations. So, the quantum gate calculation complexity of the reversible FRG gate is $Q = 1\rho + 4\sigma + 3\Omega$.

1.12 Fan-Out

Fan-out is a term that defines the maximum number of inputs in which the output of a single logic gate can be fed. The fan-out of any reversible circuit is 1.

Example 1.11 The fan-out of any reversible circuit is 1.

1.13 Self-Reversible

A gate is said to be self-reversible if its dual combination is the same as itself.

Example 1.12 In Figure 1.8, there are two Toffoli gates that are in the cascading form. If the outputs of the first Toffoli gate are fed to the input of the second Toffoli gate, then the output of the second Toffoli gate is equal to the input of the first Toffoli gate. Here the outputs of first gate are P, Q, and R, where $P = A$, $Q = B$, and $R = AB \oplus C$. Then the outputs of second gate are X, V, and Z, where $X = A$, $Y = B$, and $Z = AB \oplus AB \oplus C = 0 \oplus C = C$.

1.14 Reversible Computation

In a reversible circuit, correct output is found by applying correct input instance and controlling one or more inputs if needed. Feynman gate (FG) is already presented to illustrate

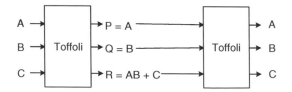

Figure 1.8 Toffoli gates as self-reversible.

the idea of garbage output, Feynman gate is 2 × 2 reversible gate where inputs are A, B, and corresponding functions are $P = A$, $Q = A \oplus B$. The Feynman gate is used here to show how to control input to produce expected output. Both the inputs A and B are used as control inputs, and their impact on output is shown below.

A as control input:

- For $A = 0$, output $P = 0$, and $Q = B$,
- For $A = 1$, output $P = 1$, and $Q = B'$.

B as control input:

- For $B = 0$, output $P = A$, and $Q = A$.
- For $B = 1$, output $P = A$, and $Q = A'$.

It is better to note that when B is used as control input and $B = 0$, both the outputs $P = B$ and $Q = A$. By controlling B, the copies of A can be created. This circuit can be easily used as a copying circuit.

1.15 Area

The area of a logic circuit is the summation of individual areas of each gate of the circuit. Suppose a reversible circuit consists of n reversible gates. Area of those n gates are $a_1, a_2, \cdots, a_n$. Then by using above definition area, denoted by A, of that circuit is

$$A = \sum_{i=1}^{n} a_i$$

The above definition for the area of a circuit can be calculated easily by obtaining area of each individual gate using CMOS 45 nm Open Cell Library and Synopsis Design Compiler.

Area of a gate can also be defined by the feature size. This size varies according to the number of quantum gates. As the basic quantum gates are fabricated with quantum dots with the size ranges from several to tens of nanometers (10^{-9} m) in diameter, the size of the basic quantum gates ranges from 50–300 Å. Quantum circuits can be implemented with the basic quantum gates and the number of quantum gates depends on the number of basic quantum gates needed to implement it. So, the area of a gate can be defined as follows: Area = Number of quantum gates × Size of basic quantum gates.

1.16 Design Constraints for Reversible Logic Circuits

The following are the important design constraints for reversible logic circuits:

- Reversible logic gates do not allow fan-outs.
- The reversible logic circuits should have minimum number of reversible gates.
- Reversible logic circuits should have minimum quantum cost.
- The design can be optimized so as to produce minimum number of garbage outputs.
- The reversible logic circuits must use minimum number of constant inputs.

- The reversible logic circuits must use a minimum logic depth or gate levels.
- Reversible logic circuits should have minimum area and power.
- The reversible logic circuits must use minimum hardware complexity and minimum quantum gate calculation complexity.

1.17 Quantum Analysis of Different Reversible Logic Gates

Calculating quantum cost of reversible circuit is always an interesting one. Quantum circuits, DNA technologies, nano-technologies and optical computing are the most common applications of quantum theory. Every reversible gate can be calculated in terms of quantum cost and hence the reversible circuits can be measured in terms of quantum cost. Reducing the quantum cost from reversible circuit is always a challenging issue and research are still going on in this area. In this section, the quantum equivalent diagram of some popular reversible gate is presented.

Property 1.17.1 The quantum cost of every 2×2 gate is the same. It can be easily assumed that 1×1 gate cost nothing, since it can always be included to arbitrary 2×2 gate that precedes or follows it. Thus, in first approximation, every permutation quantum gate will be built from 1×1 and 2×2 quantum primitives and its cost is calculated as a total sum of 2×2 gates used. All gates of the form 2×2 has equal quantum cost, and the cost is unity.

1.17.1 Reversible NOT Gate (Feynman Gate)

Example 1.13 A 2×2 Feynman gate is also called CNOT. This gate is one through because it passes one of its inputs. Every linear reversible function can be built by using only 2×2 Feynman gates and inverters. Since this is a 2×2 gate, the quantum cost is 1. Quantum equivalent circuit of the Feynman gate is shown in Figure 1.9.

1.17.2 Toffoli Gate

Figure 1.10 shows the equivalent quantum realization of three input Toffoli gate. The cost of the Toffoli gate is five 2×2 gates, or simply 5. In Figure 1.10, V is a square-root of NOT gate and V^+ is its hermitian. Thus, VV^+ creates a unitary matrix of NOT gate and $VV^+ = I$ (an identity matrix, describing just a quantum wire).

Figure 1.9 Quantum cost calculation of Feynman gate.

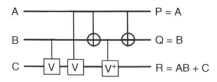

Figure 1.10 Quantum circuit of Toffoli gate.

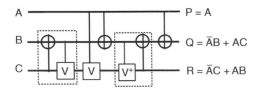

Figure 1.11 Quantum circuit of Fredkin gate.

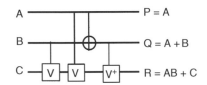

Figure 1.12 Quantum circuit of a Peres gate.

1.17.3 Fredkin Gate

The Fredkin gate costs the same as the Toffoli gate. The Toffoli gate includes a single Davio gate, while the Fredkin gate includes two multiplexers. The quantum equivalent Toffoli gate is shown in Figure 1.10. Each dotted rectangles in Figure 1.11 is equivalent to a 2×2 Feynman gate and so the cost is 1 for the particular case.

1.17.4 Peres Gate

This gate can be realized with cost 4. It is just like a Toffoli gate but without the last Feynman gate from right. This is the cheapest realization of a complete (universal) 3×3 permutation gate. Figure 1.12 shows the quantum realization of a Peres gate.

1.18 Summary

Maxwell's demon and Szilard's analysis of the demon at first suggested the connection between a single degree of freedom (one bit) and a minimum quantity of entropy. In the 1950s, this connection had been popularly interpreted to mean that computation must dissipate a corresponding minimum amount of energy during every elemental act of computation. Landauer later recognized that energy dissipation is only unavoidable when information is destroyed. Bennett and Toffoli first realized that a reversible computation, in which no information is destroyed, may dissipate arbitrarily small amounts of energy. The reversible circuits form the basic building block of quantum computers. This chapter presents some reversible gates. This chapter will help researchers/designers in designing higher complex computing circuits using reversible gates. It can further be extended toward the digital design development using reversible logic circuits, which are helpful in quantum computing, low-power CMOS, nanotechnology, cryptography, optical computing, DNA computing, digital signal processing (DSP), quantum dot cellular automata, communication, and computer graphics.

2

Reversible Adder and Subtractor Circuits

In computers, numbers are stored in straight binary format. Due to inherent characteristics of floating-point numbers and limitations on storing formats, not all floating-point numbers can be represented with desired precision. So, computing in decimal format is gaining popularity because the precision can be avoided in this format. However, hardware support for binary arithmetic allows it to be performed faster than decimal arithmetic. Faster hardware for decimal floating-point arithmetic is also imminent, as it has its importance in financial and Internet-based applications. So, faster circuits for binary coded decimal (BCD) numbers have great impact, as it is likely to be incorporated in more complex circuits like future mathematical processors.

2.1 Reversible Multi-Operand n-Digit Decimal Adder

Reversible multi-operand n-digit is capable of adding as many operands as possible, and there is no boundary for number of digit in each operand. Design of such circuit should be scalable and compact. In this circuit number of carry of each stage depends on the number of operands and equal to $\lceil log(m) \rceil$, where m is the number of operands. The basic mechanism of multi-operand n-digit decimal adder is shown in Figure 2.1.

The design of reversible multi-operand n-digit decimal adder is composed of three components: CSA (carry skip adder), CLA (carry look-ahead adder) and carry generator unit, adjust circuit and final CLA unit. Figure 2.2 shows the architecture of m-operand n-digit decimal adder. Figure 2.3 shows single i^{th} digit block diagram. The carry generator generates carry for i^{th} digit i.g., C_i. Then this C_i, C_{i-1} and sum, S_i, is used to generate adjust digit, Adj_i, which is used to adjust BCD digit. Then S_i is added with Adj_i to get adjust BCD digit. Algorithm 2.1.1 describes the whole process.

To implement this design, it requires full adder, CSA, and CLA. In the following subsections, designs of full adder, CSA, and CLA are described. An example of 3-operand n-digit decimal adder is also shown at the end of this section.

2.1.1 Full Adder

Full adder is a versatile and widely used building block in digital arithmetic processing. A full adder is a combinational circuit that performs the arithmetic sum of three input bits.

Reversible and DNA Computing, First Edition. Hafiz Md. Hasan Babu.
© 2021 John Wiley & Sons Ltd. Published 2021 by John Wiley & Sons Ltd.

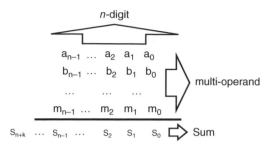

Figure 2.1 Multi-operand n-digit decimal adder.

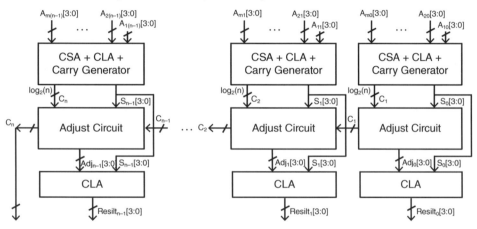

Figure 2.2 The architecture of reversible multi-operand n-digit decimal adder.

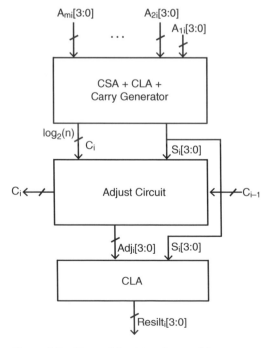

Figure 2.3 The architecture of reversible single-digit block of m-operand n-digit adder.

Algorithm 2.1.1 Reversible m-Operand n–Digit Decimal Adder, when $n \geq 1$

Take n-digit m numbers $A_m, ..., A_2, A_1$
Input: $A_m, ..., A_2, A_1$ *(all are n-digit numbers)*
Output: *Sum, Carry($log_2(n)$-bit)*

1: Begin
2: **for** $n=0$ to $n-1$ **do**
3:   Get one 3-operand BCD block, Blk
4:   Generate $S_i[3:0]$
5:   Get $C_{i+1} = generateCarry()$
6:   $Adj_i[3:0] = generateAdjustBits(S_i, C_i, C_{i+1})$;
7:   Set, $Result_i = sum(S_i[3:0], Adj_i[3:0])$
8: **end for**
9: set, $Sum = Result$
10: set, $Carry = C_n$
11: End

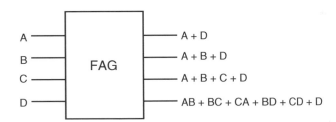

Figure 2.4 4 × 4 reversible FAG gate.

It consists of three inputs and two outputs. Three of the input variables can be defined as A, B, C_{in} and the two output variables can be defined as S for sum, C_{out} for carry.

$$S = A \oplus B \oplus C_{in}$$
$$C_{out} = AB \oplus BC_{in} \oplus C_{in}A$$

The FAG Gate
In this part, a 4 × 4 reversible gate, namely FAG gate, is described. The input vector, I_v, and output vector, O_v, of the gate are as follows:

$$I_v = a, b, c, d; \text{ and}$$
$$O_v = a \oplus d, a \oplus b \oplus d, a \oplus b \oplus c \oplus d, ab \oplus bc \oplus ca \oplus bd \oplus cd \oplus d$$

Figure 2.4 shows the diagram of the 4 × 4 reversible FAG gate and Figure 2.5 shows its quantum representation. The quantum cost of FAG gate is 6. The corresponding truth table of the gate is shown in Table 2.1. It can be verified from the truth table that the input pattern corresponding to a particular output pattern can be uniquely determined.

In this chapter, FAG gate is used to design a full adder circuit. Quantum realization of the FAG gate is shown in Figure 2.6. Here, the dotted rectangle is equivalent to a 2 × 2 CNOT gate. So, the quantum cost of the FAG gate is six.

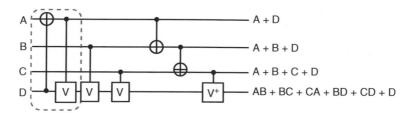

Figure 2.5 Reversibility of 4 × 4 reversible FAG gate.

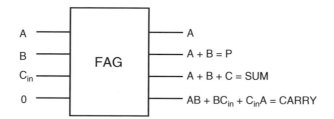

Figure 2.6 Reversible FAG gate as full adder.

Table 2.1 Reversibility of 4 × 4 Reversible FAG Gate.

INPUT				OUTPUT			
A	B	C	D	P	Q	R	S
0	0	0	0	0	0	0	0
0	0	0	1	1	1	1	1
0	0	1	0	0	0	1	0
0	0	1	1	1	1	0	0
0	1	0	0	0	1	1	0
0	1	0	1	1	0	0	0
0	1	1	0	0	1	0	1
0	1	1	1	1	0	1	0
1	0	0	0	1	1	1	0
1	0	0	1	0	0	0	1
1	0	1	0	1	1	0	1
1	0	1	1	0	0	1	1
1	1	0	0	1	0	0	1
1	1	0	1	0	1	1	1
1	1	1	0	1	0	1	1
1	1	1	1	0	1	0	0

Algorithm 2.1.1.2 Design of a Full Adder Circuit

Input: A, B, C_{in}
Output: *Sum, Carry*

1: Begin
2: Take a FAG gate, (a, b, c, d)⇔(p, q, r, s)
3: Set, FAG.a=A; FAG.b=B;
4: FAG.c=C_{in}; FAG.d=0;
5: Output, *Sum* = FAG.r and
6: *Carry* = FAG.s
7: End

Reversible FAG Gate as Full Adder

The FAG gate is designed in such a way that it can be used as full adder in efficient way. Using fourth input as constant input zero (0), propagate, sum, and carry can be generated as shown in Figure 2.4. Algorithm 2.1.1.2 shows the design procedure of the full adder.

Property 2.1.1.1   A full adder can be realized with at least six quantum cost.

Proof: Required output, sum, and carry can be generated by the following equations:

$$S = A \oplus B \oplus C_{in}$$
$$C_{out} = AB \oplus BC_{in} \oplus C_{in}A = AB \oplus C_{in}(B \oplus A)$$

In reversible logic, each logical operation requires at least one quantum gate. From the above equations, there are six logical operations in the full adder circuit; it requires at least six quantum gates. So for each operation if the minimum single unit cost is required, then it needs the minimum required cost which is six.

Example 2.1.1.1   In Figure 2.6, design of a full adder is shown. It has six logical operations.

2.1.2 Carry Skip Adder

The carry skip adder reduces the delay due to the carry computation. In this adder, when the input is a logical one, the cell will propagate the carry input to the carry output. Therefore, the ith full adder carry input, C_i, will propagate to its carry output, C_{i+1}, when $P_i = x_i \oplus y_i$. Furthermore, multiple full adders, called a block, can generate a "block" propagate signal to detour the incoming carry around to the block's carry output signal. Figure 2.8 shows the block diagram of an n-bit CSA. Each block is a small ripple carry adder producing the block's sum and carry bits. However, each block quickly calculates whether the block's carry input is propagated to its carry output.

The block carry input C_{in} is propagated as the block carry output C_{out}, if the block propagate P is one. The block propagate signal is generated with an AND gate. Figure 2.7 shows the carry skip compatible full adder constructed with FAG. The worse case delay of the CSA happens when the carry is generated in the very first full adder stages in the first block, generates the carry without using all the intermediate blocks, and ripples through the full adder stages of the last block.

Algorithm 2.1.2.1.1 Design of *m*-bit Single Chunk Carry Skip Adder Circuit

---

Input: A, B; (both are *m*-bit binary number), C_{out}(*1*-bit)
Output: Sum (*m*-bit), Carry (1-bit)

1: Begin
2: **for** $n=0$ **to** *m-1* **do**
3:     Take one FAG gate, F_i
4:     Set $F_i.a = A_i$; $F_i.b = B_i$; $F_i.c = F_{i-1}.s$; $F_i.d = 0$;
5:     Output, $Sum_i=F_i.r$;
6:     Get, $Sum_i=F_i.r$;
7: **end for**
8: set, $Carry = (F_0.q$ AND $F_1.q$ AND ... AND $F_{m-1}.q)$ OR C_{out}
9: End

2.1.2.1 Design of Carry Skip Adder

The CSA propagates the block carry input to the next block if block group propagate signal P is one. Figure 2.7 shows the *n*-bit CSA circuit. Here the *m*-bit chunk is used for carry skip. Thus, $k = \lceil m \rceil$ blocks are required. Figure 2.8 shows block diagram of the *m*-bit CSA using Algorithm 2.1.2.1.1.

The 4-bit CSA is shown in Figure 2.9. Here 4-FAG is used to generate sum and carry. In this figure, the AND-OR gate combination is used in the bottom for carry skip part. Rectangle part of that circuit is called carry skip block circuit. Design of carry skip circuit using reversible gates is shown in Figure 2.10. In this design, four Peres gates is needed.

The AND-OR carry skip generates a carry out, i.e., $C_{out} = 1$. When the most significant carry C_4 of the full adder equals to one, the block propagates P equal to one, where the carry input C_{in} signal is equal to zero.

Property 2.1.2.1.1 A reversible *n*-bit comparator requires $455 \times n \ \mu m^2$ area, where *n* is the number of bits in each operand.

Proof: For *n*-digit CSA, *n* single digit blocks are needed. From Figure 2.9, it can be shown that for single bit CSA, four full adders and one carry skip block are needed. Area of one-digit carry skip adder is

$$A = 4A_{FA} + A_{CSB}$$

The areas of A_{FA} block and A_{CSB} block are 90 μm^2, and 9 5 μm^2, respectively. So, the total area of a single bit CSA can be modeled as below:

$$A = (4 \times 90) \ \mu m^2 + 95 \ \mu m^2 \ = 455 \ \mu m^2$$

So, the area for single digit CSA is 455 μm^2. Thus, the area of *n*-digit CSA, $A = 455 \times n \ \mu m^2$.

Property 2.1.2.1.2 A reversible *n*-bit comparator requires $455 \times n \ \mu m^2$ area, where *n* is the number of bits in each operand.

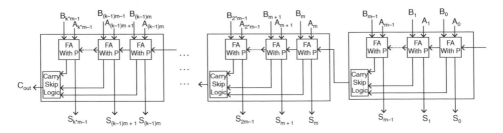

Figure 2.7 The *n*-bit carry skip adder circuit.

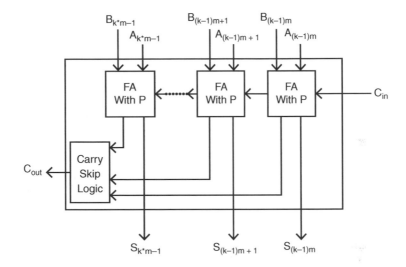

Figure 2.8 Single block carry skip adder circuit.

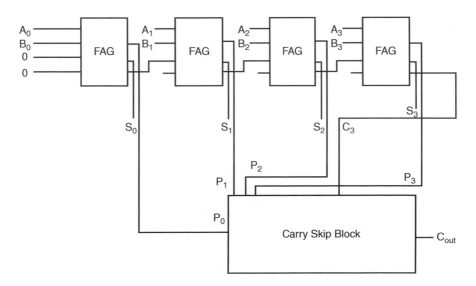

Figure 2.9 Reversible CSA for single digit.

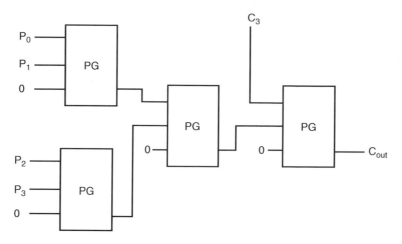

Figure 2.10 Reversible carry skip circuit.

Proof: For n-digit CSA, n single digit blocks are needed. From Figure 2.9, it can be shown that for single-bit CSA, four full adders and one carry skip block are needed. The area of one-digit CSA is

$$A = 4A_{FA} + A_{CSB}$$

The areas of A_{FA} block and A_{CSB} block are 90 μm² and 9 5 μm², respectively. So, the total area of single-bit CSA can be modeled as below:

$$A = (4 \times 90) \text{ μm}^2 + 95 \text{ μm}^2 = 455 \text{ μm}^2$$

So, the area for single-digit CSA is 455 μm². Thus, the area of n-digit CSA is $A = 455 \times n$ μm².

Property 2.1.2.1.3 A reversible n-bit comparator requires $10.6 \times n$ mW power, where n is the number of bits in each operand.

For n-digit CSA, n single digit blocks are needed. From Figure 2.9, it can be shown that for a single-bit CSA, four full adders and one carry skip block are needed. So the total required power of an n-digit CSA can be measured using following equation:

$$\text{Total power } P = n \cdot (4 \cdot P_{FA} + P_{CSB})$$

The powers obtained by full adder block and carry skip block are 3.4 mW and 7.2 mW, respectively. So, the power of a single-bit carry skip adder can be calculated as below:

$$P_{single-digit} = 3.4 + 7.2 = 10.6 \ mW$$

Thus, the total required power of n-digit carry skip adder can be modeled as below:

$$P = 10.6 \cdot nmW$$

Property 2.1.2.1.4 A reversible n-bit carry skip adder with m-bit in each block requires $\lceil n/m \rceil \cdot [\frac{1}{2^m} \cdot [T_P + T_{m-AND}] + (1 - \frac{1}{2^m}) \cdot [m \cdot T_{FA} + T_{m-AND}]]$ timing delay, where T_p be the delay of single propagate output, T_{FA} be the delay of full adder and T_{m-AND} be the delay of m AND operations.

Proof: Suppose, $T_{m-bit\ Block}$ be the delay of a single block of m-bit carry skip adder. Then, from the design, total timing delay can be modeled by

$$T_d = k \cdot T_{m-B}, \text{where } k = \lceil n/m \rceil$$

The delay of a single block depends on carry skip logic. If carry is skipped, the output C_{out} of Figure 2.9 is one, where the delay is calculated by the summation of the delay of single propagate output, denoted by T_P and the delay of m-bit AND operations, denoted by T_{m-AND}. Otherwise, delay is equal to delay of m-bit ripple carry adder, which is equal to $m \cdot T_{FA}$.

$$T_{m-B} = p(skip) \cdot [T_P + T_{m-AND}] + p(not-skip) \cdot [m \cdot T_{FA} + T_{m-AND}]$$

Here $P(skip)$ be the probability of skipping carry and $P(not\text{-}skip)$ be probability of not skipping carry. For, m-bit block $P(skip)$ is calculated by

$$P(skip) = \frac{2^m}{4^m} = \frac{1}{2^m}$$

And $P(not\text{-}skip)$ is calculated by

$$P(not-skip) = 1 - P(skip) = 1 - \frac{1}{2^m}$$

So, $T_{m-B} = \frac{1}{2^m} \cdot [T_P + T_{m-AND}] + (1 - \frac{1}{2^m}) \cdot [m \cdot T_{FA} + T_{m-AND}]$ Therefore, the total timing delay can be obtained by

$$T_d = k \cdot \frac{1}{2^m} \cdot [T_P + T_{m-AND}] + (1 - \frac{1}{2^m}) \cdot [m \cdot T_{FA} + T_{m-AND}]$$
$$= \lceil n/m \rceil \cdot [\frac{1}{2^m} \cdot [T_P + T_{m-AND}] + (1 - \frac{1}{2^m}) \cdot [m \cdot T_{FA} + T_{m-AND}]]$$

Therefore, a reversible n-bit CSA with m-bit block requires $\lceil n/m \rceil \cdot [\frac{1}{2^m} \cdot [T_P + T_{m-AND}] + (1 - \frac{1}{2^m}) \cdot [m \cdot T_{FA} + T_{m-AND}]]$ timing delay.

Property 2.1.2.1.5 A reversible n-bit CSA with m-bit in each block requires $(6 \cdot n + 4 \cdot m \cdot k)$ quantum cost.

Proof: For every bit of CSA, it requires a single FAG gate. So the quantum cost for n-bit CSA is $6 \cdot n$. For m-bit chunk, m AND operation and a single OR operation requires mPeres gate and total blocks is k. So, the required cost is $4 \cdot m \cdot k$.

Therefore, the total quantum cost for a reversible n-bit CSA with m-bit chunk requires $6 \cdot n + 4 \cdot m \cdot k$ quantum cost.

2.1.3 Carry Look-Ahead Adder

Carry look-ahead adders (CLAs) are the fastest of all adders and achieve speed through parallel carry computations. For each bit in a binary sequence to be added, the CLA logic determines whether that bit pair will generate a carry or propagate a carry. This allows the circuit to "pre-process" the two numbers being added to determine the carry ahead of time. Then, when the actual addition is performed, there is no delay from waiting for the ripple carry effect. Figure 2.12 shows the design of a carry look-ahead adder circuit using Algorithm 2.1.3.1.

The adder is based on the fact that a carry signal will be generated in two cases:

1. When both bits A_i and B_i are 1, or
2. When one of the two bits is 1 and carry-in is also 1.

Thus, it can be written as $C_{OUT} = C_{i+1} = A_i.B_i + (A_i \oplus B_i).C_i$.

The above expression can also be represented as:

$$C_{i+1} = G_i + P_i.C_i.$$

where $G_i = A_i.B_i$ and $P_i = A_i \oplus B_i, 0 \leq i < n$

$$C_i = \begin{cases} C_i; & i = 0 \\ G_{i-1} \sum_{j=0}^{i=2} \left(\prod_{k=j+1}^{i=1} p_k \right) G_j + \left(\prod_{k=j+1}^{i=1} p_k \right) c_{in}; & 1 \leq i \leq n-1 \\ C_{Out}; & i = n \end{cases}$$

Applying this to a 4-bit adder, we get

$$C_1 = G_0 + P_0.C_0$$
$$C_2 = G_1 + P_1 C_1$$
$$ = G_1 + P_1 G_0 + P_1 P_0 C_0$$
$$C_3 = G_2 + P_2 C_2$$
$$ = G_2 + P_2 G_1 + P_2 P_1 G_0 + P_2 P_1 P_0 C_0$$
$$C_4 = G_3 + P_3 C_3 = G_3 + P_3 G_2 + P_3 P_2 G_1 + P_3 P_2 P_1 G_0 + P_3 P_2 P_1 P_0 C_0$$

The sum of the signals can be calculated as follows:

$$S_i = A_i \oplus B_i \oplus C_i = P_i \oplus C_i$$

The CLA can be broken up into two modules:

1. Partial full adder (PFA): This generates G_i, P_i, C_i and S_i. Figure 2.8 shows PFA design using a Toffoli gate and a FAG gate. The quantum cost of this circuit is $5 + 6 = 11$.
2. Carry look-ahead logic (CLA): The CLA generates the carry-out bits. Figure 2.11 shows carry look-ahead logic.

Property 2.1.3.1 A reversible n-bit CLA requires $T_s + n \cdot T_{PG} + n \cdot T_c$ timing delay, where T_s be startup delay and T_{PG} be the delay for generating P_i and G_i.

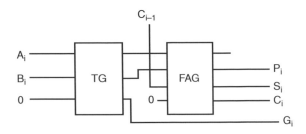

Figure 2.11 Reversible partial full adder.

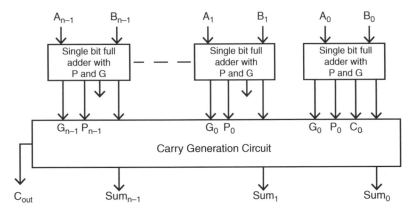

Figure 2.12 *n*-bit CLA adder circuit.

Algorithm 2.1.3.1 Design of *n*-bit Carry Look-Ahead Adder Circuit

---

Input: *A, B*; (both are *n*-bit binary numbers)
Output: *Sum(n*-bit), *Carry*(1-bit)

  1: Begin
  2: **for** i=0 **to** *n*-1 **do**
  3:     Take one FAG gate, F_i
  4:     Set $F_i.a = A_i$; $F_i.b = B_i$; $F_i.c = 0$; $F_i.d = 0$;
  5:     Output, Sum_i=$F_i.r$;
  6: **end for**
  7: **set**, *Carry* = **GenerateCarry** ($C_0, P_0, G_0, P_1, G_1, ..., P_{n-1}, G_{n-1}$) [Algorithm 2.1.3.2]
  8: End

Proof: The total timing delay is $T_d = T_s + n \cdot T_{PG} + T_{carry}$, where T_{carry} =Time for generating *carry*.

Again, $T_{carry} = n \cdot T_c$. Here, T_c is time to calculate a carry of ith bit, which needs the constant time, that is $O(1)$.

Thus, the total delay, $T_d = T_s + n \cdot T_{PG} + n \cdot T_c$. So, a reversible *n*-bit CLA requires $T_s + n \cdot T_{PG} + n \cdot T_c$ timing delay.

Algorithm 2.1.3.2 Carry Generation

Input: C_0, $P(P_0, P_1, ..., P_{n-1})$, $G(G_0, G_1, ..., G_{n-1})$;
Output: *Carry*(1-bit)

1: Begin
2: **for** $i{=}1$ **to** n **do**
3:    $C_i = G_{i-1} \mathbf{OR}\ P_{i\cdot1} \mathbf{AND}\ C_{i\cdot1}$
4: **end for**
5: **set**, $Carry = C_n$
6: End

2.2   Reversible BCD Adders

In this section, the designs for reversible BCD and carry skip reversible BCD adders have been presented. For both designs, the detail algorithms with figures and examples are shown.

Property 2.2.1   If each digit of a decimal number is represented by its equivalent straight binary code, it is known as binary coded decimal (abbreviated as BCD).

Example 2.2.1   BCD representation for decimal 801 is 1000 0000 0001. There are 10 valid digits ranging from 0 to 9. The straight binary representation for 0 and 9 is 0000 and 1001, respectively. Four bits are required to represent decimal 9 and with 4 bits, it can generate up to 16 different states, ranging from 0 to 15. Therefore, there are six forbidden states and only 10 numbers (0000 to 1001, i.e., decimal 0 to 9) are valid in BCD system.

Property 2.2.2   A BCD adder is an adder that performs addition of two BCD numbers.

Example 2.2.2   After adding two BCD numbers 0100 and 0101, BCD adder will produce their sum as 1001. It is the addition of 4 and 5 and produces the result 9, which is correct. But if we add 1000 and 0100, a simple binary addition will produce the output 1100, which is an invalid BCD number. The correct result of this BCD addition is 0001 0010. However, a combinational logic must be used to produce this correct result, which will eventually detect that the resulting number overflows. If the number is greater than the highest BCD number, 1001, and it notifies the appropriate correction logic that should be incorporated.

Property 2.2.3   A combinational circuit for BCD overflow detection is a circuit that checks whether the result of the binary addition of the two BCD numbers overflows or not, i.e., the result of the addition is greater than 1001 (Decimal 9).

A BCD number overflow occurs if the resulting number is greater than 1001. Let $A_3\,A_2\,A_1\,A_0$ and $B_3\,B_2\,B_1\,B_0$ be two BCD numbers to be added and the resulting number is represented by $T_3\,T_2\,T_1\,T_0$. The carry out is represented by C_4. C_4 is set when the resulting number is greater than binary 1111, i.e. decimal 15. Six invalid BCD numbers can be detected by the condition $(T_2{+}T_1)T_3$. So, the expression for overflow detection bit, F is $(T_2{+}T_1)T_3{+}C_4$.

Algorithm 2.2.1.1 Overflow Detection Algorithm (T)

Input: T (C_4, T_3, T_2, T_1): a 4-bit vector received from the binary adder.
Output: The vector $R=(T \cup F)$ would be the output from this algorithm, where F is the overflow detection bit (1 indicates overflow, 0 otherwise). The reversible logic design states that there must be no fan-out from any segment of the circuit. It should be noted that, the T vector is required again for correction after overflow detection, but T was fed to this detection circuit. There are numerous ways of generating copies of T vector at any level, but this detection circuit is preferred to produce T vector as well.

1: Begin
2: Overflow detection bit, $F = (T_2 T_1)T_3 \oplus C_4$. The expression shows that the resulting circuit may contain at least two blocks. The approach is similar to the following:
3: The first block takes T_1 and T_2 and produces output($T_2 + T_1$).
4: The second block takes the T_3, C_4 and output ($T_2 + T_1$) from the first block and computes the result $F = (T_2 + T_1)T_3 \oplus C_4$.
5: **return** R: $= T * F$.
6: End

However, it is easy to note that $(T_2+T_1)T_3$ and C_4 cannot be set at the same time. Therefore, a revised expression for overflow detection bit is $F = (T_2+T_1)T_3 \oplus C_4$.

If F is a set, then an overflow occurs and binary 0110 (decimal 6) is added to the partial sum $T_3 T_2 T_1 T_0$ to generate the final result.

2.2.1 Design Procedure of the Reversible BCD Adder

A reversible BCD adder consists of three components: a 4-bit parallel adder, BCD adder overflow detection logic, and BCD adder overflow correction logic. The design approach is presented here along with proper algorithms and appropriate figures. In order to design the one-digit BCD adder, three algorithms have been introduced. Algorithm 2.2.1.1 termed as overflow detection algorithm (ODA), is used to detect the overflow produced by adding two BCD digits. The overflow correction algorithm (OCA), or Algorithm 2.2.1.2 is used to correct the error generated by adding two BCD digits and finally, Algorithm 2.2.1.3 is termed as BCD adder construction algorithm, which is used to design the overall circuit.

Example 2.2.1.1 Figure 2.13 shows a direct implementation of Algorithm 2.2.1.1 where $T = (C_4, T_3, T_2, T_1)$.

Example 2.2.1.2 Figure 2.14 shows a direct implementation of Algorithm 2.2.1.2.

Example 2.2.1.3 Figure 2.15 shows a direct implementation of Algorithm 2.2.1.3.

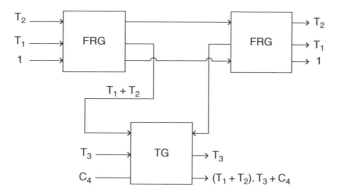

Figure 2.13 A 1-digit BCD adder's overflow detection logic.

Algorithm 2.2.1.2 Overflow Correction Algorithm (R)

Input: $R=(T \cup F)$: a 4-bit vector received from the overflow detection logic circuit.
Output: Final corrected BCD sum $S=(C_{out}, S_3, S_2, S_1, S_0)$. As T vector that was fed to the detection logic does not include T_0, it is free and intact to use as S_0. It is not mandatory to wait for the final carry out, because if F is 1, it is sure that the final carry out $C_{out} =1$, so it is not needed to propagate further to compute this carry.

1: Begin
2: The first block will take T_1 and F from the overflow detection logic circuit and generate $S_1 = T_1 \oplus F$ and $carry\_out_1 = T_1 F$.
3: The second block will take carry out of the first block, T_2 from the overflow detection circuit, and F (this F can be duplicated using numerous techniques, in the circuit first block generates F again) and generate $S_2 = T_2 \oplus F \oplus carry\_out_1$. It will also generate $carry\_out_2 = (T_2 \oplus F). carry\_out_1 \oplus T_2 F$.
4: The third block will take carry out of the second block, T_3 from the overflow detection circuit, and generate $S_3 = T_3 \oplus carry\_out_2$.
5: **return** S
6: End

2.2.1.1 Properties of the Reversible BCD Adder

In the design of reversible BCD adder, the primary concern is to keep the number of gates and number of garbage outputs as minimum as possible. As the number of gates is reduced, it is likely that delay will also be reduced. Garbage output is another important criterion. Circuits with less number of garbage outputs are always desirable. Several properties for reversible BCD adder in terms of number of gates and garbage outputs are presented in this subsection.

Property 2.2.1.1.1 A combinational circuit for BCD overflow detection can be realized by at least two reversible gates.

Proof: One Toffoli gate is needed to generate the overflow expression, $F = (T_1 + T_2)T_3 \oplus C_4$ using intermediate output. As a local application of Bennett's compute–uncompute trick,

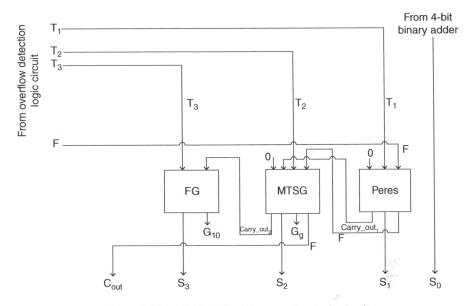

Figure 2.14 A 1-bit BCD adder correction logic circuit.

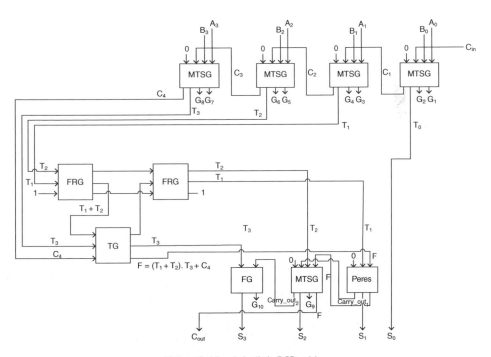

Figure 2.15 A 1-digit BCD adder.

Algorithm 2.2.1.3 BCD Adder Construction Algorithm (A, B)

Input: $A = (A_3, A_2\,A_1, A_0)$ and $B = (B_3, B_2, B_1, B_0)$ are two 4-bit input BCD vectors.
Output: Final corrected BCD sum $S = (C_{out}, S_3, S_2, S_1, S_0)$.

1: Begin
2: T: Binary adder output(A, B)
3: R: Overflow detection algorithm (T)
4: S: Overflow correction algorithm(R)
5: **return** S
6: End

---

one Fredkin gate is used to generate intermediate output $(T_1 + T_2)$ from which the second Fredkin gate restores the original input fed to the first one. But the second Fredkin gate is used to regenerate the primary input and F alone can be generated without the help of a second Fredkin gate, and hence, at least two reversible gates are needed to generate the overflow expression, $F = (T_1 + T_2)T_3 \oplus C_4$.

Property 2.2.1.1.2   A combinational circuit for BCD overflow detection can be realized by zero garbage output.

Proof: Property 2.2.1.1.2 shows that one FRG is used to generate intermediate output while other FRG stores the original input fed to the first one. A 3×3 TG also leaves two of its inputs unchanged. Constant 1 fed to the circuit is also outputted as constant 1, which is not considered as a garbage output. So, the combinational circuit for BCD overflow detection generates no garbage output.

Property 2.2.1.1.3   A reversible 4-bit parallel adder can be realized by at least eight garbage outputs.

Proof: Figure 2.15 shows that a reversible full adder circuit can be realized by at least two garbage outputs. A reversible 4-bit parallel adder consists of 4 reversible full adders. So, a reversible 4-bit parallel adder can be realized by $4 \times 2 = 8$ garbage outputs.

Property 2.2.1.1.4   Let g_{pa} be the minimum number of garbage outputs for a reversible 4-bit parallel adder, g_{od} be the minimum number of garbage output produced by the overflow detection circuit, and g_{ocl} be the minimum number of garbage outputs for overflow correction logic. Let g_{BCD} be the number of garbage outputs for a reversible BCD adder. Then $g_{BCD} >= g_{pa} + g_{od} + g_{ocl}$, where $g_{pa} >= 8, g_{od} >= 0\ g_{ocl} >= 2$.

Proof: A reversible BCD adder consists of a 4-bit reversible parallel adder, overflow detection logic and overflow correction logic circuit. A 4-bit parallel adder consists of four full adders and according to Property 2.2.1.1.3, a 4-bit parallel adder will be realized by at least eight garbage outputs. So, minimum number of garbage output for a reversible 4-bit parallel adder is $g_{pa} >= 8$.

In the overflow detection logic circuit, the overflow expression, $F = (T_1 + T_2)T_3 \oplus C_4$ is realized and according to Property 2.2.1.1.1, BCD overflow detection logic can be realized by at least zero garbage outputs. So, the minimum number of garbage output for overflow detection logic is $g_{od} >= 0$. In the overflow correction logic, overflow F is propagated. According to the improved design presented in this chapter, overflow correction logic generates only two garbage outputs. So, the minimum number of garbage outputs for overflow correction logic is $g_{ocl} >= 2$. As a result, the total number of garbage outputs for a reversible BCD adder, $g_{BCD} >= g_{pa} + g_{od} + g_{ocl}$, where $g_{pa} >= 8, g_{od} >= 0 g_{ocl} >= 2$.

Property 2.2.1.1.5 Let gt_{pa} be the minimum number of gates for a reversible 4-bit parallel adder, gt_{od} be the minimum number of gates required by overflow detection circuit and gt_{ocl} be the minimum number of gates for overflow correction logic circuit. Let gt_{BCD} be the total number of gates for a reversible BCD adder. Then $gt_{BCD} >= gt_{pa} + gt_{od} + gt_{ocl}$, where $gt_{pa} >= 4$; $gt_{od} >= 3$ and $gt_{ocl} >= 2$.

Proof: A reversible BCD adder consists of a 4-bit reversible parallel adder, overflow detection logic and overflow correction logic, where a 4-bit parallel adder consists of four full adders and a 4-bit parallel adder can be realized by at least four reversible gates. So, the minimum number of gates for a reversible 4-bit parallel adder is $gt_{pa} >= 4$.

In the overflow detection logic, the overflow expression, $F = (T_1 + T_2)T_3 \oplus C_4$ is realized and according to Property 2.2.1.1.1, BCD overflow detection logic can be realized by at least two reversible gates. So, the minimum number of gates for overflow detection logic is $gt_{od} >= 2$. In the overflow correction logic, BCD number is corrected. According to the design, the overflow correction logic can be realized with only three reversible gates. So, the minimum number of gates for overflow correction logic is $gt_{ocl} >= 2$. As a result, the total number of reversible gates for a reversible BCD adder is $gt_{BCD} >= gt_{pa} + gt_{od} + gt_{ocl}$, where $gt_{pa} >= 4$; $gt_{od} >= 3$ and $gt_{ocl} >= 2$.

2.2.2 Design Procedure of the Reversible Carry Skip BCD Adder

A carry skip reversible BCD adder consists of the following components: a 4-bit parallel adder, carry skip logic, BCD adder overflow detection logic, and BCD adder overflow correction logic. Carry skip logic may generate the carry out (C_{out}) instantaneously. These components are presented here with proper algorithms and appropriate figures.

The carry skip logic circuit is the fundamental part to this design. The carry-in, C_{in} can be propagated to the carry-out, C_{out} of the block. Let A_i and B_i be the inputs to i^{th} full adder and either of them is set. Propagation $P_i = A_i \oplus B_i$ and C_{in} to the block will propagate to the carry output of the block if the entire P_i's are set. In this way, C_{out} can be generated without waiting for it to be generated in ripple carry fashion. Let, the propagation signal for the block is denoted by P. If P is set, C_{in} will be propagated to the C_{out}. However, in the other case, C_{out} will be generated in the ripple carry fashion. So, carry skip logic bit of the block is $K = PC_{in} + tC_4$ where C_4 is the carry generated in the ripple carry fashion. K can be slightly modified to realize the logic more clearly: $K = PC_{in} \oplus PC_4$ the expression reveals the fact that if P (Propagate) is true, it does not have to wait for the generation of find carry out C_4, but if

---

Algorithm 2.2.2.1 Carry Skip BCD Adder Algorithm (A, B,C_i)

Input: $A = (A_3, A_2, A_1, A_0)$ and $B = (B_3, B_2, B_1, B_0)$ are two input vectors and C_{in} is the carry in.

Output: A BCD adder capable of performing the sum $=A+B$. The buffer vector $S = (C_{out}, S_3, S_2, S_1, S_0)$ will store the result.

1: Begin
2: Compute P (propagate bit).
3: Initially $P:=$ true
4: **for** i to $0, 1, \ldots, 3$ **do**
5:    $P:= P$ AND $(A_i B_i)$
6: **end for**
7: Compute $T:= \{C_4, T_3, T_2, T_1, T_0\}$, where $T_i:= A_i \oplus B_i \oplus C_i$ and C_i's are generated from each adder block.
8: **return** S.
9: Compute carry skip logic bit,$K:=P.C_{in} \oplus \overline{P}.C_4$
10: Add binary 011 to T if overflow detection bit F is true.
11: Compute $S:=\{C_{out}, S_3, S_2, S_1, S_0\}$, the final sum of the addition process.
12: End

$P$is false, it is needed to go in formal way, i.e., wait for C_4 to generate. The overall overflow detection bit, $F = (T_1 + T_2)T_3 \oplus K$is generated in the same way with reversible BCD adder presented earlier in this chapter. Overflow correction logic incorporated is the same as the reversible BCD adder. Algorithm 2.2.2.1 is used to design the one-digit cary skip BCD adder as shown in Figure 2.16.

Example 2.2.2.1 Figure 2.16 shows a direct implementation of Algorithm 2.2.2.1.

The FRGs in the middle of the Figure 2.16 generate the block propagation, $P$and carry skip logic bit, K. FRGs and TG on the left side perform the BCD overflow detection same as for conventional BCD adder. BCD overflow correction logic is as like as the conventional one. In this figure, all G's are garbage outputs.

2.2.2.1 Properties of the Reversible Carry Skip BCD Adder
In the design of carry skip reversible BCD adder, the primary concern is to keep the number of gates and number of garbage outputs as minimum as possible. Several properties for reversible carry skip BCD adder in terms of number of gates and garbage outputs are presented in this subsection.

Property 2.2.2.1.1 Any combinational circuit for carry skip logic can be realized by at least four reversible gates.

Proof: Block propagation, $P=P_3 P_2 P_1 P_0$ can be generated using FRGs. As one FRG is used for AND operation, three FRGs are needed to generate P. Another FRG is needed for intermediate output to generate the carry skip logic, $K=PC_{in} \oplus \overline{P}C_4$. Hence, at least $4(= 3+1)$ reversible gates are needed to generate the carry skip logic $K=PC_{in} \oplus \overline{P}C_4$.

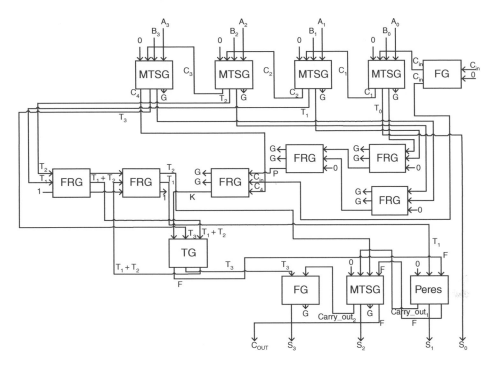

Figure 2.16 A carry skip 1-digit BCD adder.

Property 2.2.2.1.2 The combinational circuit for carry skip logic can be realized by eight garbage outputs.

Proof: As one FRG is used for AND operation, three FRGS are needed to generate P. For each AND operation, only one of three outputs is used and other two remain as garbage outputs. So, $3 \times 2 = 6$ garbage outputs are generated to produce P. Another FRG is needed for intermediate output to generate the carry skip logic $K = PC_{in} \oplus \overline{P}C_4$ using intermediate output, which generates two other garbage outputs. So, according to the design of BCD adder in this subsection, the combinational circuit generates $6 + 2 = 8$ garbage outputs for carry skip logic.

Property 2.2.2.1.3 Let g_{cpa} be the minimum number of garbage outputs for a reversible 4-bit parallel adder, g_{cod} be the minimum number of garbage outputs produced by the overflow detection circuit and g_{cocl} be the minimum number of garbage output for overflow correction logic. Let g_{cBCD} be the number of garbage outputs for a reversible carry skip BCD adder. Then $g_{cBCD} >= g_{cpa} + g_{cod} + g_{cocl}$, where $g_{cpa} >= 4, g_{cod} >= 8$ and $g_{cocl} >= 2$.

Proof: A carry skip reversible BCD adder consists of a 4-bit reversible parallel adder, overflow detection logic and overflow correction logic. A 4-bit parallel adder consists of 4 full adders. According to Property 2.2.1.1.3, a 4-bit parallel adder can be realized by at least 8 garbage outputs and from Algorithm 2.2.2.1, it is clear that four garbages from the four full adders are used to generate the propagate bit P. So, the number of garbage outputs for a reversible 4-bit parallel adder, g_{cpa} reduces to 4.

In the overflow detection logic, the overflow expression, $F=(T_1+T_2)T_3\oplus K$ is realized where $K = PC_{in}\oplus \overline{P}C_4$. According to Property 2.2.1.1.5 and Property 2.2.1.1.2, carry skip logic produces eight and BCD overflow detection logic produces at least zero garbage output. So, the minimum number of garbage outputs for overflow detection logic is $g_{cod} >= 8 + 0 = 8$. In the overflow correction logic, the overflow F is propagated. According to the design, garbage output for overflow correction logic is $g_{cocl} >= 2$. As a result, the total number of garbage outputs for a carry skip BCD adder is

$g_{cBCD} >= g_{cpa} + g_{cod} + g_{cocl}$, where $g_{cpa} >= 4, g_{cod} >= 8$ and $g_{cocl} >= 2$.

Property 2.2.2.1.4 Let gt_{cpa} be the minimum number of gates for a reversible 4-bit parallel adder, gt_{cod} be the minimum number of gates for the overflow detector and gt_{cocl} be the minimum number of gates for overflow correction logic. Let gt_{cBCD} be the number of gates for a reversible carry skip BCD adder. Then $gt_{cBCD} >= gt_{cpa} + gt_{cod} + gt_{cocl}$, where $gt_{cpa} >= 4, gt_{cod} >= 7$ and $gt_{cocl} >= 3$.

Proof: A reversible carry skip BCD adder consists of a 4-bit reversible parallel adder, overflow detection logic and overflow correction logic. A 4-bit parallel adder consists of four full adders and a 4-bit parallel adder can be realized by at least four reversible gates. So, the minimum number of gates for a reversible 4-bit parallel adder is $gt_{cpa} >= 4$.

In the overflow detection logic, the overflow expression, $F=(T_1+T_2)T_3\oplus K$ is realized where $K=PC_{in}\oplus \overline{P}C_4$. According to Property 2.2.1.1.4 and Property 2.2.1.1.1, carry skip logic can be realized by at least 4 and BCD overflow detection logic produces at least two reversible gates. So, the minimum number of garbage outputs for overflow detection logic is $gt_{cod} >= 4 + 3 = 7$. In the overflow correction logic, the BCD number is corrected. According to the design, the overflow correction logic can be realized with only three reversible gates. So, the minimum number of gate for overflow correction logic is $gt_{ocl} >= 3$.

As a result, the total number of reversible gates for a reversible BCD adder is $gt_{cBCD} >= gt_{cpa} + gt_{cod} + gt_{cocl}$, where $gt_{cpa} >= 4, gt_{cod} >= 7$ and $gt_{cocl} >= 3$.

2.3 Reversible BCD Subtractor

In this section, at first some mechanism and working procedures of BCD subtractor are shown. Then the BCD subtractor is designed as reversible fashion.

In the BCD subtraction, the nine's complement of the subtrahend is added to the minuend. In the BCD arithmetic, the nine's complement is computed by nine minus the number whose nine's complement is to be computed. This can be illustrated as the nine's complement of 5 will be 4 (9 − 5 = 4), which can be represented in BCD code as 0100. In BCD subtraction using nine's complement, there can be two possibilities.

The sum after the addition of minuend and the nine's complement of subtrahend is an invalid BCD code (an example is when 5 is subtracted from 8) or a carry is produced from

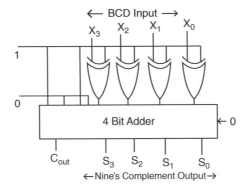

Figure 2.17   Nine's complement circuit.

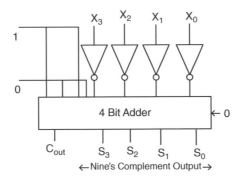

Figure 2.18   Modified nine's complement circuit.

the MSB (for example, 5 is subtracted from 8). In this case, add decimal 6 (binary 0110) and the end around carry (EAC) to the sum. The final result will be the positive number represented by the sum.

The sum of the minuend and the nine's complement of the subtrahend is a valid BCD code, which means that the result is negative and is in the nine's complement form. An example is, when 8 is subtracted from 5.

In BCD arithmetic, instead of subtracting the number from nine, the nine's complement of a number is determined by adding 1010 (Decimal 10) to the one's complement of the number. The nine's complement circuit using a 4-bit adder and XOR gates is shown in Figure 2.17. It has been realized that there is no need to use EX-OR gates in the nine's complement for complementing. The use of NOT gates will better suit the purpose and will reduce the complexity of the circuit, both in CMOS as well as reversible logic implementation. The modified design of nine's complement is shown in Figure 2.18. It replaces four EX-OR gates by four NOT gates. The one-digit BCD subtractor using the nine's complement circuit is shown in Figure 2.19. In Figure 2.19, after getting the nine's complement of the subtrahend, it is added to the minuend using the BCD adder. Then the required 1010 is added by using the complement of the output carry of the BCD adder. The sign represents whether the number stored is positive or negative (for example, 5-8 will be stored as Sign $= 1$ and Magnitude $(S_3 \cdots S_0) = 3$).

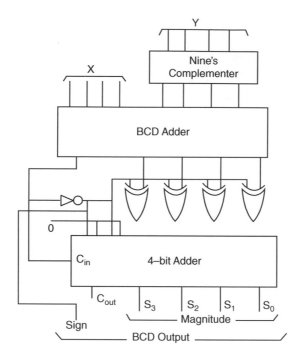

Figure 2.19 Modified conventional BCD subtractor.

2.3.1 Carry Look-Ahead BCD Subtractor

The carry look-ahead BCD subtractor can be designed by integrating the BCD adder, 4-bit adder and nine's complement components. Figure 2.20 shows the design of the carry look-ahead BCD subtractor. It is to be noted that the emphasis has been laid on improving the individual modules of the BCD subtractor to improve its overall efficiency and make it more suitable for reversible logic implementation.

2.3.2 Carry Skip BCD Subtractor

Figure 2.21 shows the design of the carry skip BCD subtractor. It is to be noted that the carry skip implementation of the nine's complement in the circuit will not be beneficial, making the carry look-ahead as the best choice for its implementation. The carry skipping property of the BCD adder can be beneficial only when its input carry $C_{in}=1$. Thus, in order to extract the benefit of the carry skip property of the BCD adder in the BCD subtractor, the LSB output (n_0) of the nine's complement has been made as input carry $'C_{in}$ of the carry skip BCD adder and passed '0' in its place for addition to the BCD adder. Therefore, the numbers passed for addition in carry skip BCD adder will be $X + (n_3 n_2 n_1' 0') + n_0$, where n_0 will work as C_{in}. The last block of the 4-bit adder in the circuit has also been designed in the carry skip fashion to further improve the efficiency of the design further. This will result in the generation of C_{out} in Figure 2.21 in carry skip fashion.

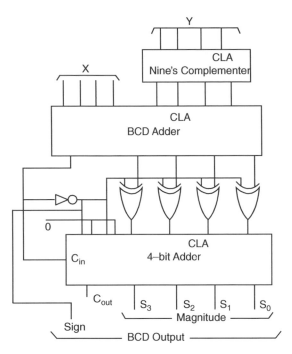

Figure 2.20 CLA BCD subtractor.

2.3.3 Design of Conventional Reversible BCD Subtractor

It is evident from Figure 2.19 that in order to design reversible BCD subtractors, the whole reversible design must be divided into three sub-modules:

1. Design of the reversible nine's complement (which, in turn, has to be designed using reversible parallel adders).
2. Design of the reversible BCD adder.
3. Integration of the modules to design the reversible BCD subtractor.

The primary goal of this section is to design reversible BCD subtractors with a minimal number of reversible gates and garbage output.

2.3.3.1 Reversible Nine's Complement

Figure 2.22 shows the reversible nine's complement using the NOT gates, New gates (NG) and the 3 × 3 Feynman gate (FG). The design is implemented with seven reversible gates and three garbage outputs. To minimize the garbage at the bottom 4-bit adder, property of regenerating the constant value at the garbage output has been utilized (the constant input "1" at the NG gate is regenerated at one of its garbage outputs and is used as input to FG. It is observed that the S_0 can be directly generated without requiring any addition circuitry (referring to Figure 2.17, it is needed to mention that the second input to the full adder is 0 and the C_{in} is 0. Further examination showed that there is no need for the full adder in the second place, third place, and fourth place of the bottom 4-bit adder. Half adders and

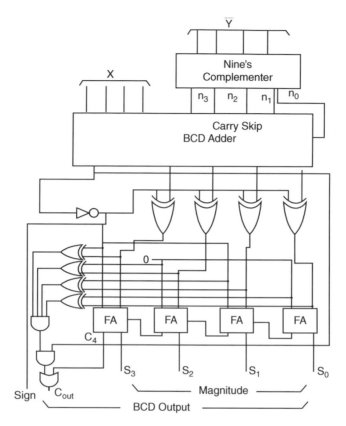

Figure 2.21 Carry skip BCD subtractor.

3-input Ex-OR gate can perform the required addition operations. The reversible half adder can be designed by New gate (NG) with only one garbage output and the 3-input Ex-OR gate can be designed using FG with only two garbage outputs. Utilizing the reversible full adder in those places would have increased the garbage, as at least two garbage output are required in a reversible full adder. Moreover, the output carry is not required in the nine's complement. Thus, the reversible full adder would have generated the output carry, leading to an increase in garbage count.

2.3.3.2 Reversible BCD Subtractor

The reversible BCD subtractor using the reversible nine's complement, reversible BCD adder, TSG, NG, and Feynman gate (FG) is shown in Figure 2.23. It has been proved that the reversible designs of the nine's complement and BCD adder are designed with a minimal number of reversible gates and garbage outputs. In order to design a more efficient, complete BCD subtractor in terms of the number of reversible gates and garbage outputs, the Feynman gate has been used for generating the Ex-OR/NOT function and copying the output (as fan-out is not allowed in reversible logic). Here Feynman gate is chosen as it can generate Ex-OR/NOT function and copy the output with minimum number of reversible gates and garbage output. This can be understood by the fact that there are exactly two outputs corresponding to the input of a Feynman gate, a 0 in the

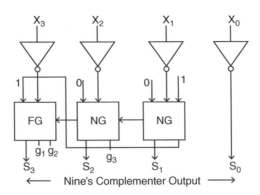

Figure 2.22 Reversible nine's complement.

second input will copy the first input in both the output of that gate. It makes the Feynman gate most suitable for a single copy of bit, as it does not produce any garbage output.

The bottom 4-bit binary adder required in BCD subtractor is also designed very efficiently to minimize the garbage output. This is achieved by carefully passing the input signal and thereby utilizing the garbage output for further computation along with identifying suitable places where reversible full adders can be replaced by reversible half adders. An inefficient approach of simply designing the 4-bit adder with the reversible full adder could lead to eight garbage outputs (at least two garbage outputs are produced in a reversible full adder). The BCD adder requires 10 reversible gates and 13 garbage outputs, as it is proved above.

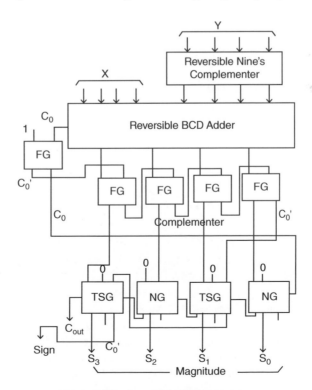

Figure 2.23 Reversible BCD subtractor.

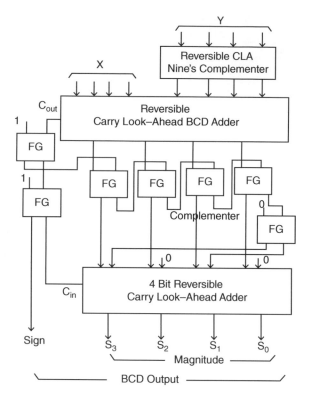

Figure 2.24   Reversible CLA BCD subtractor.

The nine's complement is designed with seven reversible gates and three garbage outputs. The generation of Ex-OR functions, copying, and NOT functions are designed in such an optimal manner that it requires five Feynman gates with zero garbage output. The bottom 4-bit reversible adder is designed with four reversible gates and four garbage outputs. Thus, the reversible BCD subtractor is designed with $10 + 7 + 5 + 4 = 26$ reversible gates with the minimum number of garbage outputs $13 + 3 + 4 = 20$.

2.3.3.3   Reversible Design of Carry Look-Ahead BCD Subtractor

After designing the individual reversible components of the carry look-ahead BCD subtractor, the components are combined together to design the complete reversible carry look-ahead BCD subtractor, as shown in Figure 2.24. It is to be noted that the same strategy of connecting the Feynman gates is used as chains for generating the Ex-OR, copying, and NOT functions with zero garbage. Thus, the architecture is designed efficiently in terms of the numbers of reversible gates and garbage outputs.

2.3.3.4   Reversible Design of Carry Skip BCD Subtractor

Figure 2.25 shows the reversible implementation of the carry skip BCD subtractor. This is an attempt to design a reversible carry skip BCD subtractor. In the reversible implementation, the reversible nine's complement can be chosen from the nine's complement that is designed in the above subsections. The other component i.e., as shown in Figure 2.25, is

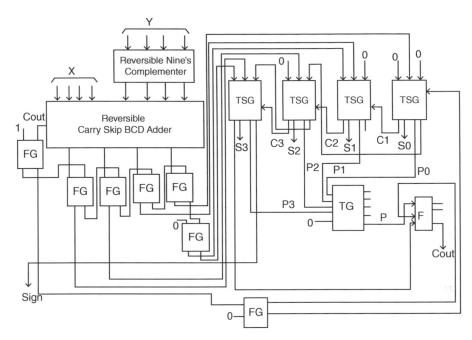

Figure 2.25 Reversible logic implementation of the carry skip BCD adder.

the reversible implementation of the 4-bit CSA block (at the bottom of the figure), which is designed with six reversible gates and seven garbage outputs. Thus, the reversible carry skip BCD subtractor is an efficient design in terms of the number of reversible gates and garbage outputs. It leads to conclude that utilizing the garbage output for regenerating the constant input like 1 and 0 will significantly help in reducing the garbage output. This can be considered as the indirect contribution of this chapter to the reversible logic community.

2.4 Summary

Adder is the basic arithmetic unit of a computer system, and it can be used in diverse areas such as complex quantum arithmetic circuits. In this chapter, a multi-operand n-digit reversible decimal adder is presented. An optimized and low quantum cost gate has been introduced. The full adder, CSA, CLA, and an optimized 3-operand n-digit decimal adder are designed using reversible logic gates. In addition, this chapter presents reversible logic implementations for binary coded decimal (BCD) adder as well as carry skip BCD adder. Furthermore, a design of conventional BCD subtractors, carry look-ahead and carry skip BCD subtractors are shown. The designs of carry look-ahead and carry skip BCD subtractors are based on the designs of carry look-ahead and carry skip BCD adders, respectively. Thus, the designs of various adders that have been shown in this chapter would help the reader to understand the next chapters clearly. In addition, the concept of designing these adder-subtractors will help in designing the reversible multiplier circuits.

3

Reversible Multiplier Circuit

Multipliers work on signed and unsigned numbers. Signed numbers can be multiplied using the Baugh-Wooley multiplication algorithm. But radix–4 Booth's algorithm can substantially reduce the number of partial products in multiplier and incur less overhead for the summation network. Moreover, a reversible multiplier will avoid the heat loss completely in comparison with the irreversible one.

3.1 Multiplication Using Booth's Recoding

Multiplication circuits are usually composed of two parts. One is the partial product generation (PPG) circuit and the other is the multi-operand addition (MOA) circuit. In a PPG circuit, partial products are generated based on the status of the multiplier bits. For each multiplier bit and its status (0 or 1), a decision must be made whether to add the multiplier with the partial product. After the generation of all the partial products, summation circuit adds them and produces the result. So, for an $\times n$ multiplier, n partial products are generated.

In this chapter, a suitable algorithm is described to optimize the complexities of $n \times n$ multipliers. Here, the Booth's algorithm is applied with radix-4. Booth's algorithm with other radix can also be applied to get $n \times n$ multiplier circuit. Higher radix representation of a number yield less number of bits. For example, a k-bit binary number can be interpreted as a $\lceil k/2 \rceil$ digit radix-4 number, a $\lceil k/3 \rceil$ digit radix-8 number, and so on. As a result, if a multiplier has a high radix representation, the number of partial products is reduced. If a k-bit number is represented using radix-4 representation, then there will be $\lceil k/2 \rceil$ partial products and less hardware will be needed to accumulate them. Booth's recoding of k-bit binary number produces a radix-4 Booth's recoded number. Table 3.1 shows Booth's recoding, and the example below shows how a 2's complement binary number is recoded using Booth's recoding:

$$(1\,0\,0\,1\,0\,1\,1\,0\,|\,0)_{2's\ complement} = (-2\,1\,2\,-2)_{four}$$

The recoding of the above binary number shows that an extra 0 has been added at the end of the bit string to get three bits to be encoded. The result of a multiplication is formed by summation of the partial products. The number of partial products is the number of bits in

Reversible and DNA Computing, First Edition. Hafiz Md. Hasan Babu.
© 2021 John Wiley & Sons Ltd. Published 2021 by John Wiley & Sons Ltd.

Table 3.1 Radix-4 Booth's Recoding.

Bit position (i+1)	Bit position (i)	Bit position (i-1)	Value to be added with multiplicand (M)
0	0	0	0×M
0	0	1	1×M
0	1	0	1×M
0	1	1	2×M
1	0	0	-2×M
1	0	1	-1×M
1	1	0	-1×M
1	1	1	-0×M

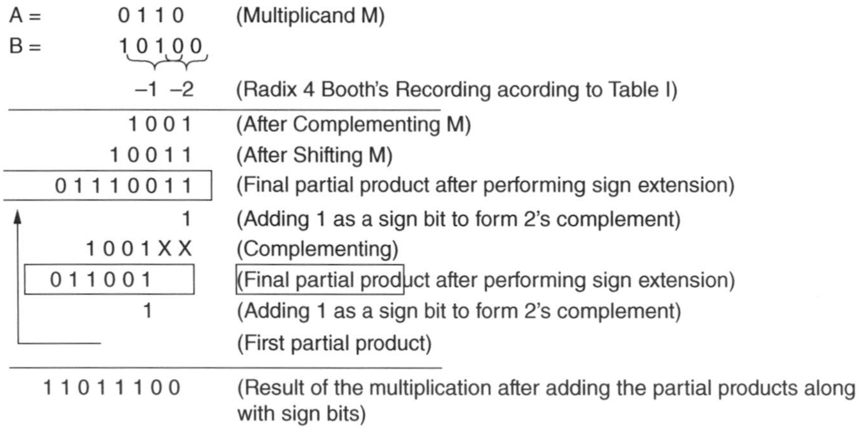

```
A =        0 1 1 0      (Multiplicand M)
B =        1 0 1 0 0
              ‿‿‿
           -1 -2         (Radix 4 Booth's Recording acording to Table I)
         ──────────
           1 0 0 1       (After Complementing M)
           1 0 0 1 1     (After Shifting M)
       0 1 1 1 0 0 1 1   (Final partial product after performing sign extension)
                   1     (Adding 1 as a sign bit to form 2's complement)
         1 0 0 1 X X     (Complementing)
       0 1 1 0 0 1       (Final partial product after performing sign extension)
                 1       (Adding 1 as a sign bit to form 2's complement)
       ──────────        (First partial product)
     1 1 0 1 1 1 0 0     (Result of the multiplication after adding the partial products along
                         with sign bits)
```

Figure 3.1 Process of 4 × 4 multiplications.

the recoded multiplier. Booth's recoding represents a multiplier with fewer bits and hence, the number of partial products is decreased.

In Booth's recoding, the multiplicand has to be multiplied by numbers in the range of [-2, 2]. This can be achieved in hardware by shifting and complementing logic. A 4-bit multiplication is shown in Figure 3.1.

3.2 Reversible Gates as Half Adders and Full Adders

Peres gate (PG) can be used as a half adder if its third input is set to 0. HNG gate can be used as a full adder. There are several other full adders too. But, HNG is the best in terms of quantum cost and hardware complexity. The quantum cost of HNG is six and its total logical calculation is as follows:

$$T = 5\alpha + 2\beta; \text{and}$$

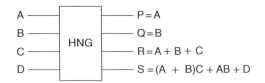

Figure 3.2   Block diagram of a reversible HNG gate.

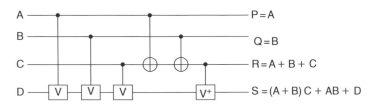

Figure 3.3   The quantum representation of a reversible HNG gate.

The quantum cost of PG is four and its total logical calculation is as follows:

$$W = 2\alpha + \beta.$$

The input and output vectors of HNG as shown in Figure 3.2 gate are given below:

$$I_v = (a, b, c, d)$$

$$O_v = (a, b, a \oplus b \oplus c, (a \oplus b)c \oplus ab \oplus d))$$

Figure 3.3 presents the quantum presentation of HNG gate.

By setting the third input as C_{in} (carry in) and fourth input as 0, HNG can work as a full adder. The third output is the sum of the addition of a, b and C_{in}, while the fourth output is the carry generated by the addition.

3.3   Some Signed Reversible Multipliers

Fast and efficient reversible multiplier circuit designing is a very interesting area of research, and different types of designs have been done by the researchers. But very few designs are considered multiplying signed numbers and numbers involving large number of bits like 16 bits, 32 bits, 64 bits, etc. The comparisons made by them are usually done on the basis of small multiplication circuits like 4 × 4 or 5 × 5. When generalizing and designing large multipliers, these designs come up with heavy number of quantum gates, garbage outputs, constant inputs and hardware complexity. The reason behind it lies in the number of partial products. If the number of partial products is not reduced, a huge summation network is required for designing large-scale multipliers.

3.4   Design of Reversible Multiplier Circuit

The design of a reversible multiplier is composed of three components: recoding cell (R cell), partial product generator (PPG) circuit, and multi-operand addition circuit (MOA).

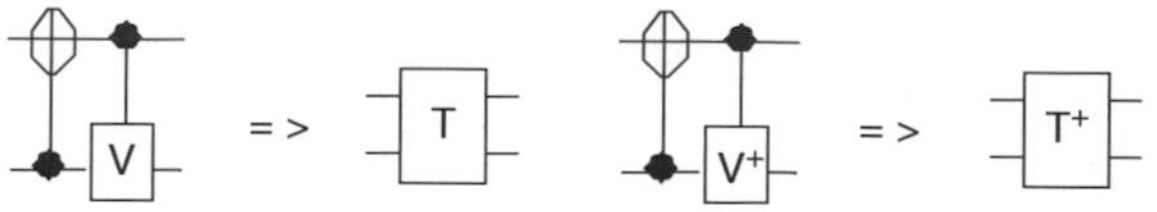

(a) Symbols of the Controlled-T Gate

(b) Symbols of the Controlled-T$^+$ Gate

Figure 3.4 Symbols of the controlled-T and controlled-T^+ gate.

3.4.1 Some Quantum Gates

In this subsection, two quantum gates, namely controlled-T and controlled-T^+ gate, are introduced. By using these gates and tensor products, any other related gates can be constructed. By cascading a CNOT gate and a controlled-V gate, a quantum gate is shown, namely controlled-T gate. Controlled-T^+ gate is also described by cascading a CNOT gate and a controlled-V^+ gate.

The matrix of a quantum circuit is derived from its quantum gate matrices using matrix multiplication. The matrix for CNOT gate (M_{CNOT}), matrix for controlled-V gate ($M_{Controlled-V}$) and matrix for controlled-V^+ gate ($M_{controlled-V+}$) are described, which are given below:

$$M_{CNOT} = \begin{pmatrix} 1 & 0 & 0 & 0 \\ 0 & 1 & 0 & 0 \\ 0 & 0 & 0 & 1 \\ 0 & 0 & 1 & 0 \end{pmatrix} \quad M_{Controlled-V} = \begin{pmatrix} 1 & 0 & 0 & 0 \\ 0 & 1 & 0 & 0 \\ 0 & 0 & \frac{i+1}{2} & \frac{1-i}{2} \\ 0 & 0 & \frac{i-1}{2} & \frac{i+1}{2} \end{pmatrix} \quad M_{Controlled-V+} = \begin{pmatrix} 1 & 0 & 0 & 0 \\ 0 & 1 & 0 & 0 \\ 0 & 0 & \frac{1-i}{2} & \frac{1+i}{2} \\ 0 & 0 & \frac{1+i}{2} & \frac{1-i}{2} \end{pmatrix}$$

In the following, matrix for controlled-T ($M_{Controlled-T}$) and controlled-T$^+$ ($M_{Controlled-T+}$) are shown. Figure 3.4 shows the block diagram of the gates.

$$M_{Controlled-T} = M_{CNOT} \times M_{Controlled-V} = \begin{pmatrix} 1 & 0 & 0 & 0 \\ 0 & 1 & 0 & 0 \\ 0 & 0 & \frac{i-1}{2} & \frac{1+i}{2} \\ 0 & 0 & \frac{1+i}{2} & \frac{1-i}{2} \end{pmatrix}$$

$$M_{Controlled-T+} = M_{CNOT} \times M_{Controlled-V+} = \begin{pmatrix} 1 & 0 & 0 & 0 \\ 0 & 1 & 0 & 0 \\ 0 & 0 & \frac{i+1}{2} & \frac{1-i}{2} \\ 0 & 0 & \frac{1-i}{2} & \frac{1+i}{2} \end{pmatrix}$$

3.4.2 Recoding Cell

The recoding cell takes three bits of the multiplier and produces three recoded bits, which are the shift enable (SE) bit, operation bit (H), and the sign bit (D). If $SE = 1$, then it can be concluded that $+2M$ or $-2M$ is to be added to the partial product, where M is the multiplicand. If $H = 1$, then it can be concluded that the combination of multiplier bits is not 0. If $D = 1$, it can be concluded that it is needed to form 2's complement and hence toggle the bits of the multiplicand. The recoded bits corresponding to the input bits are given in Table 3.2. The logical operations to generate the recoded bits are given below:

Table 3.2 Truth Table for Recoding Cell.

Bit position ($i + 1$)	Bit position (i)	Bit position ($i - 1$)	Shift enable (SE)	Operation bit (H)	Sign bit (D)
0	0	0	0	0	0
0	0	1	0	1	0
0	1	0	0	1	0
0	1	1	1	1	0
1	0	0	1	1	1
1	0	1	0	1	1
1	1	0	0	1	1
1	1	1	0	0	1

$D = X_{i+1}$(Value at bit position $i+1$)

$$H = A + B$$

$$SE = A \oplus H$$

Here, $A = X_i \oplus X_{i-1}$ and $B = X_i \oplus X_{i+1}$

$$O = \overline{SE \oplus D}$$

The last output O is needed when 2's complement generation and shifting are done. In order to develop the R cell (recoding cell), a 5×5 gate (BSJ gate) has been introduced. The input vector, I_v and output vector, O_v of the BSJ gate are as follows:

$$I_v = \{A, B, C, D, E\} \text{ and}$$
$$O_v = \{P = A, Q = A \oplus B, R = A \oplus B \oplus C \oplus E, S = C \oplus D, T = E\}$$

Figure 3.5 shows the diagram of 5×5 BSJ gate and Figure 3.6 shows its equivalent quantum representation. If the truth table of the corresponding input and output of BSJ gate is generated, it will verify the unique input–output mapping between them. Figure 3.6 shows that the quantum cost of BSJ gate is 4. The gate requires five Ex-OR operations. So, the hardware complexity of the BSJ gate is 5α. The Peres gate has been modified to construct the R cell. Figure 3.7 shows the diagram of the 3×3 modified Peres gate (MPG) and Figure 3.8 shows the quantum representation of the gate. The minimization of quantum gates of MPG is shown in Figure 3.9 and Figure 3.10. The number of quantum gates and hardware complexity of modified Peres gate are 4 and $2\alpha + \beta + 3d$, respectively. The input–output mapping and reversibility of MPG can be easily verified by generating all possible inputs and obtaining the corresponding outputs.

Figure 3.11 shows the block diagram of the R cell. Among the reversible gates, Feynman gate and F2G gate have been used in R cell. Two garbage outputs (denoted by G) are generated by R cell. The third and fourth outputs (denoted by D) of the R cell are the same. Both of these outputs are used in the PPG circuit (Figure 3.19). For this reason, the duplicate output is not considered as garbage output. The total quantum cost to realize R cell is:

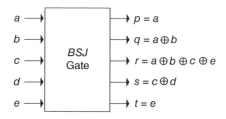

Figure 3.5 5×5 BSJ gate and its corresponding input–output mapping.

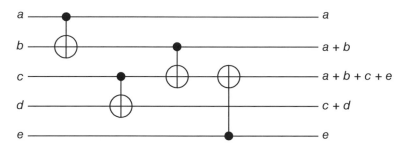

Figure 3.6 Quantum realization of 5×5 BSJ gate.

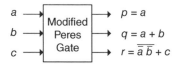

Figure 3.7 3×3 MPG and its corresponding input–output mapping.

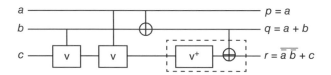

Figure 3.8 Quantum realization of 3×3 MPG.

$$Q_R = Q_{FG} + Q_{F2G} + Q_{MPG} + Q_{BSJ}$$
$$= 1 + 2 + 4 + 4$$
$$= 11$$

The total hardware complexity of R cell is:

$$H_R = H_{FG} + H_{F2G} + H_{BSJ} + H_{MPG}$$
$$= \alpha + 2\alpha + 5\alpha + (2\alpha + \beta + 3d)$$
$$= 10\alpha + \beta + 3d$$

The delay of R cell can be modeled as below:

$$D_R = D_{FG} + D_{F2G} + D_{MPG} + D_{BSJ}$$
$$= (0.1 + 0.12 + 0.15 + 0.12)ns$$
$$= 0.49ns$$

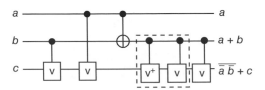

Figure 3.9 Quantum analysis of 3 × 3 MPG.

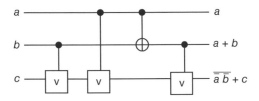

Figure 3.10 A compact quantum realization of 3 × 3 MPG.

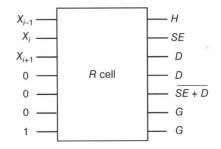

Figure 3.11 Block diagram of R cell.

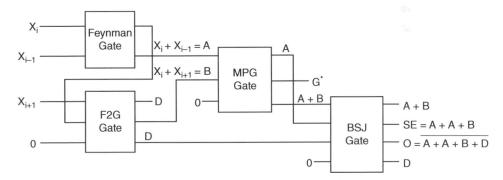

Figure 3.12 Construction of R cell.

The detailed diagram of R cell has been shown in Figure 3.12. R cell is used as a part of the PPG circuit to ensure the proper recoding of the multiplier bits. Among four constant inputs that are required for the operation of R cell, three are set to zero, while the remaining is set to one.

3.4.3 Partial Product Generation Circuit

The PPG circuit takes the multiplicand and multiplier as input and produces the partial products as output. The whole process can be divided into three steps. At first, the multiplicand bits have been taken as input and converted into recoded bits. A Feynman gate

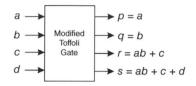

Figure 3.13 3×3 MTG and its corresponding input–output mapping.

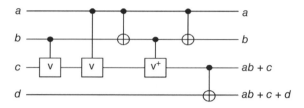

Figure 3.14 Quantum realization of 4×4 MTG.

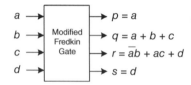

Figure 3.15 3×3 MFRG and its corresponding input–output mapping.

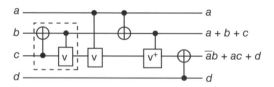

Figure 3.16 Quantum realization of 4×4 MFRG.

and two R cells have been used for this purpose. Second, the multiplier bits and the recoded bits are fed to a component, which produces some partial products. modified Toffoli gates, Fredkin gates and modified Fredkin gates have been used to construct this part. Finally, the sign extensions of partial products are generated. TS-3 gates have been used for this purpose.

The design of PPG circuit requires two types of gates: modified Toffoli gate (MTG) and modified Fredkin gate (MFRG). The block diagrams of MTG and MFRG have been shown in Figure 3.14 and Figure 3.15, respectively. Quantum realizations of MTG and MFRG have been shown in Figure 3.14 and Figure 3.16, respectively. The minimization of quantum gates of MFRG is shown in Figure 3.17. The reversibility and unique input–output mapping of MTG and MFRG can be easily proved by generating all possible inputs and obtaining the corresponding outputs.

The R cell is integrated with the PPG circuit to recode the multiplier bits as shown in Figure 3.19. TS-3 gates work as the sign extension units. The partial product array for a 16×16 multiplier is shown in Figure 3.18. The sign extension bits are denoted by the letter E, whereas the sign bits are denoted by S. The sign bit is the D output generated from the R cell. For each partial product, one Rcell is needed which recodes three bits. Algorithm 3.4.3.1

Algorithm 3.4.3.1 Partial Product Generation

Input: A, B; (both are m-bit binary number), C_{out}(1-bit)

Output: Sum(m-bit), Carry(1-bit)

1: Begin

2: **for** $i=0$ **to** n-1 **do**

3:    **if** $i=0$ **then**

4:       Apply R cell with

5:       Input := $\{0, x_0, x_1, 0, 0, 0, 1\}$

6:    **else**

7:       Input := $\{x_{2i-1}, x_{2i}, x_{2i+1}\}, 0, 0, 0, 1\}$

8:    **end if**

9:    Output := $\{H, SE, D, D, O, G, G\}$

10:    **if** $H=0$ **then**

11:      **for** j-0 **to** n-1 **do**

12:        PPG_ array[j] $= 0$

13:      **end for**

14:    **else**

15:     **if** $SE=0$ **then**

16:      **if** $D=0$ **then**

17:        **for** j-0 **to** n-1 **do**

18:          PPG_ array[j] $=$ multiplicand[j]

19:        **end for**

20:      **else**

21:        **for** j-0 **to** n-1 **do**

22:          PPG_array[j] $= \overline{multiplicand[j]}$

23:        **end for**

24:      **end if**

25:     **else**

26:      PPG_array[0] $=\overline{multiplicand[j]}$. D $\oplus$multiplicand[j]. $\overline{(SE \oplus D)}$

27:      **if** $D=0$ **then**

28:        **for** j-0 **to** n **do**

29:          PPG_array[j] $=$ PPG_ array[j] $=$ multiplicand[j-1]

30:        **end for**

31:      **else**

32:        **for** j-0 **to** n **do**

33:          PPG_array[j] $= \overline{multiplicand[j-1]}$

34:        **end for**

35:      **end if**

36:     **end if**

37:    **end if**

38:    Generate the sign extension bits E $=\overline{YH \oplus D}$

39: **end for**

40: End

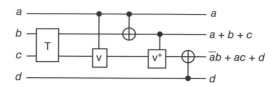

Figure 3.17 A Compact quantum realization of 4 × 4 MFRG.

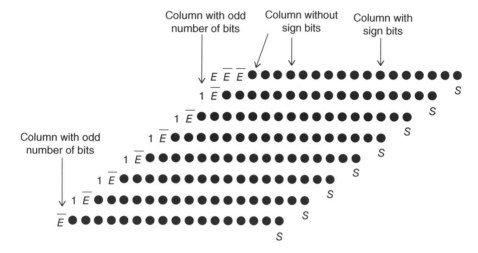

Figure 3.18 16 × 16 PPG array.

shows the partial product generation process. Figure 3.19 shows the detailed partial product array for a 4 × 4 multiplier. The generalized diagram for $n \times n$ PPG is shown in Figure 3.20.

3.4.4 Multi-Operand Addition Circuit

The multi-operand addition (MOA) circuit adds the partial products column-wise that have been generated by the PPG circuit. To add the partial products, some full adders and half adders are required. HNG gate can be used as full adder and PG gate can be used as half adder. The summation of bits are propagated downward in a column, while the carry bits are propagated from right to left most column. The construction of 4 × 4 MOA has been shown in Figure 3.21. Algorithm 3.4.4.1 describes the process of multi-operand addition.

Figure 3.22 shows the generalized representation of the MOA based on HNG and PG gates. It can be easily observed that for multiplying two n-bit numbers, where $n = 2m$, only the last row requires some PGs. There are six columns in the MOA gate that have the depth of k, where $k = \lceil n/2 \rceil$.

3.4.5 Calculation of Area and Power of $n \times n$ Multiplier Circuit

Area is an important issue to design a circuit. If the area of a circuit is very large, then the cost of that circuit will increase. Therefore, area is an important considerable matter. Algorithm 3.4.5.1 describes the area calculation of the $n \times n$ multiplier circuit.

Power is also an important issue to design a circuit. It is needed to design such a circuit, which needs less power as the world is facing the lack of power. Therefore, power is also

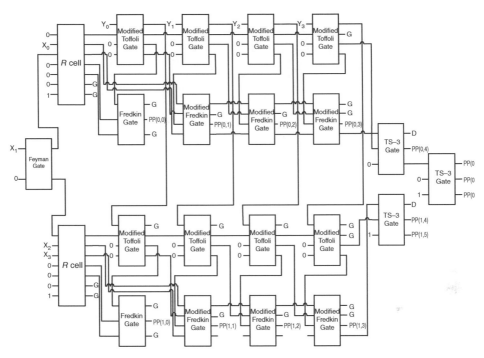

Figure 3.19 Gate level diagram of a 4 × 4 PPG for reversible Booth's multiplier.

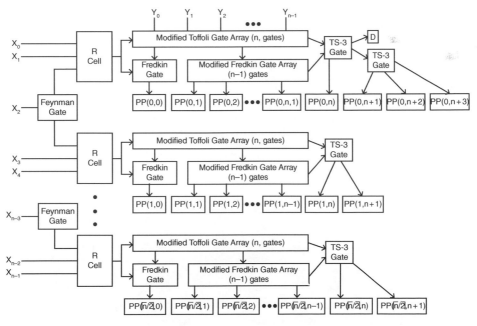

Figure 3.20 Block diagram of an $n \times n$ PPG for reversible Booth's multiplier.

Algorithm 3.4.4.1 Multi-Operand Addition

Input: A, B; (both are m-bit binary number), C_{out} (1-bit)
Output: Sum (m-bit), Carry (1-bit)

1: Begin
2: **for** each column $i = 0$ **to** $i = 2n$ in PPG Array **do**
3:    **if** Number_of_bits to add in each column is even **then**
4:       Apply (Number_of_bits/2)-1 HNG
5:       Apply 1 PG
6:    **else**
7:       Apply ⌊Number_of_bits/2⌋ HNG
8:    **end if**
9: **end for**
10: End

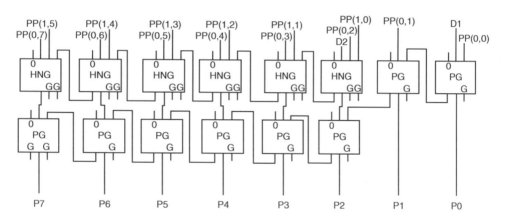

Figure 3.21 Gate level diagram of a 4 × 4 MOA for reversible Booth's multiplier.

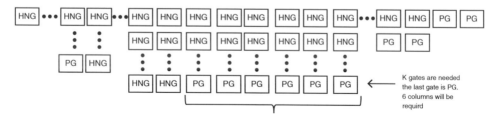

Figure 3.22 Diagram of an $n \times n$ MOA for reversible Booth's Multiplier.

an important considerable matter. Algorithm 3.4.5.2 describes the power calculation of the $n \times n$ multiplier circuit.

Complexity of a circuit is very important to analyze the cost of the circuit in terms of numbers of gates, garbage outputs, quantum costs, constant inputs, hardware complexity, quantum gates calculation complexity, delay of the reversible multiplier, delay of the quantum multiplier, area of the quantum multiplier, and power of the quantum multiplier. In this section, different complexities of the reversible $n \times n$ multiplier are represented.

Algorithm 3.4.5.1 Calculation of Area for a Quantum $n \times n$ Multiplier circuit

---

Input: A, B; (both are m-bit binary number), C_{out}(1-bit)
Output: Sum(m-bit), Carry(1-bit)

1: Begin
2: *Power* = 0, Q_{PPG} = 0, Q_{MOA} = 0
3: **for** i =1 **to** n **do**
4:    Get the number of quantum gates of PPG (partial product generation) circuit, PPG
5:    $Q_{PPG} = Q_{PPG} + Q.PPG$
6: **end for**
7: **for** i =1 **to** n **do**
8:    Get the number of quantum gates of a MOA (multi-operand addition) circuit, MOA
9:    $Q_{MOA} = Q_{MOA} + Q.MOA$
10: **end for**
11: *Power* = 142.3× ($Q_{PPG} + Q_{MOA}$)
12: **return** *Power*
13: End

Algorithm 3.4.5.2 Calculation of Power for a Quantum $n \times n$ Multiplier circuit

---

Input: A, B; (both are m-bit binary number), C_{out}(*1*-bit)
Output: Sum(m-bit), Carry(*1*-bit)

1: Begin
2: *Area* = 0, Q_{PPG} = 0, Q_{MOA} = 0
3: **for** i =1 **to** n **do**
4:    Get the number of quantum gates of PPG (partial product generation) circuit, PPG
5:    $Q_{PPG} = Q_{PPG} + Q.PPG$
6: **end for**
7: **for** i =1 **to** n **do**
8:    Get the number of quantum gates of a MOA (multi-operand addition) circuit, MOA
9:    $Q_{MOA} = Q_{MOA} + Q.MOA$
10: **end for**
11: *Area* = 50× ($Q_{PPG} + Q_{MOA}$)
12: **return** *Area*
13: End

Property 3.4.5.1   At least $(n-1)/2$ full adders are needed to add n bits if $n = 2m + 1$. For $n = 3$, at least $(n/2) - 1$ full adders and one half adder are needed to compute the summation. Thus, the number of gates required to add $n$bits is $m = \lfloor n/2 \rfloor$.

Proof: For $n = 2$, at least one PG, which works as a half adder, is needed to compute the sum. For $n = 3$, at least one HNG, which works as a full adder is needed for adding them. Now, if $n > 3$ and $n = 2m$, then after the addition of first three bits with HNG, one sum bit will be generated and it will be subsequently passed to the next level leaving bits to be added. Before the last level two bits are added along with the sum bit from the previous level. The sequence of bit addition in each level is given below:

$$N_B = 3 + 2 + 2 + \ldots + 2m - 1 + 1$$
$$= 3 + 2m - 4 + 1$$
$$N_G = 1 + (2m - 4) \div 2 + 1$$
$$= 1HNG + (m - 2)HNG + 1PG$$

Here, N_B = Total number of bits added

N_G = Number of gates needed

So, if $n = 2m$, $(n/2)$ - 1 HNG gates and one PG are needed to sum n bits. Considering when $n = 2m + 1$, the sequence of bit addition in each level is given below:

$$N_B = 3 + 2 + 2 + \ldots + 2m - 2 + 2$$
$$= 3 + 2m - 4 + 2$$
$$N_G = 1 + (2m - 4) \div 2 + 2$$
$$= 1HNG + (m - 2)HNG + 1HNG$$
$$= m$$

So, if $n = 2m$, $(n/2) - 1$ HNG gates are needed to compute the sum. It can be also verified from above that the number of gates to add n bits is $m = \lfloor n/2 \rfloor$. It is mentioned that, Property 3.4.5.1 represents the property of addition of n bits.

Example 3.4.5.1 In order to add 8 bits, $(8/2) - 1 = 3$ full adders and 1 half adder are needed, but $(9 - 1)/2 = 4$ full adders are needed to add 9 bits.

Property 3.4.5.2 An MOA circuit requires n gates to accumulate the partial products generated by PPG circuit, where n is the number of bits in the partial products except the sign bits.

Proof: The PPG generates partial products from multiplicand and recoded multiplier, which is shown in Figure 3.18. The S bits are the sign bits, while the E bits are sign extension bits. Let B_N be the number of bits in column N in the partial product array (without considering the sign bit S) and G_N be the number of gates required to sum the bits in column N. As the number of bits to be added in column N is B_N and the number of carry bits generated by the gates in column $N-1$ is G_{N-1}, the following equation can be written using Property 3.4.5.2 for columns not containing the sign bit:

$$G_N = \lfloor (B_N + G_{N-1})/2 \rfloor$$

Here, G_{N-1} is the number of gates in column N-1. So, the number of carry bits generated from column N-1 is G_{N-1}. Now, if $G_{N-1} = B_{N-1}$ is considered and it is shown that $G_N = B_N$, then $G_N = B_N$ is proved by induction. Now the above equality can be written as below:

$$G_N = \lfloor (B_N + B_{N-1})/2 \rfloor$$

Now, it can be easily verified from Figure 3.17 that the immediate previous right columns of the columns not containing the sign bit must have the same number of bits or one more bit. From this observation, the following equation can be obtained:

$$G_N = \lfloor (B_N + B_{N-1})/2 \rfloor = B_N$$

Hence, for columns not containing the sign bit, it has been proved that $G_N = B_N$. For columns containing the sign bit, the following equation can be written using Property 3.4.5.1:

$$G_N = \lfloor (B_N + G_{N-1} + 1)/2 \rfloor$$

Again by considering, $G_N = B_{N-1}$, the equation above can be written as below:

$$G_N = \lfloor (B_N + B_{N-1} + 1)/2 \rfloor$$

If a column containing sign bit and its number of bits is q, then the immediate previous column contains $q - 1$ bits. Now, $B_{N-1} = B_N - 1$. So, the following equation can be derived:

$$G_N = \lfloor (B_N + B_{N-1} + 1)/2 \rfloor = B_N$$

Therefore, the total number of gates needed to realize the MOA is the total number of bits in the partial product except the sign bits.

Example 3.4.5.2   To realize a 4×4 multiplier, $k = 4/2 = 2$ partial products are needed. The first partial product contains $4×2 = 8$ bits. The second partial product contains two bits less than the first partial product according to the rule of radix-4 Booth's multiplication. So, the second partial product has *six* bits. The total number of bits in these two partial products except the sign bits is 14. It can be easily verified from design of the MOA of the 4×4 reversible multiplier in Figure 3.16 that it has 14 gates.

Property 3.4.5.3   A Booth's recoded reversible $n×n$ multiplier requires $(3nk + 9k)$ gates, where n is the number of bits of multiplicand and multiplier and $k = \lceil n/2 \rceil$.

Proof: In radix-4 Booth's recoding, k partial products are generated by a multiplier where $k = \lceil n/2 \rceil$. For each partial product one R cell, one Fredkin Gate, $n - 1$ modified Fredkin gates, one TS-3 gate, n modified Toffoli gates are needed. Also, one TS-3 gate and $k - 1$ FG gates are required for the complete multiplier circuit. Therefore, the total number of gates required for PPG is modeled as below:

$$T_{ppg} = kRcell + (n - 1) \times kMFRG + kFRG + n \times kMTG$$
$$+ kTS - 3 + 1TS - 3 + (k - 1)FG \qquad (3.4.5.1)$$

As the R cell contains four gates, the number of gates can be written as below:

$$T_{ppg} = k \times 4 + (n - 1) \times k + k + n \times k + k + 1 + (k - 1)$$
$$= 2nk + 6k$$

With the help of Property 3.4.5.2, it can be shown that the summation circuit will contain $nk + 3k$ gates, as there are n bits in each row of partial products. Extra three bits are generated for sign extension in each row except the first and the last rows. The first row contains four bits and the last row contains two bits for sign extension as shown in Figure 3.18. This makes three extra bits per partial product on an average. Therefore, the total number of gates is:

$$T_G = T_{PPG} + T_{MOA} = 2nk + 6k + nk + 3k = 3nk + 9k \qquad (3.4.5.2)$$

Example 3.4.5.3   It can be seen from the 4 × 4 PPG in Figure 3.17 that it contains $2nk + 6k = 2 \times 4 \times 2 + 6 \times 2 = 28$ gates (Each R cell contains 4 gates inside). Here, $n = 4$ and $k = \lceil n/2 \rceil = 2$.

Property 3.4.5.4   The MOA of Booth's recoded reversible $n \times n$ n multiplier requires $(nk + 4k - 2n - 1)$ full adders and $(2n - k + 1)$ half adders, where n is the number of bits of multiplicand and multiplier, $n = 2m$, $k = \lceil n/2 \rceil$ and $n > 4$. The MOA of Booth's recoded reversible $n \times n$ multiplier requires $(nk + 4k - 2n - 1)$ full adders and $(2n - k + 1)$ half adders, where n is the number of bits of multiplicand and multiplier, $n = 2m, k = \lceil n/2 \rceil$ and $n > 4$.

Proof: In order to calculate the sum from one column in PPG array, the bits in that column and the carry bits generated from the previous column must be added. In the PPG array, there are $k - 1$ columns, for which odd number of bits have to be added and these columns do not need half adders as shown in Property 3.4.5.1. When the number of carry bits of the previous column combines with the number of bits of the current column and produces an even number of bits, the current column of the circuit requires a half adder. As the number of columns for an $n \times n$ bit multiplier is $2n$, and the column with the even number of bits (the even number is formed by the number of carry bits of the previous column and the number of bits of the current column) requires a half adder, the number of columns for half adders is:

$$\text{Number of half adders } (N_h) = 2n - (k - 1) = 2n - k + 1 \qquad (3.4.5.3)$$

According to Property 3.4.5.3, the total number of half and full adders required in MOA is $nk + 3k$. Hence, the total number of full adders is:

$$\text{Number of full adders } (N_f) = nk + 3k - (2n - k + 1)$$
$$= nk + 4k - 2n - 1 \qquad (3.4.5.4)$$

Example 3.4.5.4   In Figure 3.23 (a) and Figure 3.23 (b), H denotes HNG gate i.e., reversible full adder gate, and PG denotes Peres gate, which is a reversible half-adder gate. It can be seen from Figure 3.23 (a) that a 6x6 MOA contains 6x3 + 3x 3 = 27 gates (according to Equation 3.2). Morever, from Figure 3.23 (b), it can be observed that an 8x8 MOA contains 8x4 + 3x4 = 44 gates (according to Equation 3.4.5.2). The number of half

```
H HHHHHHHHH P P       H HHHHHHHHHHHHHH P P
   H HHHHHH P PP    H HHHHHHHHH P P
   P  PPPPP            H HHHHHH P P
                         P PPPPP
       (a)                 (b)
```

Figure 3.23   (a) 6 × 6 MOA (b) 8 × 8 MOA.

adders in Figure 3.23 (a) is 2×6 − 3 + 1 = 10 (according to Equation 3.4.5.3), where the number of full adders is 6×3 + 4×3 − 2×6 − 1 = 17. The number of half adders in Figure 3.23 (b) is 2×8 − 4 + 1 = 13 (according to Equation 3.4.5.3)), where the number of full adders is 8×4 + 4×4 − 2×8 − 1 = 31 (according to Equation 3.4.5.4).

Property 3.4.5.5 A Booth's recoded reversible $n \times n$ multiplier requires $(17nk + 34k − 4n − 1)$ quantum gates, where n is the number of bits of multiplicand and multiplier, $k = \lceil n/2 \rceil, n = 2m$.

Proof: According to Equation 3.4.5.3) in Property 3.4.5.3, the number of quantum gates of PPG can be modeled using FG, R cell, MFRG, FRG, MTG, and TS-3 gates as follows:

$$QG_{ppg} = k \times 11 + (n − 1) \times k \times 5 + k \times 5 + n \times k \times 6 + k \times 2 + 2 + (k − 1) \times 1$$
$$= 11nk + 14k + 1$$

The number of half adders to realize the MOA is $2n − k + 1$ for $n = 2m$, when $k > 2$. For $k < 2$, no half adder is required.

$$T_{MOA} = (nk + 3k − 2n + k − 1)HNG + (2n − k + 1)PG$$
$$QG_{MOA} = (nk + 4k − 2n − 1) \times 6 + (2n − k + 1) \times 4$$
$$= 6nk + 20k − 4n − 2$$

So, the total number of quantum gates of the multiplier can be modeled as follows:

$$QG_M = QG_{PPG} + QG_{MOA}$$
$$= 17nk + 34k − 4n − 1$$

Property 3.4.5.6 A Booth's recoded reversible $n \times n$ multiplier generates $(3nk + 12k − n − 1)$ garbage outputs, where n is the number of bits of multiplicand and multiplier, $k = \lceil n/2 \rceil$ and $n = 2m$.

Proof: To generate one partial product, the R cell, FRG, MFRG and MTG gates produce 2, 2, n, and 1 garbage outputs, respectively. MTG gates produce extra $(n − 1)$ garbage outputs while generating the last partial product. As k partial products are generated, the total number of garbage outputs generated by the PPG circuit can be modeled as follows:

$$G_{ppg} = 2k + 2k + nk + k + (n − 1)$$
$$= nk + 5k + n − 1$$

In the MOA circuit, HNG gates generate two garbage outputs and PG gates generates one garbage output. The leftmost PG in the last row of MOA generates one more garbage as shown in Figure 3.17. As there are $nk + 4k − 2n − 1$ HNG and $2n − k + 1$ PG, the total number of garbage outputs generated by the MOA circuit can be modeled as follows:

$$G_{MOA} = (nk + 4k − 2n − 1) \times 2 + (2n − k + 1) \times 1 + 1$$
$$= 2nk + 7k − 2n$$

Therefore, the total number of garbage outputs generated by the multiplier is given below:

$$G_M = 3nk + 12k - n - 1$$

Property 3.4.5.7 A Booth's recoded reversible $n \times n$ multiplier requires $(3nk + 10k - 1)$ constant inputs, where n is the number of bits of multiplicand and multiplier, $k = \lceil n/2 \rceil$ and $n = 2m$.

Proof: In order to generate a partial product, the R cell, MTG and TS-3 require 4, $2n$, and 1 constant input respectively. Moreover, $k - 1$ Feynman gates require $k - 1$ constant inputs and to generate the first partial product two extra constant inputs are needed for the TS-3 gate in the first row of PPG as shown in Figure 3.19. So, the total number of constant inputs required by the PPG circuit can be modeled as follows:

$$C_{PPG} = (4 + 2n + 1) \times k + (k - 1) + 2$$
$$= 2nk + 6k + 1$$

Each of the HNG gates and Peres gates in the MOA requires one constant input. Also, it can be seen from the structure of partial products that each of them has a "1" in their most significant bit as shown in Figure 3.18, except the first and last partial products. So, $k - 2$ constant inputs are also added. The total number of constant inputs required by the PPG circuit can be modeled as follows:

$$C_{MOA} = (nk + 4k - 2n - 1) \times 1 + (2n - k + 1) \times 1 + (k - 2)$$
$$= nk + 4k - 2$$

Therefore, the total number of constant inputs required by the multiplier is given below:

$$C_M = C_{PPG} + C_{MOA}$$
$$= 3nk + 10k - 1$$

Property 3.4.5.8 A Booth's recoded reversible $n \times n$ multiplier requires $[\alpha(12nk + 29k - 6n - 2) + \beta(6nk + 10k - 2n - 1) + d(nk + 4k)]$ logical calculation or hardware complexity, where n is the number of bits of multiplicand and multiplier, $k = \lceil n/2 \rceil$ and $n = 2m$.

Proof: The hardware complexities of R cell, MTG, MFRG, TS-3, FG, and FRG are $10\alpha + \beta + 3d$, $3\alpha + 2\beta$, $4\alpha + 2\beta + d$, 2α, α and $2\alpha + 4\beta + 2d$, respectively. According to Equation 3.4.5.1 in Property 3.4.5.3, the hardware complexity of PPG can be modeled as below:

$$H_{PPG} = k(10\alpha + \beta + 3d) + (n - 1)k(4\alpha + 2\beta + d) + k(2\alpha + 4\beta + 2d) +$$
$$nk(3\alpha + 2\beta) + k(2\alpha) + 2\alpha + (k - 1)\alpha$$
$$= \alpha(7nk + 11k + 1) + \beta(4nk + 3k) + d(nk + 4k)$$

The hardware complexities of HNG and PG gates are $5\alpha + 2\beta$ and $2\alpha + \beta$, respectively. Using Property 3.4.5.4, it can be shown that $nk + 4k - 2n - 1$ HNG and $2n - k + 1$ PG gates are required to realize the MOA. Therefore, the hardware complexity of MOA can be written as below:

$$H_{MOA} = (nk + 4k - 2n - 1)(5\alpha + 2\beta) + (2n - k + 1)(2\alpha + \beta)$$
$$= \alpha(5nk + 18k - 6n - 3) + \beta(2nk + 7k - 2n - 1)$$

Therefore, the total hardware complexity of the multiplier can be calculated as below:

$$H_M = H_{PPG} + H_{MOA}$$
$$= \alpha(12nk + 29k - 6n - 2) + \beta(6nk + 10k - 2n - 1) + d(nk + 4k)$$

Property 3.4.5.9 A Booth's recoded reversible $n \times n$multiplier requires $(0.53n + 0.25k + 0.53)$ ns delay, where n is the number of bits of multiplicand and multiplier, $k = \lceil n/2 \rceil$ and $n = 2m$.

Proof: The critical path, shown in Figure 3.24, shows that the PPG circuit requires one Feynman gate, one R cell, $k + 1$ modified Toffoli gates, $n - 1$ modified Fredkin gates, and one TS-3 gate. The critical path is shown in Figure 3.24, where FG, R Cell, MTG, FRG, MFRG, and TS-3 are indicated by node 1, 2, 3, 4, 5 and 6, respectively. So, the delay of the PPG circuit can be modeled as below:

$$D_{PPG} = D_{FG} + D_R + (k + 1)D_{MTG} + (n - 1)D_{MFRG} + D_{TS-3}$$
$$= (0.1 + 0.49 + (k + 1) \times 0.15 + (n - 1) \times 0.23 + 0.12) \ ns$$
$$= (0.23n + 0.15k + 0.63) \ ns$$

From Figure 3.25, it can be observed that the critical path delay is the delay obtained from the gates in the lowest layer of the MOA. The lowest layer contains $(2n - k + 1)$ PG and $k - 1$ HNG. So, the delay of the MOA can be calculated as below:

$$D_{MOA} = ((k - 1) \times 0.25 + (2n - k + 1) \times 0.15) \ ns$$
$$= (0.30n + 0.1k - 0.1) \ ns$$

So, the delay of the multiplier can be modeled as below:

$$D_M = D_{PPG} + D_{MOA}$$
$$= (0.53n + 0.25k + 0.53) \ ns$$

Property 3.4.5.10 The area of a Booth's recoded quantum $n \times n$multiplier is $(850nk + 1700k - 200n - 50)$Å, where n is the number of bits of multiplicand and multiplier, $k = \lceil n/2 \rceil$, $n = 2m$, and Å is the unit of measuring area.

Proof: From Property 3.4.5.5, it is found that the number of quantum gates of PPG can be modeled using FG, R cell, MFRG, FRG, MTG, and TS-3 gates as below:

$$QG_{ppg} = k \times 11 + (n - 1) \times k \times 5 + k \times 5 + n \times k \times 6 + k \times 2 + 2 + (k - 1) \times 1$$
$$= 11nk + 14k + 1$$

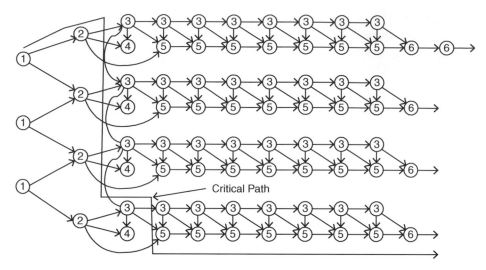

Figure 3.24 Critical path for an 8 × 8 PPG for reversible Booth's multiplier.

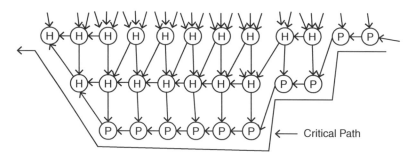

Figure 3.25 Critical path for a 6 × 6 MOA for reversible Booth's multiplier.

The number of half adders to realize the MOA is $2n - k + 1$ for $n = 2m$, when $k > 2$. For $k < 2$, no half adder is required.

$$T_{MOA} = (nk + 3k - 2n + k - 1)HNG + (2n - k + 1)PG$$
$$QG_{MOA} = (nk + 4k - 2n - 1) \times 6 + (2n - k + 1) \times 4$$
$$= 6nk + 20k - 4n - 2$$

So, the total number of quantum gates of the multiplier can be modeled as below:

$$QG_M = QG_{PPG} + QG_{MOA}$$
$$= 17nk + 34k - 4n - 1$$

Therefore, the total area of the quantum multiplier circuit can be modeled as below:

$$A_M = 50 \times QG_M$$
$$= 50 \times (17nk + 34k - 4n - 1)\text{Å}$$
$$= (850nk + 1700k - 200n - 50)\text{Å}$$

Property 3.4.5.11 The power of a Booth's recoded quantum $n \times n$ multiplier is $(2419.1nk + 4838.2k - 569.2n - 142.3)meV$, where n is the number of bits of multiplicand and multiplier, $k = \lceil n/2 \rceil$, $n = 2m$, and meV is the unit of measuring power.

Proof: From Property 3.4.5.5, it is found that the number of quantum gates of PPG can be modeled using FG, R cell, MFRG, FRG, MTG, and TS-3 gates as below:

$$QG_{ppg} = k \times 11 + (n-1) \times k \times 5 + k \times 5 + n \times k \times 6 + k \times 2 + 2 + (k-1) \times 1$$
$$= 11nk + 14k + 1$$

The number of half adders to realize the MOA is $2n - k + 1$ for $n = 2m$, when $k > 2$. For $k < 2$, no half adder is required.

$$T_{MOA} = (nk + 3k - 2n + k - 1)HNG + (2n - k + 1)PG$$
$$QG_{MOA} = (nk + 4k - 2n - 1) \times 6 + (2n - k + 1) \times 4$$
$$= 6nk + 20k - 4n - 2$$

So, the total number of quantum gates of the multiplier can be modeled as follows:

$$QG_M = QG_{PPG} + QG_{MOA}$$
$$= 17nk + 34k - 4n - 1$$

Therefore, the total power of the quantum multiplier circuit can be modeled as follows:

$$P_M = 142.3 \times QG_M = 142.3 \times (17nk + 34k - 4n - 1)meV$$
$$= (2419.1nk + 4838.2k - 569.2n - 142.3)meV$$

Property 3.4.5.12 A Booth's recoded quantum $n \times n$ multiplier requires $(17nk + 34k - 4n - 1)\Delta$ delay, where n is the number of bits of multiplicand and multiplier, $k = \lceil n/2 \rceil$, $n = 2m$, and Δ is the unit delay.

Proof: From Property 3.4.5.5, it is found that the number of quantum gates of PPG can be modeled using FG, R cell, MFRG, FRG, MTG, and TS-3 gates as follows:

$$QG_{ppg} = k \times 11 + (n-1) \times k \times 5 + k \times 5 + n \times k \times 6 + k \times 2 + 2 + (k-1) \times 1$$
$$= 11nk + 14k + 1$$

The number of half adders to realize the MOA is $2n - k + 1$ for $n = 2m$, when $k > 2$. For $k < 2$, no half adder is required.

$$T_{MOA} = (nk + 3k - 2n + k - 1)HNG + (2n - k + 1)PG$$
$$QG_{MOA} = (nk + 4k - 2n - 1) \times 6 + (2n - k + 1) \times 4$$
$$= 6nk + 20k - 4n - 2$$

So, the total number of quantum gates of the multiplier can be modeled as below:

$$QG_M = QG_{PPG} + QG_{MOA}$$
$$= 17nk + 34k - 4n - 1$$

Therefore, the total delay of the quantum multiplier circuit can be modeled as below:

$$D_{QG} = (17nk + 34k - 4n - 1)\Delta$$

Property 3.4.5.13   A Booth's recoded quantum $n \times n$ multiplier requires $[\sigma(18k + 8nk - 2n) + \Omega(16k + 10nk - 2n - 1)]$ quantum gate calculation complexity, where n is the number of bits of multiplicand and multiplier, $k = \lceil n/2 \rceil$, $n = 2m$, σ is a CNOT gate calculation complexity and Ω is a controlled-V or controlled-V^+ gate calculation complexity.

Proof: The quantum gate calculation complexities of R cell, MTG, MFRG, TS-3, FG, and FRG are $7\sigma + 3\Omega$, $3\sigma + 3\Omega$, $3\sigma + 3\Omega$, 2σ, σ and $4\sigma + 3\Omega$ respectively. According to Equation 3.4.5.1) in Property 3.4.5.3, the quantum gate calculation complexity of PPG can be modeled as below:

$$
\begin{aligned}
QGC_{PPG} &= k(7\sigma + 3\Omega) + (n - 1)k(3\sigma + 3\Omega) + k(4\sigma + 3\Omega) \\
&\quad + nk(3\sigma + 3\Omega) + k(2\sigma) + 2\sigma + (k - 1)\sigma \\
&= \sigma(11k + 6nk + 1) + \Omega(3k + 6nk)
\end{aligned}
$$

The quantum gate calculation complexities of HNG and PG gates are $2\sigma + 4\Omega$ and $\sigma + 3\Omega$, respectively. Using Property 3.4.5.4, it can be shown that $nk + 4k - 2n - 1$ HNG and $2n - k + 1$ PG gates are required to realize the MOA. Therefore, the quantum gate calculation complexity of MOA can be written as below:

$$
\begin{aligned}
QGC_{MOA} &= (nk + 4k - 2n - 1)(2\sigma + 4\Omega) + (2n - k + 1)(\sigma + 3\Omega) \\
&= \sigma(2nk + 7k - 2n - 1) + \Omega(4nk + 13k - 2n - 1)
\end{aligned}
$$

Therefore, the total quantum gate calculation complexity of the multiplier can be calculated as below:

$$
\begin{aligned}
QGC_M &= QGC_{PPG} + QGC_{MOA} \\
&= \sigma(11k + 6nk + 1) + \Omega(3k + 6nk) + \sigma(2nk + 7k - 2n - 1) \\
&\quad + \Omega(4nk + 13k - 2n - 1) \\
&= \sigma(18k + 8nk - 2n) + \Omega(16k + 10nk - 2n - 1)
\end{aligned}
$$

3.5   Summary

This chapter presents the design of a reversible signed multiplier, which is based on Booth's recoding. The multiplier has been designed in three steps. First, a recoding cell (R cell) has been designed to produce the recoded bits. Second, a partial product generation (PPG) circuit has been constructed to generate the partial products and finally, a multi-operand addition circuit (MOA) has been developed to add the partial products. To realize a compact design, the reversible and quantum gates have been introduced. A generalized architecture of the $n \times n$ multiplier has been presented, where n is the number of bits of multiplicand and multiplier. The algorithms are described to calculate area and power of the quantum multiplier. The design methodology can be integrated as a part of reversible central processing unit (CPU), reversible signal processing and reversible arithmetic logic unit (ALU) optimized in terms of ancillary inputs and garbage outputs.

4

Reversible Division Circuit

Division is a complex operation in the computer arithmetic. Nowadays, people use a hardware module divider to implement division algorithm. The division circuit can be used in the arithmetic unit of a processor. Conventionally sequential circuits are used to implement the divider.

4.1 The Division Approaches

In a fixed-point division, inputs are divisor V and dividend D, and outputs are quotient Q and remainder R such that $D = Q \cdot V + R$. In every step i, the divisor V is shifted i bits to the right, which represents $2^{-i}V$. If $2^{-i}V$ is less than partial remainder R_i, then quotient bit q_i is set to 1; otherwise, it is set to 0. Then a partial remainder R_{i+1} is calculated as follows: $R_{i+1} = R_i - q_i \times 2^{-i} \times V$ which is equivalent to $R_{i+1} = 2R_i - q_i \times V$. There are two approaches to perform division operation, namely, restoring division and nonrestoring division.

4.1.1 Restoring Division

In restoring approach, at every step, the operation $R_{i+1} = 2R_i - V$ is performed. If subtraction result is negative, the partial remainder is restored to the correct value by performing the addition operation $R_{i+1} = R_{i+1} + V$.

4.1.2 Nonrestoring Division

In nonrestoring approach, if subtraction result is negative, the partial remainder is not restored immediately. It is based on the inspection that a restoring step of the form $R_i = R_i + V$ followed by the next partial remainder calculation step $R_{i+1} = 2R_i - V$ can be combined into a single operation $R_{i+1} = 2R_i + V$.

Thus, if the quotient bit $q_i = 1$, then the next partial remainder is calculated by performing a subtraction. On the other hand, if the quotient bit $q_i = 0$, instead of restoring the partial remainder, the next step is calculated by performing the addition of the divisor to the partial remainder. If the last quotient bit is 0, then the partial remainder is negative due to trial subtraction. Therefore, an essential correction step is to restore the remainder by adding the divisor.

Reversible and DNA Computing, First Edition. Hafiz Md. Hasan Babu.
© 2021 John Wiley & Sons Ltd. Published 2021 by John Wiley & Sons Ltd.

4.2 Components of Division Circuit

In order to design reversible sequential division hardware, reversible multiplexers (MUXs), registers, parallel-in parallel-out (PIPO) left-shift registers, and parallel adder are required. Each component is described in the following subsections.

4.2.1 Reversible MUX

Using Fredkin gates, the Figure 4.1 shows a 2-input n-bit reversible MUX where S is the select input, $A_1 A_2 A_3 \ldots A_n$ and $B_1 B_2 B_3 \ldots B_n$ are two inputs. If $S = 0$, then $(Z_1 \ Z_2 \ Z_3 \ldots Z_n) = (A_1 \ A_2 \ A_3 \ldots A_n)$ or if $S = 1$, then $(Z_1 \ Z_2 \ Z_3 \ldots Z_n) = (B_1 \ B_2 \ B_3 \ldots B_n)$. This reversible MUX requires n Fredkin gates, generates n garbage outputs, and needs $5n$ quantum cost.

4.2.2 Reversible Register

Figure 4.2 shows the implementation of reversible clocked D flip-flop. The reversible D flip-flops are used to implement an n-bit reversible register, which is shown in Figure 4.3. It requires $2n$ gates, produces $n + 1$ garbage outputs and needs $6n$ quantum cost.

4.2.3 Reversible PIPO Left-Shift Register

In PIPO shift register, all data bits are loaded into the register at once with the next clock pulse. After shift operation all data bits appear on the parallel outputs immediately. A reversible PIPO right-shift register was found in the literature. The authors make it compatible for implementing left-shift register. The control inputs $(HOLD, E)$ select the operation of the register according to the function entries in Table 4.1.

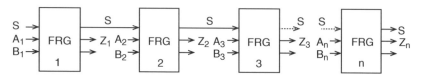

Figure 4.1 Two-input n-bit reversible MUX.

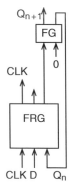

Figure 4.2 A clocked D Flip-Flop.

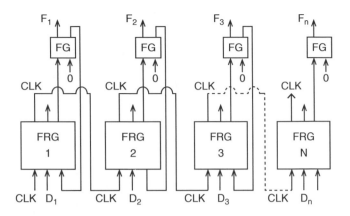

Figure 4.3 An *n*-bit reversible D flop-flop.

Table 4.1 Function Table for Reversible PIPO Left-Shift Register.

HOLD	E	Final Output Q_i^+	
0	0	Q_{i-1} (Left shift)	
0	1	I_i	(Parallel load)
1	Don't care	Qi	(No change)

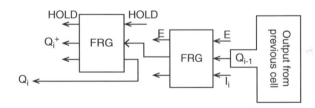

Figure 4.4 Implementation of the characteristic function of Equation (4.2.3.1).

From Table 4.1, when both *HOLD* and *E* are low, the shift register performs the left-shift operation. When *HOLD* is low and *E* is high, the inputs $I_1, I_2, I_3 ..., I_n$ are loaded in parallel into the register coincident with the next clock pulse. The outputs $O_1, O_2, O_3 ..., O_n$ are available in parallel from the Q output of the flip-flops. When *HOLD* is high, present value of flip-flop is returned to the *D* input of that flip-flop. In other words, the register is inactive when *HOLD* is high, and the contents are stored indefinitely.

The characteristic function of Q_i^+ (shown in Figure 4.4) can be obtained from Table 4.1.

$$Q_i^+ = (HOLD)'.E.I_i + (HOLD)'.E.Q_{i-1} + HOLD.Q_i \qquad (4.2.3.1)$$

For the first stage Q_{i-1} is the serial input (*SI*) and for last stage Q_i is the serial output (*SO*). Figure 4.5 and Figure 4.6 show the basic cell of the reversible PIPO left-shift register and its block diagram. By cascading *n* basic cells, reversible PIPO left-shift register can be implemented as shown in Figure 4.7. An *n*-bit reversible PIPO shift register can be implemented by 5*n* reversible gates and 3*n* + 3 garbage outputs with quantum cost of *18n*.

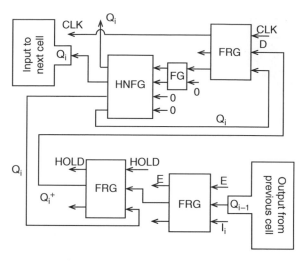

Figure 4.5 The structure of the basic cell for the reversible PIPO left-shift register.

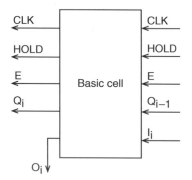

Figure 4.6 The block diagram of the basic cell for the reversible PIPO left-shift register.

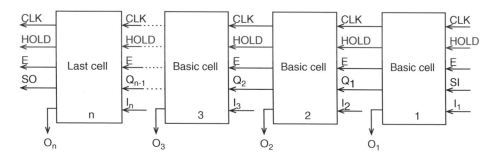

Figure 4.7 An n-bit reversible PIPO left-shift register.

4.2.4 Reversible Parallel Adder

Reversible parallel adder using MTSG gate and TS-3 gate is shown in this subsection. As carry-out of the parallel adder is ignored in the division circuit, implementation of $(n + 1)$-bit parallel adder requires n full adders and the last bit position of the adder requires

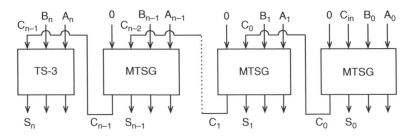

Figure 4.8 An $(n + 1)$-bit parallel adder (carry-out ignored).

computing two Ex-OR operations. Thus, it can be implemented using n MTSG gates and one TS-3 gate as shown in Figure 4.8.

4.3 The Design of Reversible Division Circuit

The reversible design of nonrestoring division circuit for positive integers is shown in Figure 4.9. It has two PIPO reversible left-shift registers: one is $n+2$ bits named as S_1. S.A and other is n bits named as Q. It also contains an n-bit reversible register to store the divisor. Initially, $S = 0$, A $(a_{n-1} a_{n-2} \ldots a_0) = 0$, D $(d_{n-1} d_{n-2} \ldots d_0)$ = dividend, V $(v_{n-1} v_{n-2} \ldots v_0)$ = divisor, and $SIG = 0$. When the division operation is completed, register Q $(q_{n-1} q_{n-2} \ldots q_0)$ contains the quotient and A $(a_{n-1} a_{n-2} \ldots a_0)$ contains the remainder. If $SELECT = 1$, then two-input $(n + 1)$-bit MUX selects $S = 0$ and A $(a_{n-1} a_{n-2} \ldots a_0) = 0$, and

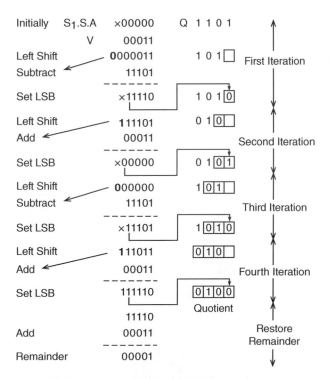

Figure 4.9 Illustration of the division circuit.

n-bit MUX selects dividend D (d_{n-1} d_{n-2} ... d_0). During the clock pulse when $E = 1$ and $HOLD_2 = 0$, the input $S_1 = 1$ and output data from $(n + 1)$-bit MUX are loaded into $S_1.S.A$, and when $HOLD_1 = 0$, outputs from n-bit MUX are loaded into Q in parallel. When $E = 0$, both $S_1.S.A$ and Q act as left-shift registers. Initially, the value of S_1 is not important, it is important only after the left shift of $S_1.S.A.Q$ ($S_1.S.A.Q$ means SO of register Q is connected to SI of $S_1.S.A$); thus, the value of S is shifted to S_1, which is used to select the operation to be performed on $S.A$ and V. If S_1 is 1, then $S.A + V$ is performed; otherwise, $S.A-V$ is computed. Addition or subtraction is performed using $(n + 1)$-bit reversible parallel adder. The complement of the most significant bit (MSB) of the sum is loaded into q_0 bit position of register Q and $(n + 1)$-bit sum is loaded into $S.A$ during next clock pulse when $SELECT$ is 0. It requires $2n + 1$ clock pulses to store the value of quotient into register Q.

After $2n + 1$ clock pulses, control outputs a high signal SIG. This signal is connected to $HOLD_1$ input of Q register. Thus, Q stores the quotient indefinitely. AND operation of (S, SIG) that can be implemented using the Peres gate is connected to $HOLD_2$ input. If S is 0, then remainder restoration is not required, $HOLD_2$ will be high and A will store the remainder indefinitely. If S is 1, then remainder restoration is necessary. During $2n + 1$ clock pulse, as S_1 is 1, restoration is performed by adding V with $S.A$. During the next clock pulse, the correct value of remainder is loaded into $S.A$ when E is 1. After remainder restoration, S must be 0. This results $HOLD_2$ to be high and A will store the remainder indefinitely. The working principle of the reversible design is described in Algorithm 4.3.0.1.

The example in Figure 4.9 will clarify the architecture of a division circuit. It is assumed that dividend = 1101 and divisor = 0011. Initially, $S = 0$, A ($a_3 a_2 a_1 a_0$) = 0, V ($v_3 v_2 v_1 v_0$) = divisor = 0011, and Q ($q_3 q_2 q_1 q_0$) = dividend = 1101. After the division operation, Quotient Q is ($q_3 q_2 q_1 q_0$) = 0100 and remainder A is ($a_3 a_2 a_1 a_0$) = 0001.

Algorithm 4.3.1 Reversible Division

Inputs: $S = 0$, A (a_{n-1} a_{n-2} ... a_0) = 0, D (d_{n-1} d_{n-2} ... d_0) = dividend and V (v_{n-1} v_{n-2} ... v_0) = divisor

Outputs: Q (q_{n-1} q_{n-2} ... q_0) = quotient and A (a_{n-1} a_{n-2} ... a_0) = remainder

The design is evaluated through the generation of a theorem that gives the necessary lower bounds.

Begin
Division (A, D, V)
$SIG := 0$
$SELECT := 1$
$count := 0$
Rest of the algorithm at the end.

Property 4.3.0.1 The n-bit reversible division circuit can be realized by at least $18n + 17$ gates and $11n + 18$ garbage outputs with quantum cost of $61n + 50$.

while TRUE **do**

  E := 1

  if CLK is High AND E=1 **then**

    if SELECT =1 **then**

      Inputs $(S1 = 1, S = 0, A(a_{n-1}a_{n-2}...a_0) = 0$ and D $(d_{n-1}d_{n-2}...d_0)$= dividend) are loaded into $S_1.S.A.Q$

      $SELECT := 0$

    else

      $S_1 = 1$ and outputs from previous step are loaded into $S.A.Q$

      count := count +1

    end if

    $E := 0$

    if *count* = 2n+1 **then**

      break

    end if

    if CLK is high **then**

      Occurrence of left shift of $S_1.S.A.Q$, thus S_1 gets the value of S

      count := count +1

      if $S1 = 0$ **then**

        The values of $S.A$ and 2's complement of divisor are the inputs of Adder

      end if

      if the MSB of sum is 0 **then**

        q_0 will be 1 during the next CLK

      else

        Q_0 will be 0 during the next CLK

      end if

    end if

  end if

  if count = 2n+1 **then**

    $SIG = 1$, $HOLD_1$ will be 1, and Q will store quotient indefinitely

    if AND operation of (S', SIG) is 0 **then**

      The values of $S.A$ and divisor are the inputs of adder

      $E := 1$

      Restored remainder is loaded into A during next CLK

      S will be 0 which results $HOLD_2$ to be 1 and A will store remainder indefinitely

    end if

  else

    $HOLD_2$ will be 1 and A will store remainder indefinitely

  end if

end while

End

Proof: The division circuit has following properties:

- Two-input $(n + 1)$-bit reversible MUX requires $n + 1$ gates, $n + 1$ garbage outputs and $5(n + 1)$ quantum cost.
- Two-input n-bit reversible MUX requires n gates, n garbage outputs and $5n$ quantum cost.
- $(n + 2)$-bit reversible PIPO shift register requires $5(n + 2)$ gates, $3(n + 2) + 3$ garbage outputs, and $18(n + 2)$ quantum cost.
- n-bit reversible PIPO shift register requires $5n$ gates, $3n + 3$ garbage outputs, and $18n$ quantum cost.
- n-bit register to store divisor requires $2n$ gates, $n + 1$ garbage outputs, and $6n$ quantum cost.
- $3n + 3$ Feynman gates to avoid fan-out and to invert input bit require $(3n + 3)$ quantum cost.
- nMTSG gates and one TS-3 gate for $(n + 1)$ bit adder require $n + 1$ gates, $2n + 2$ garbage outputs, and $6n + 2$ quantum cost.
- Peres gate for AND operation produces two garbage outputs and four quantum cost.
- One NOT gate that has no quantum cost.

Thus, an n-bit reversible division circuit can be realized by at least $n + 1 + n + 5(n + 2) + 5n + 2n + 3n + 3 + n + 1 + 1 + 1 = 18n + 17$ gates and $n + 1 + n + 3(n + 2) + 3 + 3n + 3 + n + 1 + 2n + 2 + 2 = 11n + 18$ garbage outputs with quantum cost of $5(n + 1) + 5n + 18(n + 2) + 18n + 6n + 3n + 3 + 6n + 2 + 4 = 61n + 50$.

4.4 Summary

This chapter presents a design of sequential division circuit using reversible logic. The hardware has its application in the design of a reversible arithmetic logic unit. An algorithm is also provided to describe the division module. Lower bound of the design was established by providing relevant theorem. The contents of this chapter form an important step in the building of complex reversible circuits for quantum computers.

5

Reversible Binary Comparator

Comparison of two binary numbers finds its wide application in general purpose microprocessors, communication systems, encryption device, sorting networks etc. In this chapter,
an n-bit comparator is presented that has fewer gates; it also produces less garbage outputs, quantum cost, and delay. With the help of properties, the efficiency of reversible logic
synthesis of an n-bit comparator has also been proved in this chapter.

5.1 Design of Reversible n-Bit Comparator

In this section, a compact and improved version of reversible n-bit comparator has been
designed. A binary comparator compares the magnitude of two binary numbers to determine whether they are equal or one is greater/less than the other. To construct the optimized
n-bit comparator, here two reversible gates are introduced, namely Babu-Jamal-Saleheen
(BJS) and Hasan-Lafifa-Nazir (HLN) gate.

5.1.1 BJS Gate

In this subsection, BJS gate is used to design the 1-bit comparator and the MSB comparator.
Figure 5.2 shows the equivalent quantum realization of the BJS gate. Here, each dotted
rectangle is equivalent to a 2×2 CNOT gate. Hence, the quantum cost of the BJS gate is six.

A 3×3 reversible gate is also introduced in this subsection, which is named as HLN gate,
as shown in Figure 5.3. The corresponding truth table of the gate is shown in Table 5.2. The
truth table shows that the input–output combinations preserve the one-to-one mapping
between them. The HLN gate can implement all Boolean functions. When $C = 1$ and $C = 0$,
it implements EX-OR $(Q = A \oplus B)$ and AND $(Q = AB)$ functions, respectively. When $B = 1$,
it implements OR function $(R = A + B)$. When $B = 0$ and $C = 0$, it depicts NOT function
$(Q = A')$.

In this subsection, HLN gate is used to design the single-bit greater or equal comparator
cell. Figure 5.4 shows the equivalent quantum realization of HLN gate. The quantum cost
of the HLN gate is five.

Reversible and DNA Computing, First Edition. Hafiz Md. Hasan Babu.

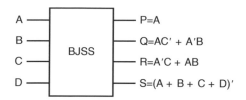

Figure 5.1 Reversible BJS gate.

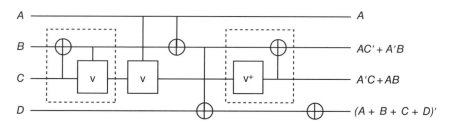

Figure 5.2 Quantum realization of the BJS gate.

Table 5.1 Truth Table of the BJS Gate.

A	B	C	D	P	Q	R	S
0	0	0	0	0	0	0	1
0	0	0	1	0	0	0	0
0	0	1	0	0	0	1	0
0	0	1	1	0	0	1	1
0	1	0	0	0	1	0	0
0	1	0	1	0	1	0	1
0	1	1	0	0	1	1	1
0	1	1	1	0	1	1	0
1	0	0	0	1	1	0	0
1	0	0	1	1	1	0	1
1	0	1	0	1	0	0	1
1	0	1	1	1	0	0	0
1	1	0	0	1	1	1	1
1	1	0	1	1	1	1	0
1	1	1	0	1	0	1	0
1	1	1	1	1	0	1	1

5.1.2 Reversible 1-Bit Comparator Circuit

The 1-bit comparator compares two 1-bit binary numbers and determines the result among $ALB(A < B)$, $AEB(A = B)$ and $AGB(A > B)$. The truth table of 1-bit comparator is shown in Table 5.3. From the truth table, it is obvious that $ALB = A'B$, $AEB = (A \oplus B)'$ and $AGB = AB'$. The BJS gate can be used as a 1-bit comparator. The gate takes two 1-bit binary numbers

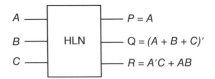

Figure 5.3 Reversible HLN gate.

Table 5.2 Truth Table of the BJS Gate.

A	B	C	P	Q	R
0	0	0	0	1	0
0	0	1	0	0	1
0	1	0	0	0	0
0	1	1	0	1	1
1	0	0	1	0	0
1	0	1	1	1	0
1	1	0	1	1	1
1	1	1	1	0	1

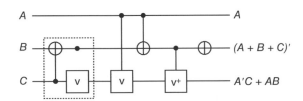

Figure 5.4 Quantum realization of the HLN gate.

Table 5.3 Truth Table of the 1-Bit Binary Comparator.

Inputs		Outputs		
A	B	ALB	AEB	AGB
0	0	0	1	0
0	1	1	0	0
1	0	0	0	1
1	1	0	1	0

as inputs A and B and two ancilla inputs. The outputs of the gate produce one garbage output and three consecutive outputs AGB, ALB, and AEB as shown in Figure 5.5.

5.1.3 Reversible MSB Comparator Circuit

The MSB comparator circuit takes MSB of two binary numbers A_n, B_n, and sets two ancilla inputs as 0. It produces three outputs based on the bits present at the input level. $ALB(R_n), AGB(Q_n)$, and $AEB(P_n)$ are the three outputs produced from this circuit. Later on

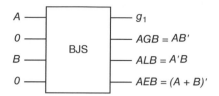

Figure 5.5 BJS gate works as reversible 1-bit comparator.

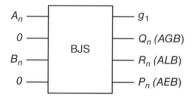

Figure 5.6 BJS gate works as reversible MSB comparator.

these outputs are fed into the next level, where $(n-1)^{th}$ bits have also been compared. This MSB comparator circuit shown in Figure 5.6 consists of only one BJS gate, which produces one garbage output and has the quantum cost of six.

5.1.4 Reversible Single-Bit Greater or Equal Comparator Cell

In this subsection, the design of a reversible single-bit comparator cell is shown, which consists of one HLN gate and two Peres gates (PGs). It takes $(n-1)^{th}$ bits of two binary numbers A and B and two more inputs P_n and Q_n from previous comparison result. Together they work to produce two outputs $AEB(P_n-1)$ and $AGB(Q_n-1)$, which indicates whether the given numbers A and B are equal to each other or greater than the other. Figure 5.7

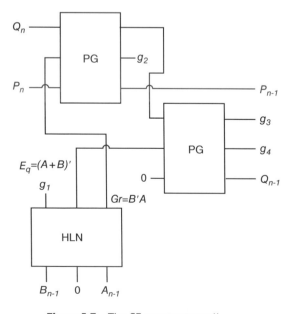

Figure 5.7 The GE comparator cell.

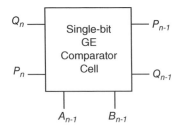

Figure 5.8 Block diagram of the single-bit GE comparator cell.

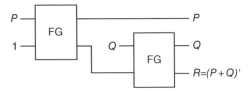

Figure 5.9 The LT comparator cell.

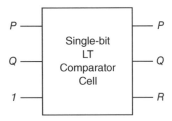

Figure 5.10 Block diagram of the single-bit LT comparator cell.

and Figure 5.8 show the circuit and the block diagram of the reversible single-bit greater or equal comparator cell, respectively.

5.1.5 Reversible Single-Bit Less Than Comparator Cell

The circuit described in previous subsection produces two outputs $P_n - 1(AEB)$ and $Q_n - 1(AGB)$. If $A_n - 1 < B_n - 1$, it can be calculated using the equation, $ALB = (AEB \oplus AGB)'$. The design of this circuit and its block diagram are shown in Figure 5.9 and Figure 5.10, respectively. As the circuit requires two FGs, the quantum cost of the circuit is two.

5.1.6 Reversible 2-Bit Comparator Circuit

A reversible 2-bit binary comparator consists of a reversible MSB comparator, a single-bit GE comparator, and a single-bit LT comparator circuit. The reversible design of 2-bit binary comparator is shown in Figure 5.11.

5.1.7 Reversible n-Bit Comparator Circuit

The reversible n-bit binary comparator is shown in Figure 5.12, which consists of one MSB comparator, $(n - 1)$ single-bit GE comparators and a single-bit LT comparator circuit. The algorithm for constructing reversible n-bit binary comparator is given in Algorithm 5.1.7.1.

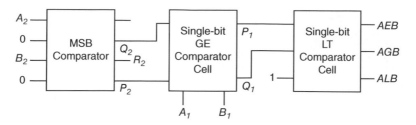

Figure 5.11 Reversible 2-bit comparator.

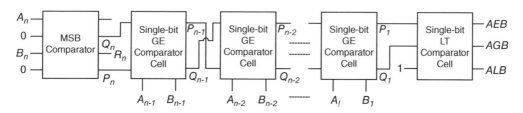

Figure 5.12 Reversible *n*-bit comparator.

An MSB comparator circuit requires one BJS gate. A single-bit GE comparator cell is constructed with one HLN gate and two PGs. A single-bit LT comparator cell requires two FGs. A 2-bit comparator circuit is constructed with one MSB comparator circuit, one single-bit GE comparator cells and one single-bit LT comparator cell as shown in Figure 5.11. Hence, the total number of gates required to construct 2-bit binary comparator (NOG_{2-bit}) is

$$NOG_{2-bit} = NOG_{MSB} + NOG_{SB-GE} + NOG_{SB-LT}$$
$$= 1 + (1 + 2) + 2$$
$$= 6$$
$$= 3 \times 2$$

Hence, the statement holds for the base case $n = 2$.

Assume that, the statement holds for $n = k$. Hence, a reversible k-bit binary comparator can be realized with $3k$ reversible gates.

A $(k + 1)$-bit binary comparator requires one MSB comparator circuit, k single-bit GE comparator cells and one single-bit LT comparator cell. Hence, the total number of gates required to construct $(k + 1)$-bit binary comparator
($NOG_{(k+1)-bit}$) is

$$NOG_{(k+1)-bit} = NOG_{MSB} + k \times NOG_{SB-GE} + NOG_{SB-LT}$$
$$= 1 + k(1 + 2) + 2$$
$$= 3k + 3$$
$$= 3(k + 1)$$

Thus, the statement holds for $n = k + 1$.

Therefore, a reversible n-bit binary comparator can be realized with $3n$ reversible gates.

Algorithm 5.1.7.1 Multi-Operand Addition

1: Begin
2: Take on MSB comparator circuit I_n. Output of this block are considered to be of level L_n.
3: $I_n[I_1] = A_n, I_n[I_2] = 0, I_n[I_3] = B_n, I_n[I_4] = 0$
4: **if** $I_n[I_1] < I_n[I_2]$ **then**
5: $\quad I_n[O_n] = R_{Ln} = 1$
6: **else if** $I_n[I_1] > I_n[I_2]$ **then**
7: $\quad I_n[O_2] = Q_{Ln} = 1$
8: **else**
9: $\quad I_n[O_4] = P_{Ln} = 1$
10: **end if**
11: For each single-bit comparator circuit, level of inputs and outputs are considered to be of L_n and L_{n-1} respectively.
12: **for** j=n − 1 **to** 1 **do**
13: $\quad$ Take one single-bit GE comparator cell C_j
14: $\quad$ **if** j=n-1 **then**
15: $\quad\quad C_j[I_1] = I_n[O_2] = Q_{Ln}, C_j[I_2] = I_n[O_4] = P_{Ln}, C_j[I_3] = A_{n-1}, C_j[I_4] = B_{n-1}$
16: $\quad$ **else**
17: $\quad\quad C_j[I_1] = C_{j-1}[O_2] = Q_{Ln-1}, C_j[I_2] = C_{j-1}[O_1] = P_{Ln-1}, C_j[I_3] = A_j, C_j[I_4] = B_j$
18: $\quad$ **end if**
19: **end for**
20: Take one single-bit LT comparator cell, L
21: $L[I_1] = C_1[O_1], L[I_2] = C_1[O_2], L[I_3] = 1, L[O_1] = C_1[O_1], L[O_2] = C_1[O_2], L[O_3] = C_1[O_1] \oplus C_1[O_2] \oplus 1$
22: End

Property 5.1.7.1 A reversible n-bit comparator requires $(0.15n − 0.03)$ ns timing delay, where n is the number of data bits.

Proof: The n-bit comparator design has a serial architecture that has latency of O(n). An n-bit comparator requires one MSB comparator circuit, $(n − 1)$ GE comparator cells, and one single bit LT comparator cell. Hence, the delay of the comparator circuit is

$$\text{Total delay } (T) = T_{MSB} + (n − 1) \times T_{SB-GE} + T_{SB-LT} \qquad (5.1.7.1)$$

The critical path for single-bit GE comparator cell contains one HLN gate and two PGs. The delays of HLN gate and PG are 0.07 ns and 0.04 *ns*, respectively. Hence, the delay of single-bit GE comparator cell can be calculated as follows:

$$T_{SB-GE} = T_{HLN} + 2T_{PG}$$
$$= (0.07 + 2 \times 0.04) \text{ ns}$$
$$= 0.15ns$$

Property 5.1.7.2   A reversible n-bit comparator requires $(42.5n - 14.5)$ µm² area, where n is the number of data bits.

Proof: An n-bit comparator is constructed with one MSB comparator circuit, $(n - 1)$ single-bit GE comparator cells and one single bit LT comparator cell. An MSB comparator circuit requires one BJS gate. A single-bit GE comparator cell is constructed with one HLN gate and two PGs. A single-bit LT comparator cell requires two FGs. Hence, the total number of gates is

$$NOG = NOG_{MSB} + (n - 1) \times NOG_{SB-GE} + NOG_{SB-LT} \qquad (5.1.7.2)$$
$$= 1 \times BJS + (n - 1)(HLN + 2 \times PG) + 2 \times FG$$

The areas of BJS gate, HLN gate, PG, and FG are 20, 17.5, 12.5 and 7 µm² respectively, which are obtained using Algorithm 5.1.7.2. Hence, the total area of an n-bit comparator can be modeled as follows:

$$A = A_{MSB} + (n - 1) \times A_{SB-GE} + A_{SB-LT}$$
$$= \{20 + (n - 1)(17.5 + 2 \times 12.5) + 7\}\text{µm}^2$$
$$= (42.5n - 42.5 + 27)\text{µm}^2$$
$$= (42.5n - 14.5)\text{µm}^2$$

Algorithm 5.1.7.2 Multi-Operand Addition

1: Begin
2: Take one MSB comparator circuit I_n.
3: **Area**=$area(I_n)$.
4: **for** j =$n - 1$ **to** 1 **do**
5:    Take one single-bit GE comparator cell C_j
6:    **Area=Area+**$area(C_j)$.
7: **end for**
8: Take on single-bit LT comparator cell, L
9: **Area=Area+**$area(L)$
10: End

Property 5.1.7.3   A reversible n-bit comparator requires $(117.76n - 32.94)\mu W$ power, where n is the number of data bits.

Proof: One MSB comparator circuit, $(n - 1)$ single-bit GE comparator cells and one single-bit LT comparator cell are required to realize an n-bit comparator. Hence, the total required power can be measured by using following equation:

$$\text{Total Power } (P) = P_{MSB} + (n - 1) \times P_{SB-GE} + P_{SB-LT} \qquad (5.1.7.3)$$

The single-bit GE comparator cell contains one HLN gate and two PGs. The power requirement of HLN gate and PG are 58.88 μW and 29.44 μW, respectively. Hence, the power requirement of a single-bit GE comparator cell, PSB-GE, can be calculated as follows:

$$P_{SB-GE} = (58.88 + 2 \times 29.44)\mu W$$
$$= 117.76\mu W$$

The power requirement of MSB and LT modules are obtained as 37.72 μW and 47.1 μW, respectively. Hence, the total power of an n-bit comparator can be modeled as below according-ing to

$$P = \{37.72 + (n-1) \times 117.76 + 47.1\}\mu W$$
$$= (117.76n - 117.76 + 84.82)\mu W$$
$$= (117.76n - 32.94)\mu W$$

Property 5.1.7.4 A reversible n-bit binary comparator ($n \geq 2$) produces at least $4n - 3$ garbage outputs, where n is the number of data bits.

Proof: A 2-bit comparator circuit consists of one MSB comparator, one single-bit GE comparator cell and one single-bit LT comparator cell. The MSB comparator circuit produces one garbage output and the single-bit GE comparator cell produces four garbage outputs. The single-bit LT comparator cell does not produce any garbage output. Hence, the total number of garbage outputs produced by a 2-bit comparator is at least

$$G = {}_{2\text{-}bit} = G_{MSB} + G_{SB-GE} + G = {}_{SB-LT}$$
$$= 1 + 4 + 0$$
$$= 5$$
$$= 4 \times 2 - 3$$

Hence, the statement holds for the base case $n = 2$.

Assume that, the statement holds for $n = k$. Hence, a reversible k-bit binary compara-tor produces at least $4k - 3$ garbage outputs. A $(k + 1)$-bit binary comparator requires one MSB comparator circuit, k single-bit GE comparator cell and one single-bit LT comparator cell. So, the total number of garbage outputs produced by the $(k + 1)$-bit binary comparator $(G(k + 1)\text{-bit})$ is

$$G_{(k+1)-bit} = G_{MSB} + k \times G_{SB-GE} + G_{SB-LT}$$
$$= 1 + k \times 4 + 0$$
$$= 4k + 1$$
$$= 4(k + 1) - 3$$

Thus, the statement holds for $n = k + 1$. Therefore, a reversible n-bit binary comparator produces at least $4n - 3$ garbage outputs.

Property 5.1.7.5 A reversible n-bit binary comparator ($n \geq 2$) can be realized with $13n - 5$ quantum cost; where n is the number of data bits.

Proof: An MSB comparator circuit requires one BJS gate. A single-bit GE comparator cell is constructed with one HLN gate and two PGs. A single-bit LT comparator cell requires two FGs. A 2-bit comparator circuit is constructed with one MSB comparator circuit, one single-bit GE comparator cell, and one single-bit LT comparator cell. The quantum cost of BJS gate, HLN gate, PG and FG are six, five, four, and one, respectively. Hence, the total quantum cost of the 2-bit comparator circuit is:

$$QC_{2\text{-}bit} = QC_{MSB} + QC_{SB-GE} + QC_{SB-LT}$$
$$= 6 + (5 + 4 \times 2) + 2$$
$$= 21$$
$$= 13 \times 2 - 5$$

Hence, the statement holds for the base case $n = 2$.

Assume that, the statement holds for $n = k$. Hence, a reversible k-bit binary comparator can be realized with $13k - 5$ quantum cost.

A $(k + 1)$-bit binary comparator requires one MSB comparator circuit, k single-bit GE comparator cell and one single-bit LT comparator cell. Hence, the quantum cost of the $(k + 1)$-bit binary comparator is $(QC_{(k+1)-bit})$ is

$$QC_{(k+1)-bit} = QC_{MSB} + k \times QC_{SB-GE} + QC_{SB-LT}$$
$$= 6 + k(5 + 4 \times 2) + 2$$
$$= 13k + 8$$
$$= 13(k + 1) - 5$$

Thus, the statement holds for $n = k + 1$.

Therefore, a reversible n-bit binary comparator can be realized with $13n - 5$ 5 quantum cost.

Property 5.1.7.6 A reversible n-bit binary comparator ($n \geq 2$) requires at least three ancilla inputs, where n is the number of data bits.

Proof: An n-bit comparator circuit ($n \geq 2$) consists of one MSB comparator, $(n - 1)$ single-bit GE comparator cells and one single-bit LT comparator cell. The MSB comparator circuit requires two ancilla inputs and the single-bit LT comparator cell requires one ancilla input. The $(n - 1)$ single-bit GE comparator cells do not require any ancilla input. Hence, the total number of ancilla inputs required for an n-bit comparator is at least

$$\text{Ancilla inputs } (AI) = AI_{MSB} + (n - 1) \times AI_{SB-GE} + AI_{SB-LT}$$
$$= 2 + (n - 1) \times 0 + 1$$
$$= 3$$

Example 5.1.7.1 The reversible 3-bit comparator requires one MSB comparator circuit, two single-bit GE comparator cells, and one single-bit LT comparator cell. To construct the comparator, it requires $3 \times 3 = 9$ reversible gates, which produce $(4 \times 3 - 3) = 9$ garbage outputs. It requires three ancilla inputs. The quantum cost is $(13 \times 3 - 5) = 34$, the required area is $(42.5 \times 3 - 12.5) = 115 \ \mu m^2$, and the required power is $(117.76 \times 3 - 32.94) = 320.34 \ \mu W$.

5.2 Summary

In this chapter, the design of a reversible low power n-bit binary comparator is discussed. An algorithm is presented for constructing a compact reversible n-bit binary comparator circuit. Two reversible gates, namely, Babu-Jamal-Saleheen (BJS) and Hasan-Lafifa-Nazir (HLN) gates, are also presented to optimize the comparator. In addition, several properties on the numbers of gates, garbage outputs, quantum cost, ancilla input, power, delay, and area of the reversible n-bit comparator have been presented. The reversible binary comparator circuit can be used in designing different reversible circuits such as reversible processing unit, complex arithmetic circuits, communication systems, etc. Thus, the concept of designing reversible binary comparator will help in designing the reversible subtractor, multiplier, divider, and various adder circuits.

6

Reversible Sequential Circuits

The sequential circuits considered in this chapter are reversible designs of latches, such as D latch, T latch, JK latch, SR latch, and their corresponding master–slave flip-flops. Negative enable reversible D latch is also introduced to be used in master–slave flip-flops. Because of the negative enable D latch, it does not require the inversion of the clock for use in the slave latch. Further, this reduces the quantum cost of the master–slave flip-flops. A strategy is also introduced of using the Fredkin gate at the outputs of a reversible latch to make it an asynchronous set/reset latch. This strategy is used to design a Fredkin gate based on asynchronous set/reset D latch and the master–slave D flip-flop.

6.1 An Example of Design Methodology

The output equations of the reversible gates are used as the templates for mapping the characteristic equation of the latch into an equivalent reversible design. For example, in the characteristic equation of a latch, suppose an expression as $A \cdot B \oplus C$, it can be easily matched with the template of the output equation of the reversible Peres gate, and hence the Peres gate can be used to synthesis this expression. The design methodology is illustrated here with the design of the reversible JK latch as an example circuit.

Step 1. Make templates of the output equations of the basic reversible gates as follows:

1. Peres gate: $PG(A, B, C) = A \cdot B \oplus C$
2. Fredkin gate: $F(A, B, C) = A'B + AC$ or $F(A, B, C) = A'C + AB$
3. Toffoli gate (4 inputs, 4 outputs): $TG(A, B, C, D) = A \cdot B \cdot C \oplus D$
4. Feynman gate: $FG(A, B) = A \oplus B$
5. NOT gate: $NOT(A) = A'$

Step 2. Derive the characteristic equation of the latch. For example, the characteristic equation of the JK latch can be derived as $Q^+ = (J \cdot \overline{Q} + \overline{K} \cdot Q \cdot E + \overline{E} \cdot Q)$ where J and K are the inputs to JK latch, Q is the previous output, E is the enable or clock signal, and Q^+ is the current output. Please note that in further discussions E represents the enable or clock signal.

Step 3. Derive the minimum number of garbage outputs needed to convert the characteristic equation as a reversible function. For example, as evident from the characteristic equation of the JK latch for the eight inputs combinations ($E = 0, J = 0, K = 0,$

Reversible and DNA Computing, First Edition. Hafiz Md. Hasan Babu.
© 2021 John Wiley & Sons Ltd. Published 2021 by John Wiley & Sons Ltd.

$Q = 0$), $(E = 0, J = 0, K = 1, Q = 0)$, $(E = 0, J = 1, K = 0, Q = 0)$, $(E = 0, J = 1, K = 1, Q = 0)$, $(E = 1, J = 0, K = 0, Q = 0)$, $(E = 1, J = 0, K = 1, Q = 0)$, $(E = 1, J = 0, K = 1, Q = 1)$, and $(E = 1, J = 1, K = 1, Q = 1)$, the output Q^+ is 0. Thus, it will require at least three garbage outputs to have eight distinct output combinations when Q^+ is 0.

Step 4. Rewrite the characteristic equation by replacing the functions in the parenthesis by variables to have the modified characteristic equation with fewer variables. For example, the JK latch characteristic equation can be rewritten as $Q^+ = M \cdot E + \overline{E} \cdot Q$, where M is the new variable substituted for $J \cdot \overline{Q} + \overline{K} \cdot Q$.

Step 5. From the templates in Step 1, find the template that exactly maps the equation in Step 4 with minimum quantum cost. For example, the equation $Q^+ = M \cdot E + \overline{E} \cdot Q$ maps on the Fredkin gate with minimum quantum cost of 5 where E is equivalent of input A to the Fredkin gate. This step is illustrated in Figure 6.1.

Step 6. Find the template matching for the functions represented by variables in Step 4. For example, the variable M substituted for can be mapped $J \cdot \overline{Q} + \overline{K} \cdot Q$ on the Fredkin gate with the Fredkin gate inputs as $A = Q, B = J$ and $C = \overline{K}$. Hence, the quantum cost is 6 (quantum cost of 1 Fredkin gate + quantum cost of 1 NOT gate). This step is illustrated in Figure 6.2.

Step 7. Avoid the fan-out in the derived reversible circuit by properly using the Feynman gates. Further maintain the lower bound in terms of garbage outputs by carefully utilizing the outputs. For example, by avoiding the fan-out using Feynman gates and carefully reutilizing the unused outputs, the derived design of the reversible JK latch is shown in Figure 6.3.

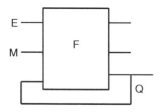

Figure 6.1 Mapping of JK latch (Equation $Q^+ = M \cdot E + \overline{E} \cdot Q$) on the Fredkin gate.

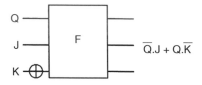

Figure 6.2 Mapping of variable M (Equation $J \cdot \overline{Q} + \overline{K} \cdot Q$) on the Fredkin gate.

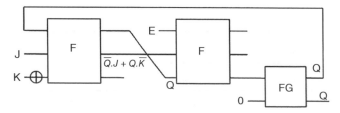

Figure 6.3 Reversible design of JK latch with minimal garbage outputs.

6.2   The Design of Reversible Latches

In this section, the designs of reversible latches are described that are optimized in terms of quantum cost, delay, and the number of garbage outputs.

6.2.1   The SR Latch

The SR latch can be designed using two cross-coupled NOR gates or two cross-couple NAND gates. The S input sets the latch to 1, while the R input resets the latch to 0. As an example, the NAND gates-based design is shown in Figure 6.4. The Toffoli gate and the Fredkin gate-based SR latches require two Toffoli gates and two Fredkin gates, respectively, and hence it will have a quantum cost of 10 and delay of 10Δ. It is possible to decrease the quantum cost and the delay of the cross-coupled reversible SR latch by using two Peres gates working as cross-coupled NAND gates as in Figure 6.5. Since the Peres gate has a quantum cost of 4 and the delay of 4Δ, the design will have a propagation delay of 8Δ and the quantum cost of 8.

In order to design a reversible SR latch that can work for all possible input combinations with minimum number of garbage outputs, the characteristic equation of the SR latch is studied, which is given as $Q^+ = S + \overline{R} \cdot Q$. From the characteristic equation, it is observed that for five input combinations $(S = 0, R = 0, Q = 1), (S = 1, R = 0, Q = 0)$, $(S = 1, R = 0, Q = 1), (S = 1, R = 1, Q = 0)$, and $(S = 1, R = 1, Q = 1)$, we get $Q^+ = 1$; thus, it requires at least three garbage outputs to have the reversible implementation of the SR latch. Further, the SR latch enters in the unstable condition when $S = 1$ and $R = 1$. In order to design a reversible SR latch that can work for all input combinations and is optimized in terms of garbage outputs, the output Q is modified for the selective input combinations as follows: When $(S = 1, R = 1, Q = 0)$ the output Q^+ is assigned the value

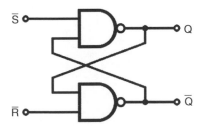

Figure 6.4   Conventional cross-coupled SR latch.

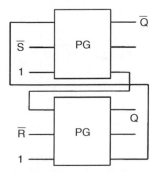

Figure 6.5   Peres gate based SR latch without enable.

of 0; and when $(S = 1, R = 1, Q = 1)$ the output Q^+ is assigned the value of 1. Thus, when $S = 1$ and $R = 1$, it is got $Q^+ = Q$. After the modifications in the output for these two selective input combinations, it is observed that now for only four input combinations $(S = 0, R = 0, Q = 1), (S = 1, R = 0, Q = 0), (S = 1, R = 0, Q = 1)$, and $(S = 1, R = 1, Q = 1)$, it is got $Q^+ = 1$. Thus, it will require only two garbage outputs to realize the reversible SR latch. Thus, the garbage outputs in the reversible SR latch will be one less than the number of garbage outputs in the conventional irreversible SR latch.

Furthermore, this removes the unstable condition in the SR latch when $S = 1$ and $R = 1$. The modified truth table of the SR latch is shown in Table 6.1. From Table 6.1, a characteristic equation as $Q^+ = (S \oplus Q) \cdot (S \oplus R) \oplus Q$ is derived for the reversible SR latch. This equation can be easily matched with the template of the Peres gate considering $A = S \oplus Q, B = S \oplus R$, and $C = Q$ where A, B, C are the inputs of the Peres gate. The functions $S \oplus Q$ and $S \oplus R$ match with the template of the Feynman gate and hence Feynman gates can be used to generate these functions. The design of the reversible SR latch is shown in Figure 6.6. The reversible SR latch has the quantum cost of 7, delay of 7Δ and has the bare minimum of 2 garbage outputs. In order to design the gated reversible SR latch, a characteristic equation of the gated reversible SR latch is derived as $Q^+ = (E \cdot S \oplus Q) \cdot (S \oplus R)) \oplus Q$. From the characteristic equation, it can be easily seen that it requires three garbage outputs for the reversible realization of the gated reversible SR latch. Further, using the design strategy considering $A = E, B = S \oplus Q), C = S \oplus R$, and $D = Q$ where A, B, C, D are the inputs of the four input Toffoli gate, the characteristic equation can be easily matched with the template of the Toffoli gate (TG). The variables $B = (S \oplus Q)$ and $C = S \oplus R$ needed above

Table 6.1 Modified Truth Table of the SR Latch

S	R	Q	Q^+
0	0	0	0
0	0	1	1
0	1	0	0
0	1	1	0
1	0	0	1
1	0	1	1
1	1	0	0
1	1	1	1

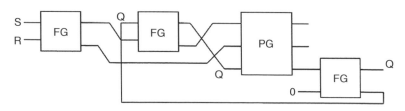

Figure 6.6 Reversible SR latch based on modified truth table.

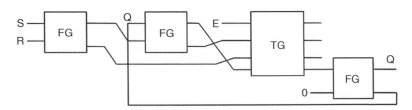

Figure 6.7 Reversible gated SR latch based on modified truth table.

match with the template of the Feynman gate. The gated reversible design of SR latch is shown in Figure 6.7 and has the quantum cost of 16 (the quantum cost of four inputs Toffoli gate is 13) and has the delay of 16Δ. The design has three garbage outputs, which is minimal.

6.2.2 The D Latch

The characteristic equation of the D latch can be written as $Q^+ = D \cdot E + \overline{E} \cdot Q$. When the enable signal (clock) is 1, the value of the input D is reflected at the output that is $Q^+ = D$. When $E = 0$, the latch maintains its previous state, that is $Q^+ = Q$. The characteristic equation of the D latch can be mapped to the Fredkin gate (F) as it matches the template of the Fredkin gate. Figure 6.8 shows the realization of the reversible D latch using the Fredkin gate and the Feynman gate (fan-out is not allowed in reversible logic and the role of Feynman gate is to avoid the fan-out). It is observed that it cannot be further optimized in terms of quantum cost and the garbage outputs. This can be understood as follows: From the characteristic equation of the D latch $Q^+ = D \cdot E + \overline{E} \cdot Q$, it is seen that for the four inputs combinations $(E = 0, D = 0, Q = 0), (E = 0, D = 1, Q = 0), (E = 1, D = 0, Q = 0)$, and $(E = 1, D = 0, Q = 1)$, the output Q^+ is 0. The addition of one garbage output can resolve only two output positions since one bit can produce only two distinct output combinations. Since $2^2 = 4 > 3$, thus it will require at least two garbage outputs to have four distinct output combinations when Q^+ is 0. The propagation delay of the Fredkin gate-based D latch is also shown. Since in this design, one Fredkin gate and one Feynman gate in series, its propagation delay is 6Δ, which is the summation of 5Δ propagation delay of the Fredkin gate and 1Δ propagation delay of the Feynman gate.

6.2.2.1 The D Latch with Outputs Q and $\overline{Q}$
The design shown in Figure 6.8 does not produce the complement output $\overline{Q}$, which is required often in sequential circuits. In this subsection, a design of the D latch is shown that

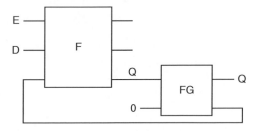

Figure 6.8 Fredkin gate-based D latch with one Feynman gate.

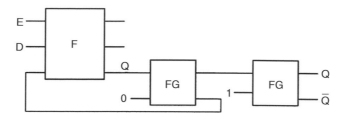

Figure 6.9 Fredkin gate-based D latch with two Feynman gates.

has both the outputs Q and $\overline{Q}$ and is designed with one Fredkin gate and two Feynman gates as shown in Figure 6.9. In the design, the Feynman gate is used to generate the complement of the output Q. The design has the quantum cost of 7 (quantum cost of the Fredkin gate + quantum cost of the two Feynman gates) and has bare minimum of two garbage outputs. The propagation delay of this design is 7Δ.

6.2.2.2 The Negative Enable Reversible D Latch

In this subsection, a design of the reversible D latch is introduced that will pass the input D to the output Q when $E = 0$; otherwise maintains the same state. The characteristic equation of such a negative enable D latch can be written as $Q^+ = D \cdot \overline{E} + E \cdot Q$. This characteristic equation of the negative enable reversible D latch matches with the template of the Fredkin gate. Thus, it can be mapped on the second output of the Fredkin gate as shown in Figure 6.10. The Feynman gate used in the design plays the role of avoiding the fan-out of more than one. In this design, the negative enable D latch is designed with a special purpose of utilizing it in master–slave flip-flops. This is because it will help to design master–slave flip-flops in which no clock inversion is required. The design shown in Figure 6.11 does not have the output $\overline{Q}$, which can be generated as shown in Figure 6.11.

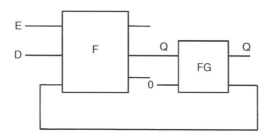

Figure 6.10 Fredkin gate-based negative enable reversible D latch with only output Q.

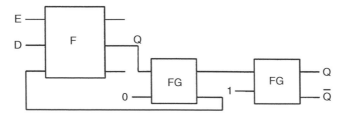

Figure 6.11 Fredkin gate-based negative enable reversible D latch with outputs Q and $\overline{Q}$.

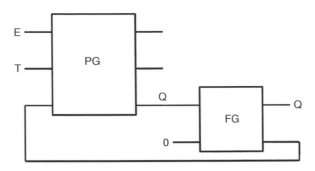

Figure 6.12 Peres gate-based *T* latch.

6.2.3 T Latch

The characteristic equation of the T latch can be written as $Q^+ = (T \cdot Q \cdot E + \overline{E} \cdot Q)$. But the same result can also be obtained from $Q^+ = (T \cdot E) \oplus Q$. The T (toggle) latch is a complementing latch that complements its value when $T = 1$, that is when $T = 1$ and $E = 1$, it is $Q^+ = \overline{Q}$. When $T = 0$, the T latch maintains its state and it will have no change in the output. The T latch characteristic equation can be directly mapped to the Peres gate and the fan-out at output Q can be avoided by cascading the Feynman gate (it can be seen that $T \cdot E \oplus Q$ matches with the template of the Peres gate). The design is shown in Figure 6.12. The design has the quantum cost of 5 (quantum cost of one Peres gate + one Feynman gate), delay of 5Δ, and requires two garbage outputs. The reversible T latch design has the minimum garbage outputs as from the characteristic equation of the T latch $Q^+ = (T \cdot E) \oplus Q$, it is seen that for the four inputs combinations $(E = 0, T = 0, Q = 0), (E = 0, T = 1, Q = 0), (E = 1, D = 0, Q = 0)$ and $(E = 1, T = 1, Q = 1)$, the output Q^+ is 0. Thus, it will require at least two garbage outputs to have four distinct output combinations when Q^+ is 0.

In this subsection, a design of the T latch based on the Peres and the Feynman gates has been introduced that has both the outputs Q and $\overline{Q}$. The design is shown in Figure 6.13 and has the quantum cost of 6, delay of 6Δ, and produces two garbage outputs. The design uses the Feynman gate to generate the complement of the output Q. The reversible T latch with outputs Q and $\overline{Q}$ achieves an significant improvement.

6.2.4 The JK Latch

The design of the reversible JK latch has been discussed in Section 6.1 and is shown in Figure 6.3. The JK latch has the capability to set the output, reset, the output or complement the output depending on the value of J and K. When E (clock) is 1, the J input can set the

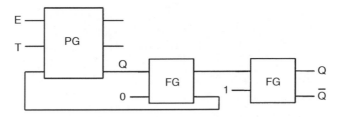

Figure 6.13 Reversible T latch with outputs Q and $\overline{Q}$.

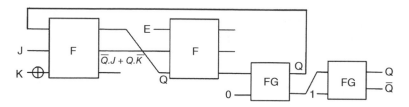

Figure 6.14 Reversible JK latch with outputs Q and $\overline{Q}$.

output to 1 (when $J = 1$), the reset input can reset the output to 0 (when $K = 1$), and when both J and K are set to 1 the output Q is complemented. The reversible JK latch design has the quantum cost of 12, delay of 12Δ, and produces three garbage outputs. Here, a design of the JK latch is illustrated in Figure 6.14, which produces both the output Q and its complement $\overline{Q}$. The design has the quantum cost of 13, delay of 13Δ, and produces three garbage outputs. In this design, the Feynman gate is also used to generate the complement of the output Q.

6.3 The Design of Reversible Master–Slave Flip-Flops

The section presents a strategy to design reversible master–slave flip-flops and the goal is to optimize the quantum cost and the delay of the flip-flops along with the garbage outputs. The master–slave flip-flops designs have a special characteristic of not requiring the clock to be inverted for use in the slave latch. This is because as they use the negative enable D latch as the slave latch, thus no clock inversion is required.

Figure 6.15 shows the design of the master–slave D flip-flop in which positive enable Fredkin gate-based D latch is used, which is shown in Figure 6.8 as the master latch, while the slave latch is designed from the negative enable Fredkin gate-based D latch, which is shown earlier in Figure 6.10. Figures 6.16, Figure 6.17, and Figure 6.18 show the designs of master–slave T flip-flop, JK flip-flop, and SR flip-flop, respectively. It is to be noted that in the master–slave flip-flop designs, the master is designed using the positive enable corresponding latch, while the slave is designed using the negative enable Fredkin gate-based D latch. For example, in the master–slave JK flip-flop, the master is designed using the positive enable JK latch, while the slave is designed with the negative enable D latch. The use of negative enable D latch makes sure that it does not need to invert the clock, which saves the

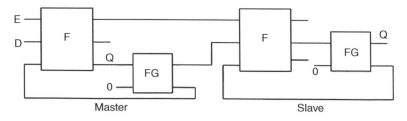

Figure 6.15 Reversible master–slave D flip-flop.

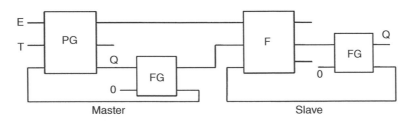

Figure 6.16   Reversible master–slave T flip-flop.

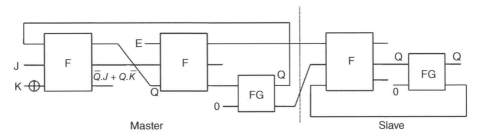

Figure 6.17   Reversible master–slave JK flip-flop.

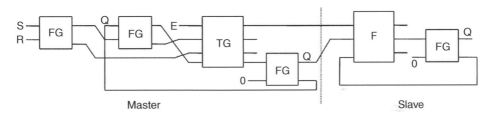

Figure 6.18   Reversible master–slave SR flip-flop.

NOT gate generally used in the designs for inversion of clock. This decreases the quantum cost and the propagation delay of the design.

6.4   The Design of Reversible Latch and the Master–Slave Flip-Flop with Asynchronous SET and RESET Capabilities

In this section, a design of the reversible D latch is introduced that can be asynchronously set and reset according to the control inputs. Before discussing the design of the asynchronously set/reset D latch, the properties of the Fredkin gate that will be helpful in understanding the design are discussed as follows:

1. As shown in Figure 6.19, the Fredkin gate can be used to avoid the fan-out of a signal by assigning that signal to its input A. The other two inputs B and C of the Fredkin gate are assigned the values as $B = 0$ and $C = 1$. This will result in copying of the input A of the Fredkin gate at the outputs P and Q, thus avoiding the fan-out problem.

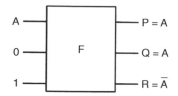

Figure 6.19 Application of the Fredkin gate to avoid the fan-out.

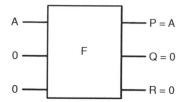

Figure 6.20 Asynchronous reset of the Q and R outputs of the Fredkin gate.

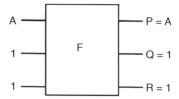

Figure 6.21 Asynchronous set of the Q and R outputs of the Fredkin gate.

2. As shown in Figure 6.20, by assigning the value 0 to the inputs B and C of the Fredkin gate it can reset the outputs Q and R to 0. This is a very useful property to asynchronously reset the Q and R outputs of the Fredkin gate.

3. As shown in Figure 6.21, by assigning the value 1 to the inputs B and C of the Fredkin gate it can set the outputs Q and R to 1. This is a very useful property to asynchronously set the outputs Q and R of the Fredkin gate.

The design of the asynchronously set/reset D latch is shown in Figure 6.22. The design has two Fredkin gates and one Feynman gate. It can be observed that the first Fredkin gate maps the D latch characteristic equation, while the second Fredkin gate helps in asynchronous set/reset of the output Q. The fan-out is avoided by the use of the Feynman gate. The design has two control inputs, C_1 and C_2. When $C_1 = 0$ and $C_2 = 1$, the design works in normal mode, implementing the D latch characteristic equation. When $C_1 = 0$ and $C_2 = 0$,

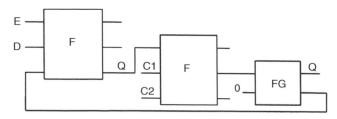

Figure 6.22 Fredkin gate-based asynchronous set/reset D latch.

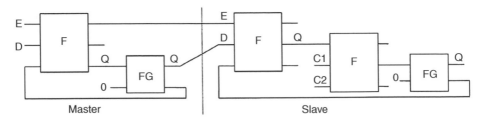

Figure 6.23 Reversible asynchronous set/reset master–slave D flip-flop.

the second Fredkin gates will reset the output Q to 0. When $C_1 = 1$ and $C_2 = 1$, the design will be set to $Q = 1$. Thus, the control inputs help the design to work in various modes. The design has the quantum cost of 11, delay of 111, and has four garbage outputs.

The reversible design of the master–slave D flip-flop with asynchronous set/reset is shown in Figure 6.23. The design contains positive enable D latch as the master latch and negative enable asynchronous set/reset D latch as the slave latch. The design has the quantum cost of 17, delay of 171, and five garbage outputs. The designs of the reversible T latch, the reversible JK latch, and the reversible SR latch can also be conditioned similarly as asynchronous set/reset designs. For example, asynchronous set/reset design of the T latch can be designed by replacing the Feynman gate in Figure 6.12 with one Fredkin gate and one Feynman gate. Fredkin gate will control signals C_1 and C_2 while the Feynman gate will avoid the fan-out.

6.5 Summary

In this chapter, the designs of reversible latches and flip-flops are presented. The designs of reversible sequential circuits are also optimized in terms of number of reversible gates and garbage outputs. In addition, the designs of reversible D latch and D flip-flop are discussed with asynchronous set/reset capability.

7

Reversible Counter, Decoder, and Encoder Circuits

A counter is a sequential circuit capable of counting the number of clock pulses that have arrived at its clock input. This chapter describes a design of n-bit reversible counter. The efficiency of the design is proved with the help of properties and algorithms. Decoders are a vigorous part of any recent digital computing device. They are used for addressing memories and caches, and they are used in aggregation with counters in multiphase clock generators. An n-to-2^n decoder takes an n-bit input combination and affirms the output line addressed by that input combination. Each output line resembles exactly one input combination. Decoders can have an *enable* line which functions as follows: If enable is activated, the decoder's outputs behave as usual, that is, exactly one is activated at any time. If the enable input is not active, all outputs of the decoder are deactivated. In this chapter, reversible decoder without enable input is described. The design procedure of reversible encoder is also described in this chapter.

7.1 Synthesis of Reversible Counter

This section describes the design for n-bit counter. Designs for both the asynchronous counter and the synchronous counter are presented here. While designing counter, this chapter also describes the design of reversible T flip-flop, which is the building block of the counter circuit although the T flip-flop was designed in a different way in the previous chapter. This section presents the design of T flip-flop, gated T flip-flop, and master–slave T flip-flop.

7.1.1 Reversible T Flip-Flop

The characteristic equation of a T flip-flop is $T Q' \oplus Q'T = T \oplus Q$. A T flip-flop can be realized by a single Feynman gate. The T flip-flop is shown in Figure 7.1. The T flip-flop with Q output has only one gate, quantum cost one, delay one, and no garbage outputs.

7.1.2 Reversible Clocked T Flip-Flop

The characteristic equation of a clocked T flip-flop is $Q = (T \oplus Q) \cdot CLK \oplus (CLK \cdot Q)'$. The equation can be simplified as $Q = (T \cdot CLK) \oplus Q$. This clocked flip-flop is realized by a Peres

Reversible and DNA Computing, First Edition. Hafiz Md. Hasan Babu.
© 2021 John Wiley & Sons Ltd. Published 2021 by John Wiley & Sons Ltd.

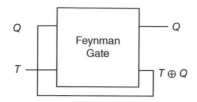

Figure 7.1 Reversible *T* flip-flop.

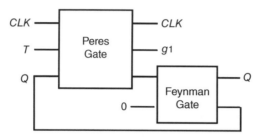

Figure 7.2 Reversible clocked T flip-flop for synchronous counter.

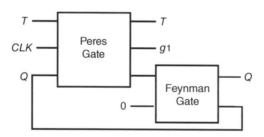

Figure 7.3 Reversible clocked T flip-flop for asynchronous counter.

gate and a Feynman gate. Two designs are shown here. Figure 7.2 and Figure 7.3 show the design of a clocked T flip-flop. Figure 7.2 is used for synchronous counter and Figure 7.3 is used for asynchronous counter. Both of the designs have quantum cost five, delay five, and garbage output one.

7.1.3 Reversible Master–Slave T Flip-Flop

To implement master–slave T flip-flop, it needs one flip-flop working as master and another as slave. The same strategy is followed here. For master flip-flop, Peres gate is modified. The input vector I_V and output vector O_V of a 3×3 modified Peres gate, MPG are defined as follows, $I_V = (A, B, C)$ and $O_V = (P = A', Q = A \oplus B, R = AB \oplus C)$. The quantum cost of MPG is 4. The block diagram and equivalent quantum representation for 3×3 MPG are shown in Figure 7.4 and Figure 7.5, respectively.

The modification is required to produce negative clocked pulse without generating any additional gate cost. The master–slave T flip-flop is shown in Figure 7.6. The design has quantum cost 10, delay 10, and garbage output 2.

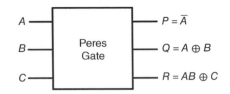

Figure 7.4   Block diagram of 3×3 MPG gate.

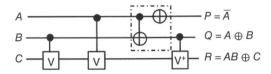

Figure 7.5   Quantum representation of 3×3 MPG gate.

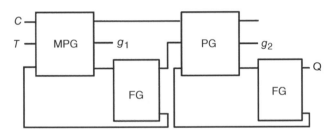

Figure 7.6   Reversible master–slave T flip-flop.

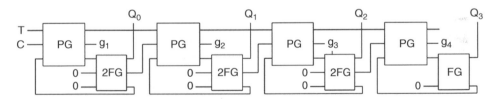

Figure 7.7   4-bit reversible asynchronous counter.

7.1.4   Reversible Asynchronous Counter

In asynchronous counter, the T flip-flops are arranged in such a way that output of one flip-flop is connected to the clock input of the next higher order flip-flop. The output of a flip-flop triggers the next flip-flop. The flip-flop holding the least significant bit receives the incoming count pulse. The 4-bit asynchronous counter is shown in Figure 7.7. The counter is realized by four Peres gates and some Feynman gates. The 4-bit asynchronous counter has quantum cost 23, delay 23, and garbage output four.

Property 7.1.4.1   If g is the total total number of gates required to design the counter producing $b$numbers of garbage outputs, then $g \geq 2n$ and $b \geq n$ to construct n–bit asynchronous counter circuit.

Proof: Each flip-flop consists of two gates, where n-bit counter requires n numbers of flip-flops. No additional gates are required to interconnect each other. So, the total number of gates required to design the counter is $2n$, hence $g \geq 2n$. Similarly, every flip-flop produces only one garbage output. No garbage output is produced while interconnection among flip-flops. So, the total number of garbage output is n. Hence $b \geq n$.

Property 7.1.4.2   The quantum cost of an n-bit asynchronous counter is $Q_n \geq 6n - 1$.

Proof: For $n = 1$, only one Peres gate and one Feynman gate are required to construct the counter. The quantum cost of Peres gate is 4 and quantum cost of Feynman gate is 1. So, the total cost is $4 + 1 = 5$.

Now for $n > 1$, one Peres gate and one double Feynman gate are required for each flip-flop in the counter circuit, except the last one, which requires one Feynman gate instead of double Feynman gate. So it needs n Peres gates, $(n - 1)$ double Feynman gate, and one Feynman gate for n-bit asynchronous counter. The quantum cost of double Feynman gate is two. So, the total quantum cost is $4 \times n + 2(n - 1) + 1 = 6n - 1$. Hence $Q_n \geq 6n - 1$.

7.1.5   Reversible Synchronous Counter

Synchronous counter is different from asynchronous counter in that clock pulses are applied to the inputs of all the flip-flops at a time. A flip-flop is complemented, depending on the input value T and the clock pulse. The flip-flop in least significant position is completed with every clock pulse. A flip-flop in other position is complemented only when all the outputs of preceding flip-flops produce one. The same strategy is followed here to implement the synchronous counter. Figure 7.8 shows the 4-bit synchronous counter. The 4-bit synchronous counter has quantum cost 32, delay 32, and garbage output four.

Property 7.1.5.1   If g is the total number of gates required to design the counter producing b numbers of garbage outputs, then $g \geq 4n - 4$ and $b \geq n$ to construct n-bit synchronous counter.

Proof: Each flip-flop consists of two gates; n-bit counter requires n numbers of flip-flops. For $n = 3$, one Toffoli gate and one Feynman gate are required to carry out all the outputs to the next higher positioned flip-flop. So, the total number of gates is $3 \times 2 + 2 = 8$.

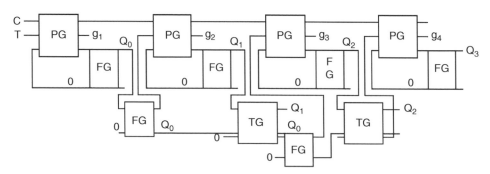

Figure 7.8   4-bit reversible synchronous counter.

For $n > 3$, $2n$ number of gates are required for the flip-flops and $2(n - 2)$ numbers of gates are required to carry out all the lower outputs to the next higher outputs. So, the total number of gates required gates is $2n + 2(n - 2) = 4n - 4$. Every flip-flop produces only one garbage output. No garbage output is produced while interconnection is made among flip-flops and also no garbage output is generated to carry out outputs to next higher flip-flop. So, the total number of garbage out is n. Hence $b \geq n$.

Property 7.1.5.2 The quantum cost of an $n(\geq 3)$-bit synchronous counter is $Q_n \geq 11n - 12$.

Proof: For $n(\geq 3)$-bit synchronous counter, it requires n flip-flops. Each flip-flop consists of one Peres gate and one Feynman gate. Additional $(n - 2)$ Toffoli gates and $(n - 2)$ Feynman gates are required to carry out all the outputs to the next higher flip-flop. So, it requires n number of Peres gate, $(n - 2)$ number of Toffoli gates, and $n + (n - 2) = 2n - 2$ numbers of Feynman gates. It is noted that quantum cost of Peres gate is four, the quantum cost of Toffoli gate is five, and quantum cost of Feynman gate is one. So, total quantum cost is $4 \times n + 5 \times (n - 2) + 1 \times (2n - 2) = 11n - 12$. Hence $Q_n \geq 11n - 12$.

7.2 Reversible Decoder

The quantum implementation of a reversible circuit is shown in Figure 7.9, which is configured as a reversible 2–*to*–4 decoder. Figure 7.10 illustrates the measurement of the quantum delay for the reversible 2–*to*–4 decoder unit. In this figure, a visual explanation of the quantum delay measurement methodology is applied to the reversible unit. Because the merged qubit state pairs in stages three and six only contribute unit quantum cost and they are considered to contribute unit quantum delay as well. Table 7.1 contains the truth table for this reversible 4-to-2 decoder unit. The input variable columns are permuted in order to group the constant inputs B and D, and generate the outputs of the reversible decoder with inputs A and C. The highlighted rows show the required decoding functionality.

In general, implementing a logical AND function or logical OR function reversibly is very costly, especially when the number of inputs increases. A 2–*to*–4 decoder must generate the four logical AND functions $A'B', A'B, AB'$, and AB. The design of a reversible 2–*to*–4 decoder reversibly generates all four necessary AND functions using a single Fredkin gate

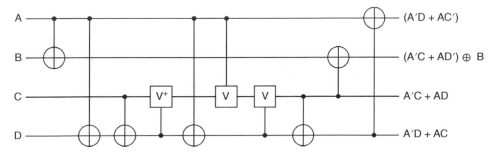

Figure 7.9 Quantum implementation of a reversible 2–*to*–4 decoder.

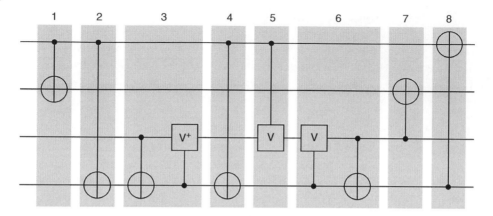

Figure 7.10 Measurement of the quantum delay for the reversible 2–*to*–4 decoder unit.

Table 7.1 Truth Table for the Reversible 2–*to*–4 Decoder

B	D	A	C	O_0	O_1	O_2	O_3
0	0	0	0	0	0	0	0
0	0	0	1	0	1	1	0
0	0	1	0	1	1	0	0
0	0	1	1	0	1	0	1
0	1	0	0	1	0	0	1
0	1	0	1	1	1	1	1
0	1	1	0	1	0	1	0
0	1	1	1	0	0	1	1
1	0	0	0	0	1	0	0
1	0	0	1	0	0	1	0
1	0	1	0	1	0	0	0
1	0	1	1	0	0	0	1
1	1	0	0	1	1	0	1
1	1	0	1	1	0	1	1
1	1	1	0	1	1	1	0
1	1	1	1	0	1	1	1

and minimal supplementary logic. The quantum cost of this unit is eight, which has two constant inputs, and zero garbage outputs, as shown in Figure 7.11.

7.2.1 Reversible Encoder

Due to the logical reversibility of the design, it can also be configured to operate "in reverse" as a reversible encoder. Simply reflecting the reversible decoder design horizontally is shown in Figure 7.9. It yields a gate that can reversibly implement a 4-to-2 encoder. This

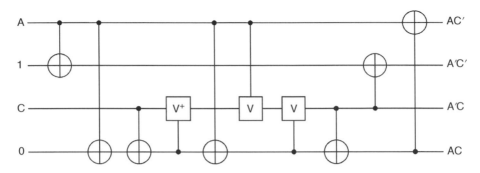

Figure 7.11   Reversible 2−to−4 decoder.

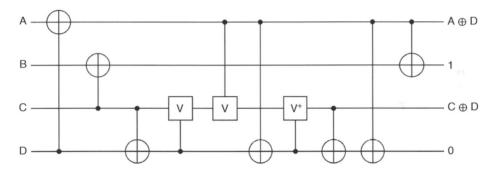

Figure 7.12   Reversible 4−to−2 encoder.

reversible 4-to-2 encoder design assumes – like traditional encoders – that exactly one of its four inputs will have an input value of logical 1 at any time. When functioning this way as a reversible 4-to-2 encoder, the unit has zero constant inputs and two garbage outputs as shown in Figure 7.12. Because the reversible 4-to-2 encoder design consists entirely of the same hardware as the reversible 2-to-4 decoder simply in a different (reflected) configuration. Its quantum cost and quantum delay are eight, which are equal to the decoder configuration.

Table 7.2 shows the truth table of the design of a reversible 4−to−2 encoder with the required encoder functionality highlighted. In the truth table, the output columns are permuted in order to partition them into garbage outputs (O_2 and O_0) and the two encoder outputs (O_3 and O_1). The rows are also permuted in order to group the possible input combinations. The xs in the remaining 12 rows represent don't-care conditions in the sense that when using the reversible 4−to−2 encoder design, none of those 12 input combinations should be expected. They are not don't-care conditions in the sense that here these are specifying the output functions in the truth table; indeed, the reversible logic implementation shown in Figure 7.12 specifies all output functions completely. In Table 7.2, it is shown the outputs for the 12 impossible input combinations for clarity of presentation.

Table 7.2 Truth Table for the Reversible 4−to−2 Encoder

I_3	I_2	I_1	I_0	O_3	O_1	O_2	O_0
1	0	0	0	1	1	1	0
0	1	0	0	0	0	1	0
0	0	1	0	0	1	1	0
0	0	0	1	1	0	1	0
0	0	0	0	x	x	x	x
0	0	1	1	x	x	x	x
0	1	0	1	x	x	x	x
0	1	1	0	x	x	x	x
0	1	1	1	x	x	x	x
1	0	0	1	x	x	x	x
1	0	1	0	x	x	x	x
1	0	1	1	x	x	x	x
1	1	0	0	x	x	x	x
1	1	0	1	x	x	x	x
1	1	1	0	x	x	x	x
1	1	1	1	x	x	x	x

7.3 Summary

In this chapter, reversible design for both n-bit synchronous and asynchronous counters are described. This chapter also describes an algorithm for synthesis the design of reversible counter that optimizes the counter in terms of quantum cost, delay, and garbage outputs. Appropriate algorithms and properties are presented to clarify the design and to establish its efficiency. In addition, of all the reversible logical unit designs, notably little attention has been paid to the implementation of reversible decoder and encoders. In this chapter, designs of reversible decoder and encoder circuit are described. The analysis in terms of its quantum cost, garbage outputs, constant inputs, and quantum delay is also shown. Reversible counters and decoder/encoder have obvious advantages of low power dissipation. So it can be applied to wireless sensors very well, in digital signal processing, in parallel circuits, etc. In addition, the designs that have been shown in this chapter would help the reader to understand the next chapter better in a better way.

8

Reversible Barrel Shifter and Shift Register

A barrel shifter is a combinational circuit that has n-input and n-output and m select lines that controls bit shift operation. Since multiple and variable bit shifting are more desirable than single bit operations, several irreversible barrel shifters have been presented in that consequence. Barrel shifters can be unidirectional, which performs the shift/rotate operation to left (or right) or bidirectional that are capable of shifting/rotating to both directions. In addition, the shift register is one of the most extensively used functional devices in digital systems. A shift register consists of a group of flip-flops connected together so that information bits can be shifted from one position to either right or left depending on the design of the device.

8.1 Design Procedure of Reversible Bidirectional Barrel Shifter

In this section, a design of an n-bit reversible bidirectional barrel shifter is described, where n is an integer power of two. The bidirectional barrel shifter can shift at most $(n-1)$ bits using log n bits select input. The reversible barrel shifter is divided into three components namely 2's complement generator, swap condition generator and right rotator. A generalized approach is presented for each component to realize the whole architecture.

This section presents the architecture of reversible barrel shifter, which performs rotate operation in both directions. If the data input is of n-bit and the maximum shift amount is b-bit, the barrel shifter is defined as (n, b) barrel shifter. The select input is of log n-bit, which represents the shift amount. There is one control input dir that represents the direction of the rotation (left or right). When control input dir is 1, a left rotation is performed. Otherwise, a right rotation is performed. If an n-bit data input is right rotated by b-bit, it is equivalent to an $(n-b)$-bit left rotation. Besides, if n is an integer power of 2, then $(n-b)$ is the 2's complement of b. Based on this concept, when $dir=1$, select input is 2's complemented and a right rotation is performed based on 2's complemented select input. When $dir = 0$, the select input is kept unchanged. A 1-bit ($S_1 = 0, S_0 = 1$) right rotation is equivalent to a 3-bit (2's complement of 1) left rotation. This statement is true for all other values of select input as long as number of bits of data input is a power of two.

The data inputs are divided into pairs, where bits in each pair swaps for a particular condition. The conditions for swapping are calculated from the select input. So, the bidirectional

Reversible and DNA Computing, First Edition. Hafiz Md. Hasan Babu.

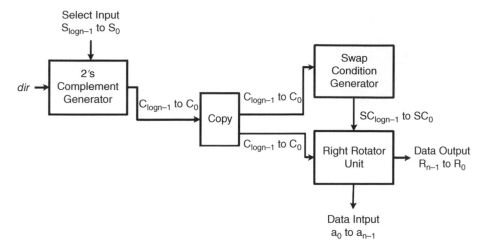

Figure 8.1 Simple block diagram of the barrel shifter.

barrel shifter can be divided into three components: 2's complement generator, swap condition generator, and right rotator. A simplified block diagram is shown in Figure 8.1 to realize the bidirectional barrel shifter. The detail circuit of the reversible bidirectional barrel shifter is presented in later subsections.

8.1.1   Reversible 3 × 3 Modified BJN Gate

A 3 × 3 reversible modified BJN gate (MBJN) is introduced here to minimize the quantum cost of the reversible 2's complement generator. This gate is a modified version of the reversible BJN gate. The gate maps three inputs (A, B, C) to three outputs $(A, A \oplus B, (A+B) \oplus C)$ and maintains a one-to-one relationship between them. Figure 8.2 and Figure 8.3 represent the block diagram and quantum realization of reversible MBJN gate, respectively. The MBJN gate requires quantum cost of 4.

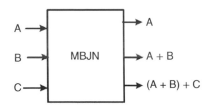

Figure 8.2 Block diagram of the reversible MBJN gate.

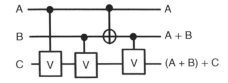

Figure 8.3 Quantum realization of the reversible MBJN gate.

8.1.2 Reversible 2's Complement Generator

The 2's complement of a binary number can be calculated in two steps: (i) find 1's complement by complementing each bit, and (ii) add 1 to the least-significant-bit (lsb) position. Here 2's complement of the select input is generated without using any adder circuitry. Algorithm 8.1.2.1 is presented to generate 2's complement of log n-bit select input of an n-bit barrel shifter.

Algorithm 8.1.2.1 2's Complement Generator

Input: Select input $S_{\log n-1}$ to S_0, Control input *dir*
Output: Modified select input $C_{\log n-1}$ to C_0

1: Begin
2: $C_0 = S_0$
3: $C_1 = S_0 \oplus S_1$
4: **for** $i=2$ **to** log n-1 **do**
5: $\quad C_i = S_i \oplus (S_{i-1}$ OR S_{i-2} OR ... OR $S_0)$
6: **end for**
7: **if** *dir=1* **then**
8: $\quad$ **return** $C_{\log n-1}$ to C_0
9: **else**
10: $\quad$ **return** $S_{\log n-1}$ to S_0
11: **end if**
12: End

Let 2's complement of select input S_i be C_i, where $0 \leq i \leq \log n - 1$. According to Algorithm 8.1.2.1, the lsb (S_0) of the select input is equivalent to it's 2's complement. If index i is one, the 2's complement of S_i is the Ex-OR of the lsb and the bit itself. To generate the bits at index $i >= 2$, the corresponding bit is Ex-ORed with the OR of all bits at lower positions. For example, the 2's complement of bit S_2 is $S_2 \oplus (S_1 + S_0)$. If control input *dir* is zero, select input S_i is returned. Otherwise, 2's complement of select input is returned, where, 0 <= i <= log n - 1. A 2-bit and 3-bit 2's complement generator as shown in Figure 8.4 and

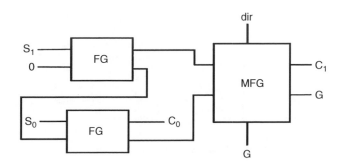

Figure 8.4 Reversible 2-bit 2's complement generator.

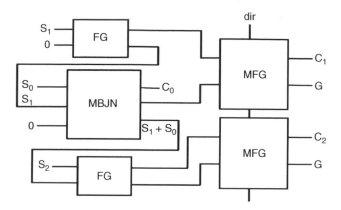

Figure 8.5 Reversible 3-bit 2's complement generator.

Figure 8.5, respectively. The 2- and 3-bit reversible 2's complement generators need total 3 and 5 gates with quantum cost of 6 and 14, respectively.

Property 8.1.2.1 Let GC be the number of gates, QC be the quantum cost, and GO be the number of garbage outputs of a log n-bit reversible 2's complement generator of an n-bit barrel shifter. Then, $GC = 2\log n - 1$, $QC = 8\log n - 10$, and $GO = \log n$.

8.1.3 Reversible Swap Condition Generator

This unit generates swap conditions to swap bits in a pair of data inputs in $n(> 4)$-bit right rotator unit. The outputs of 2's complement generator are fed to swap condition generator as inputs as shown in Algorithm 8.1.3.1. An n-bit data input can be divided into $n/2$ number of pairs. Since each pair needs a particular swap condition, $n/2$ number of conditions are needed. A methodology to implement a swap condition generator is explained with the following example.

A 3-bit and 4-bit reversible swap condition generator is shown in Figure 8.6 and Figure 8.7.

Property 8.1.3.1 Let GC be the number of gates, QC be the quantum cost and GO be the number of garbage outputs of a log n-bit reversible swap condition generator

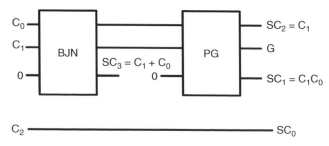

Figure 8.6 Reversible 3-bit swap condition generator.

Algorithm 8.1.3.1 2's Complement Generator

Input: Outputs of 2's complement generator $C_{\log n-1}$ to C_0.
Output: Swap conditions $SC_{(n/2)-1}$ to SC_0.

1: Begin
2: Swap condition for $pair_0$ is the msb of the select input and swap condition for $pair_{n/4}$ is the second msb of the select input. For example, a 16-bit data input needs eight swap conditions where $SC_0 = C_3$, $SC_4 = C_2$.
3: For swap conditions that are greater than zero and less than $n/4$, perform $C_{\log n-2}$ AND outputs of $(\log n - 1)$-bit swap condition generator that are greater than zero. For example, an 8-bit data input has 3-bit select input and swap conditions are $SC_0 = C_2$, $SC_1 = (C_1$ AND $C_0)$, $SC_2 = C_1$, $SC_3 = (C_1$ OR $C_0)$. Similarly, a 16-bit data input has 4-bit select input and its swap condition is less than four, which are $SC_1 = C_2$ AND $(C_1$ AND $C_0)$, $SC_2 = C_2$ AND C_1, and soon.
4: For swap conditions that are greater than $n/4$, perform $C_{\log n-2}$ OR outputs of $(\log n-1)$-bit swap condition generator that are greater than zero. A 16-bit data input has 4-bit select input and its swap condition is greater than four, which are $SC_5 = C_2$ OR $(C_1$ AND $C_0)$, $SC_6 = (C_2$ OR $C_1)$, and so on.
5: End

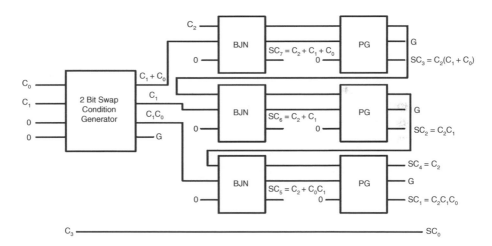

Figure 8.7 Reversible 4-bit swap condition generator.

of an n-bit reversible barrel shifter. Then, $GC = n\text{-}2\log n$, $QC = 9((n/2) - \log n)$ and $GO = (n/2) - \log n$.

8.1.4 Reversible Right Rotator

An n-bit $(n > 4)$ right rotator takes an n-bit data input, outputs of 2's complement generator, and outputs of swap condition generator to rotate at most $b(= n - 1)$ bits in right direction. In this subsection, first a $(4, 3)$ reversible right rotator is presented and then a generalized approach is developed.

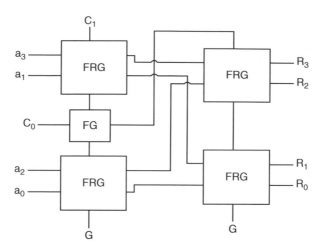

Figure 8.8 A (4, 3) reversible right rotator.

8.1.4.1 (4, 3) Reversible Right Rotator

A (4, 3) reversible right rotator takes 4-bit data input, 2-bit select input from 2's comple-ment generator and rotates at most three bits, as shown in Figure 8.8. The swap condition generator is not needed for a (4, 3) reversible right rotator. The circuit needs five gates with 21 quantum cost and it generates two garbage outputs. The (4, 3) reversible right rotator is used to realize an n-bit reversible right rotator.

8.1.4.2 Generalized Reversible Right Rotator

A methodology to implement a right rotator is explained with the following example.

Input: Data inputs a_{n-1} to a_0, outputs of 2's complement generator $C_{\log n-1}$ to C_0, outputs of swap condition generator $SC_{(n/2)-1}$ to SC_0.

Output: Right rotated outputs R_{n-1} to R_0.

- Step 1: Partition n-bit data input into two groups and take one bit from the first partition and one bit from the second partition to form a pair.
- Step 2: Call the algorithm swap condition presented in Algorithm 8.1.3.1. Bits in a pair are swapped if Ex-OR of corresponding swap condition and msb of the select input is 1.
- Step 3: Take the first bit of all pairs to form partition first and second bit of all pairs to form partition 2. The bits in the first partition and select input except the msb are fed into an $n/2$-bit right rotator. Similarly, the bits in the second partition and select input except the msb are fed into an $n/2$-bit right rotator. An $n/2$-bit right rotator is called recursively until n becomes 4. When n becomes 4, a 4-bit right rotator is needed, which is shown in Figure 8.8. The first rotator generates outputs R_{n-1} to $R_{n/2}$ and the second rotator produces outputs $R_{(n/2)-1}$ to R_0.

Property 8.1.4.2.1 Let GC be the number of gates, QC be the quantum cost, and GO be the number of garbage outputs of an n-bit reversible right rotator unit. Then, $GC = (3(n/4) - 7)logn + (5(n/2) + 3)$, $QC = (3n - 29)logn + 19(n/2) + 17$ and $GO = ((n/2 - 3)logn + (n/2) + 2$.

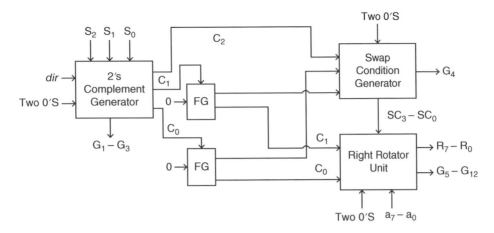

Figure 8.9 Block diagram of (8, 7) reversible bidirectional barrel shifter.

8.1.5 Reversible Bidirectional Barrel Shifter

In this subsection, the whole architecture of the reversible bidirectional barrel shifter is presented. The three reversible components discussed in previous sections are combined to implement the reversible bidirectional barrel shifter. The architecture of the reversible (8, 7) bidirectional barrel shifter is depicted in Figure 8.9.

The reversible circuits described in the previous section can be combined with additional FG gates to realize a generalized reversible bidirectional barrel shifter. The property of the n-bit reversible bidirectional barrel shifter is presented in Property 8.1.5.1 to calculate the cost parameters such as gate count, quantum cost, and garbage outputs.

Property 8.1.5.1 Let GC be the number of gates, QC be the quantum cost and GO be the number of garbage outputs of an n-bit reversible bidirectional barrel shifter. Then, $GC = (3(n/4) - 7)logn + 7(n/2) + 2$, $QC = (3n - 30)logn + 14n + 7$ and $GO = ((n/2) - 3)logn + n + 2$.

8.2 Design Procedure of Reversible Shift Register

To synthesize reversible memory circuits, reversible sequential circuits such as shift registers are very essential. This section describes a unique concept on reversible sequential circuit design that includes serial-in serial-out (SISO), serial-in parallel-out (SIPO), parallel-in serial-out (PISO), parallel-in parallel-out (PIPO) and universal shift registers.

8.2.1 Reversible Flip-Flop

A flip-flop is a bistable electronic circuit that has two stable states and can be used as a 1-bit memory device. The characteristic equation of D flip-flop is $Q+ = D.CLK + Q.CLK'$, which can be mapped onto the Fredkin gate. Figure 8.10 and Figure 8.11 show the structure and block diagram of the reversible D flip-flop, respectively. To avoid a fan-out problem, a Feynman gate is used to copy the output.

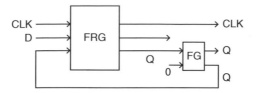

Figure 8.10 Structure of the reversible clocked D flip-flop.

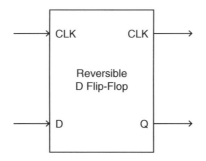

Figure 8.11 Block diagram of the reversible clocked D flip-flop.

The shift register is one of the most extensively used functional devices in digital systems. A shift register consists of a group of flip-flops connected together so that information bits can be shifted from one position to either right or left depending on the design of the device. This subsection introduces several types of shift registers including SISO, SIPO, PISO, PIPO, and universal shift registers.

8.2.1.1 Reversible SISO Shift Register

SISO shift register is the simplest shift register that contains only flip-flops. In right shift register, output of a given flip-flop is connected to the data input of the next flip-flop at its right. Each clock pulse shifts the contents of the register one bit position to the right. The serial input is provided to the leftmost flip-flop and the serial output is the output of the rightmost one. Figure 8.12 shows the n-bit reversible SISO shift register built from n reversible clocked D flip-flops, as shown in Figure 8.11. Thus, it requires n Fredkin gates and n Feynman gates, a total of $2n$ gates, and produces $n + 1$ garbage outputs with $6n$ quantum cost.

8.2.1.2 Reversible SIPO Shift Register

A SIPO shift register is similar to a SISO shift register. It is different in that it makes all the stored bits available as outputs. Reversible implementation of a SIPO shift register using clocked D flip-flops is shown in Figure 8.13. The serial data are entered to the SI input

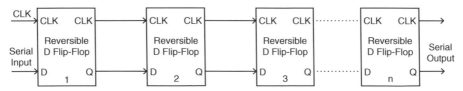

Figure 8.12 n-bit reversible SISO shift register.

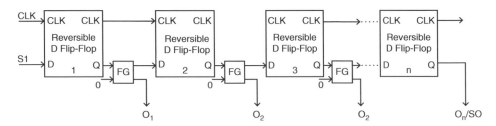

Figure 8.13 *n*-bit reversible SIPO shift register.

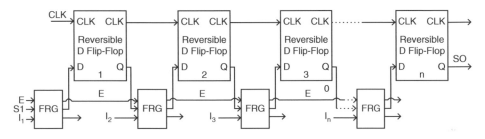

Figure 8.14 *n*-bit reversible PISO shift register.

of the reversible left-most flip-flop while the outputs O_1, O_2, O_3,...., O_n are available in parallel from the Q output of the flip-flops. It requires n reversible clocked D flip-flops and $n - 1$ Feynman gates. Thus, it requires a total of $3n - 1$ gates and produces $n + 1$ garbage outputs with quantum cost $7n - 1$.

8.2.1.3 Reversible PISO Shift Register

PISO shift register takes the data from parallel inputs and shifts it to the next flip-flop when the register is clocked. Figure 8.14 shows the reversible implementation of PISO shift register using clocked D flip-flops. The operations are controlled by the enable signal E. When E is high, the inputs $I_1, I_2, I_3, \dots, I_n$ are loaded in parallel into the register coincident with the next clock pulse. Again when E is low, the Q output of the flip-flop is shifted to the right by means of a Fredkin gate. It allows accepting data n bits at a time on n lines and then sending them one bit after another on one line. It requires n reversible clocked D flip-flops and n Fredkin gates. Thus, it requires a total of $3n$ gates and produces $2n + 2$ garbage outputs with quantum cost $11n$.

8.2.1.4 Reversible PIPO Shift Register

A PIPO shift register combines the functions of the PISO and SIPO shift registers. The control inputs (*HOLD, E*) select the operation of the register according to the function entries in Table 8.1.

From Table 8.1, when both *HOLD* and E are low, the shift register performs the shift-right operation. When *HOLD* is low and E is high, the inputs $I_1, I_2,..., I_n$ are loaded in parallel into the register coincident with the next clock pulse. The outputs $O_1, O_2, O_3,..., O_n$ are available in parallel from the Q output of the flip-flops. When *HOLD* is high, present value of flip-flop is returned to the D input of that flip-flop. In other words, the register is inactive when *HOLD* is high, and the contents are stored indefinitely.

Table 8.1 Truth Table of Reversible PIPO Shift Register

HOLD	E	Final Output Q_i^+	
0	0	Q_{i-1}(Right shift)	
0	1	I_i	(Parallel Load)
1	Don't care	Q_i	(No Change)

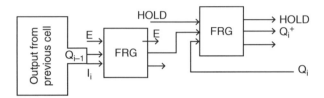

Figure 8.15 Implementation of the characteristic function of Equation (8.2.1.4.1).

The characteristic function of Q_i^+ can be obtained from Table 8.1.

$$Q_i^+ = (HOLD)'.E.I_i + (HOLD)'.E'.Q_{i-1} + HOLD.Q_i \qquad (8.2.1.4.1)$$

Where "." repersents the multiplication sign. For the first stage, Q_{i-1} is the serial input (*SI*) and for last stage, Q_i is the serial output (*SO*).

Property 8.2.1.4.1 The characteristic function of Q_i of reversible PIPO shift register can be obtained by two gates and four garbage outputs with 10 quantum cost.

Proof: The characteristic function of Q_i^+ is $Q_i^+ = (HOLD)'.E.I_i + (HOLD)'.E'.Q_{i-1} + HOLD.Q_i = ((HOLD)'.(E.I_i + E'.Q_{i-1}) + HOLD.Q_i$

Thus, this function can be implemented by only two Fredkin gates as shown in Figure 8.15. From this figure, it is clear that it generates four garbage outputs. The quantum cost of a Fredkin gate is 5, thus the implementation of characteristic function of Q_i^+ requires 10 quantum cost.

Figure 8.16 shows the basic cell of the reversible PIPO shift register and its block diagram is shown in Figure 8.17. By cascading *n* basic cells, reversible PIPO shift register can be implemented as shown in Figure 8.16.

Property 8.2.1.4.2 The *n*-bit reversible PIPO shift register using clocked *D* flip-flops can be implemented by at least $5n$ reversible gates and $3n + 3$ garbage outputs.

Proof: Each basic cell has two parts: generation of Q_i^+ and *D* flip-flop. From Property 8.2.1.4.2, generation of Q_i^+ requires two gates and produces four garbage outputs. But in *n*-bit PIPO shift register as shown in Figure 8.18, *HOLD* and *E* outputs of each basic cell are connected to the *HOLD* and *E* inputs of the next basic cell, respectively. Thus, it

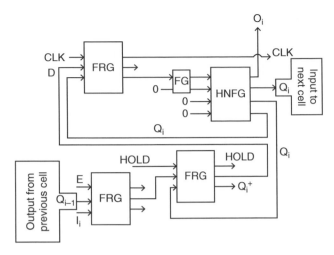

Figure 8.16 Basic cell for the reversible PIPO shift register.

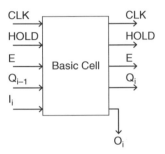

Figure 8.17 Block diagram for the reversible PIPO shift register.

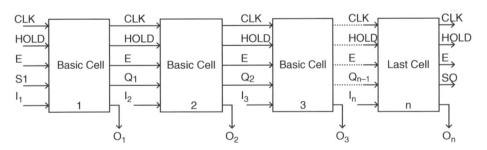

Figure 8.18 n-bit reversible PIPO shift register.

reduces two garbage outputs from each basic cell in the generation of Q_i^+ except the last cell. In other words, generation of Q_i^+ ($1 \leq i < n$) produces only two garbage outputs and Q_n^+ (i.e., SO) produces four garbage outputs.

In the basic cell, each D flip-flop (Figure 8.16) requires three gates and has four Q_i outputs. One of the Q_i outputs is used to the generation of Q_i^+, the second one is used to generate Q_{i+1}^+, one is considered as parallel output, and other is fed to its own D flip-flop's Fredkin gate. CLK output of each basic cell is connected to the CLK input of the next basic cell.

Thus, each of the basic cells (except the last cell) produces only one garbage output from each D flip-flop. D flip-flop in last cell produces two garbage outputs.

Thus, each basic cell requires $2 + 3 = 5$ gates. Each of the first $n - 1$ basic cells produces $2 + 1 = 3$ garbage outputs. And the last cell produces $4 + 2 = 6$ garbage outputs. Therefore, an n-bit reversible PIPO shift register has n basic cells and requires $5n$ gates and produces $(n - 1) \times 3 + 6 = 3n + 3$ garbage outputs.

Property 8.2.1.4.3 The quantum cost of an n-bit reversible PIPO shift register using D flip-flops is at least $18n$.

Proof: From Property 8.2.1.4.1, generation of Q_i^+ requires 10 quantum cost. Quantum costs of the Feynman gate, Fredkin gate, and HNFG gate are one, five, and two, respectively. D flip-flop of each basic cell requires one Fredkin gate, one Feynman gate and one HNFG gate, a total quantum cost of $5 + 1 + 2 = 8$. Thus, an n-bit reversible PIPO shift register requires $10n + 8n = 18n$.

8.2.1.5 Reversible Universal Shift Register

The universal shift register has both shifts (left and right) and parallel load capabilities. The selection inputs (s_0, s_1) control the mode of operation of the register according to the function entries in Table 8.2.

When both s_0 and s_1 are low, the present value of the register is applied to the D inputs of the flip-flops. This condition forms a path from the output of each flip-flop into the input of the same flip-flop. The next clock pulse transfers the previous value into each flip-flop, and thus no change of the state occurs. When s_0 is low and s_1 is high, the register acts as a shift-right register. When s_0 is high and s_1 is low, the register acts as a shift-left register. When both s_0 and s_1 are high, information on the parallel input lines is transferred into the register simultaneously during the next clock pulse.

The characteristic function of Q_i^+ can be obtained from Table 8.2.

$$Q_i^+ = s_0'.s_1'.Q_i + s_0'.s_1.Q_{i-1} + s_0.s_1'.Q_{i+1} + s_0.s_1.I_i \qquad (8.2.1.5.1)$$

Where "." repersents the multiplication sign. For the first stage, Q_{i-1} is the serial input for shift-right (SIR) and for last stage Q_{i+1} is the serial input for shift-left (SIL).

Property 8.2.1.5.1 The characteristic function of Q_i^+ of reversible universal shift register can be obtained by three gates and five garbage outputs with quantum cost of 15.

Table 8.2 Function Table for Reversible Universal Shift Register

s_0	s_1	Final Output Q_i^+
0	0	Q_i(No change)
0	1	Q_{i-1}(Right shift)
1	0	Q_{i+1}(Left shift)
1	1	I_i(Parallel load)

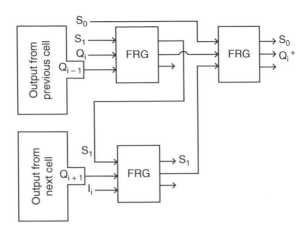

Figure 8.19 Implementation of the characteristic function of Equation (8.2.1.5.1).

Proof: The characteristic function of Q_i^+ is

$$Q_i^+ = s_0'.s_1'.Q_i + s_0'.s_1.Q_{i-1} + s_0.s_1'.Q_{i+1} + s_0.s_1.I_i$$
$$= s_0'.(s_1'.Q_i + s_1.Q_{i-1}) + s_0.(s_1'.Q_{i+1} + s_1.I_i)$$

Thus, this function can be implemented using three Fredkin gates as shown in Figure 8.19. From this figure, it is clear that it generates five garbage outputs. The quantum cost of a Fredkin gate is 5; therefore, the implementation of characteristic function of Q_i^+ requires only $3 \times 5 = 15$ quantum cost.

Figure 8.20 shows the basic cell of the universal shift register and its block diagram is shown in Figure 8.21. By cascading n basic cells, reversible universal shift register can be implemented as shown in Figure 8.22.

Property 8.2.1.5.2 The n-bit reversible universal shift register can be implemented by at least $7n - 2$ reversible gates and $4n + 3$ garbage outputs.

Proof: Each basic cell of the reversible universal shift register has two parts: generation of Q_i^+ and clocked D flip-flop.

From Property 8.2.1.5.1, generation of Q_i^+ requires three gates and produces five garbage outputs. But in n-bit universal shift register, s_0 and s_1 outputs of each basic cell are connected to the s_0, s_1 inputs of the next basic cell, respectively. Thus, it reduces two garbage outputs from each basic cell in the generation of Q_i^+ except the last cell. In other words, generation of Q_i^+ ($1 \leq i < n$) produces only three garbage outputs and Q_n^+ produces five garbage outputs.

Each of the basic cells (except the last cell) produces only one garbage output from each D flip-flop which is shown in Figure 8.20. D flip-flop in last cell produces two garbage outputs. Thus, each cell requires $3 + 3 = 6$ gates. Each of the first $n - 1$ basic cells produces $3 + 1 = 4$ garbage outputs. And the last cell produces $5 + 2 = 7$ garbage outputs. An n-bit reversible universal shift register has n basic cells and $n - 2$ Feynman gates to avoid fan-out. Therefore, the design requires $6n + n - 2 = 7n - 2$ gates and produces $4(n - 1) + 7 = 4n + 3$ garbage outputs.

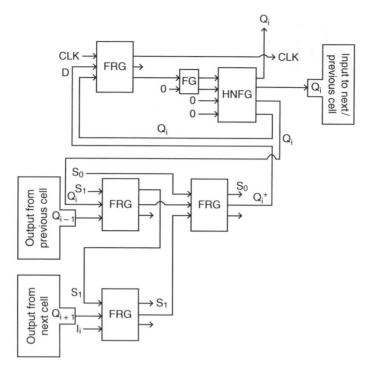

Figure 8.20 Basic cell for the reversible universal shift register.

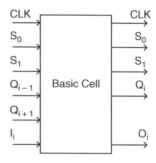

Figure 8.21 Block diagram for the reversible universal shift register.

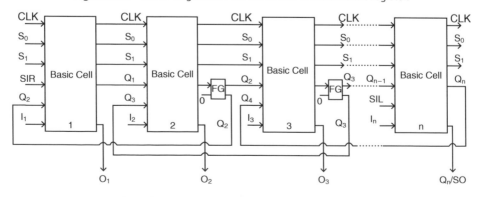

Figure 8.22 n-bit reversible universal shift register.

Property 8.2.1.5.3   The quantum cost of an n-bit reversible universal shift register is at least $24n - 2$.

Proof: From Property 8.2.1.5.1, generation of Q_i^+ requires 15 quantum cost. Quantum costs of a Feynman gate, Fredkin gate, and HNFG gate are one, five, and two, respectively. In D flip-flop, each of the basic cells requires one Fredkin gate, one Feynman gate and one HNFG gate. Therefore, the total quantum cost is $5 + 1 + 2 = 8$. To avoid fan out, the design also requires $n - 2$ Feynman gates. Thus, an n-bit reversible universal shift register requires $15n + 8n + n - 2 = 24n - 2$.

8.3   Summary

This chapter is divided into two parts, namely reversible barrel shifter and reversible shift register. In the first part of this chapter, the design methodology of a reversible barrel shifter is presented. In the way of presenting the design, a 2's complement generator is implemented using reversible logic, which can be used in arithmetic circuits for signed number. Two algorithms have been presented to realize a cost-efficient reversible bidirectional barrel shifter. In the second part of this chapter, reversible logic synthesis was carried out for SISO shift register. The key contribution of this part is the reversible realization of SIPO, PISO, PIPO, and universal shift registers. Lower bounds of the designs are established by providing relevant properties. The designs have the applications to perform serial-to-parallel and parallel-to-serial conversions. The design forms an important step in the building of complex reversible sequential circuits for quantum computers.

9

Reversible Multiplexer and Demultiplexer with Other Logical Operations

A multiplexer (mux) is a device that selects one of several input signals and forwards the selected input into a single line. A multiplexer of $2n$ inputs has n select lines, which are used selecting input line to send to the output. Multiplexers are mainly used to increase the amount of data that can be sent over the network within a certain amount of time and bandwidth. A multiplexer is also called a data selector. Multiplexers can also be used to implement Boolean functions of multiple variables. Conversely, a demultiplexer (demux) is a device taking a single input signal and selecting one of many data output lines, which is connected to the single input. A multiplexer is often used with a complementary de-multiplexer on the receiving end.

9.1 Reversible Logic Gates

Various types of reversible gates are used in the related works such as multiplexer and demultiplexer. In the following subsections, some related reversible logic gates are described with their applications.

9.1.1 RG1 Gate

Figure 9.1 shows the reversible gate 1 (RG1 gate). It has three inputs (A, B, and C) and three outputs (P, Q, and R). The output logic expressions are defined as $P = B'$, $Q = AB' + BC$, and $R = A \oplus C$. This gate can be used as a 2:1 multiplexer, and also can be used as a data copier.

9.1.2 RG2 Gate

Figure 9.1 shows the reversible RG2 gate. It has three inputs (A, B, and C) and three outputs (P, Q, and R). The outputs can be defined as $P = A'B' \oplus C$, $Q = A' \oplus B'$, and $R = A$. This gate can be used as a 2:1 multiplexer, and also can be used as a data copier.

Reversible and DNA Computing, First Edition. Hafiz Md. Hasan Babu.
© 2021 John Wiley & Sons Ltd. Published 2021 by John Wiley & Sons Ltd.

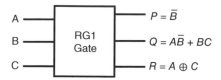

Figure 9.1 Reversible gate 1 (RG1).

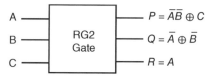

Figure 9.2 Reversible gate 2 (RG2).

9.2 Designs of Reversible Multiplexer and Demultiplexer with Other Logical Operations

Reversible logic is a computing paradigm in which one-to-one mapping is maintained between the input and the output vectors. The multiplexer is a circuit that produces only one output from many inputs. On the other hand, a demultiplexer is a circuit that produces many outputs from a single input. In the following subsections, reversible R-I and R-II gates are used to construct reversible multiplexer and demultiplexer circuits.

9.2.1 The R-I Gate

Figure 9.3 shows the R-I reversible gate. It has three inputs (A, B, and C) and three outputs (P, Q, and R). The outputs are defined as $P = B$, $Q = AB' + BC$, and $R = AB \oplus C$. The R-I reversible logic gate can be realized as a 2:1 multiplexer, 1:2 de-multiplexer, AND, and XOR gates. The truth table for this gate is shown in Table 9.1, which maintains the reversible principle. Pass transistors are used to realize this reversible gate at circuit level, since CMOS architectures are not reversible in nature. The transistor level realization of R-I gate is shown in Figure 9.4. The PMOS and NMOS transistors are labeled as MP1-MP4 and MN1-MN5, respectively. The MP3-MN4 and MP4-MN5 are two CMOS inverters connected back to back to maintain the reversibility principle. Rests of transistors are configured as pass transistors. Further, the realizations of XOR, AND, multiplexer, and demultiplexer are shown in Figure 9.4 to Figure 9.8.

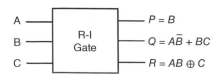

Figure 9.3 Reversible R-I gate.

Table 9.1   Truth Table of the B-I Gate

	Input			Output	
A	B	C	P	Q	R
0	0	0	0	0	0
0	0	1	0	0	1
0	1	0	1	0	0
0	1	1	1	1	1
1	0	0	0	1	0
1	0	1	0	1	1
1	1	0	1	0	1
1	1	1	1	1	0

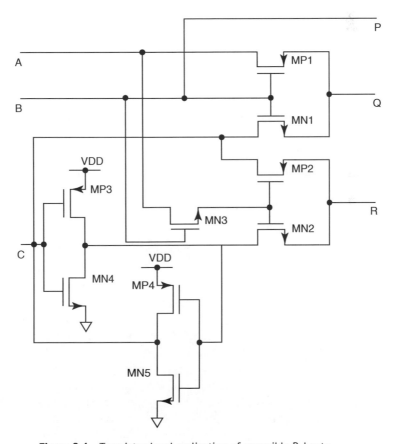

Figure 9.4   Transistor level realization of reversible R-I gate.

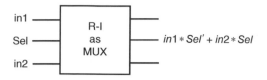

Figure 9.5 Realization of 2:1 multiplexer using reversible R-I gate.

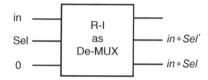

Figure 9.6 Realization of 1:2 demultiplexer using reversible R-I gate.

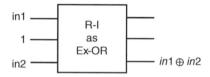

Figure 9.7 Realization of two-input XOR using reversible R-I gate.

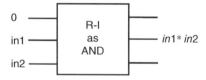

Figure 9.8 Realization of two-input AND gate using reversible R-I gate.

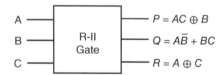

Figure 9.9 Reversible R-II gate.

9.2.2 The R-II Gate

Figure 9.9 shows the R-II reversible gate. It has three inputs (A, B, and C) and three outputs (P, Q, and R). The outputs are defined by $P = AC \oplus B$, $Q = AB' + BC$, and $R = A \oplus C$. This gate can be used as a multiplexer, XOR, etc. The truth table for this gate is shown in Table 9.2, which maintains the reversible principle. The transistor level realization of R-II gate is shown in Figure 9.10. The PMOS and NMOS transistors are labeled as MP1-MP7 and MN1-MN7, respectively. The MP4-MN4 and MP5-MN5 are two CMOS inverters connected back to back to maintain the reversibility principle. Similarly, MP6-MN6 and MP7-MN7 are another set of inverters connected back-to-back to maintain the reversibility of the logic gate. The rest of the transistors are configured as pass transistors. Further, the realization of logic functions like multiplexer, XOR, half adder, and AND gate are shown in Figure 9.11 to Figure 9.13, respectively.

Table 9.2   Truth Table of the R-II Gate

Input			Output		
A	B	C	P	Q	R
0	0	0	0	0	0
0	0	1	0	0	1
0	1	0	1	0	0
0	1	1	1	1	1
1	0	0	0	1	1
1	0	1	1	1	0
1	1	0	1	0	1
1	1	1	0	1	0

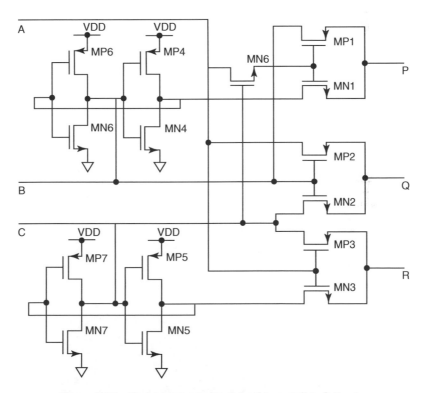

Figure 9.10   Transistor level circuit for the reversible R-II gate.

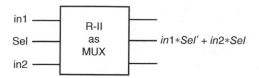

Figure 9.11 Realization of 2:1 multiplexer using reversible R-II gate.

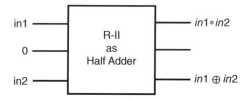

Figure 9.12 Realization of two-input XOR and half adder using reversible R-II gate.

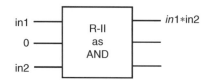

Figure 9.13 Realization of two-input AND gate using reversible R-II gate.

9.3 Summary

This chapter introduced two reversible gates, named as R-I gate and R-II gate, for realizing reversible combinational logic circuits. These two gates can be used for realization of basic logical functions such as AND, XOR, and MUX. Besides these functions, other advantage of the R-I gate is that it can be used as a 1:2 de-multiplexer without requiring any extra logic circuits and the R-II gate can be used as a half adder circuit.

10

Reversible Programmable Logic Devices

A field programmable gate array (FPGA) is a semiconductor device that can be configured by the customer or designer after manufacturing hence it is named as "field pro-grammable." FPGAs can be used to implement any logical function that an application specific integrated circuit (ASIC) could perform, but the ability to update the functionality after shipping offers advantages for many applications. FPGAs, which contain pro-grammable logic components are called "logic blocks," and a hierarchy of reconfigurable interconnects that allow the blocks to be "wired together" somewhat like a one-chip pro-grammable breadboard. Logic blocks can be configured to perform complex combinational functions, or merely simple logic gates like AND Ex-OR. In most FPGAs, the logic blocks also include memory elements, which may be simple flip-flops or more complete blocks of memory.

Array logic was introduced based on AND, OR, and NOT synthesis to implement SOP or POS, whereas reversible logic prefers Ex-OR operation as well as exclusive sum-of-product (ESOP) synthesis. ESOP synthesis gives out better result than SOP realization where many useful methods are shown for minimizing multi-output Boolean functions into ESOP. This chapter has introduced an approach of designing reversible programmable logic arrays.

10.1 Reversible FPGA

By designing the logic elements of FPGA in reversible logic can reduce the power dissi-pation. In this section, the reversible implementation of the internal architecture of logic elements of Plessey FPGA is described. Different components of logic elements of Plessey FPGA are also shown. The different components are an 8-to-2 reversible MUX, a reversible D latch, a reversible RAM, and a reversible NAND gate. A reversible 4-to-1 MUX design is introduced. Any *4n-to-n* MUX can be designed using this reversible 4-to-1 MUX. A 4×4 reversible gate named BSP gate is also described to design the 4–to–1 MUX. The block dia-gram along with the quantum analysis of reversible NH gate is also shown in this section. To design the reversible RAM, reversible write enable master–slave flip-flops are needed. So, a design of reversible write enable master–slave flip-flop with fewer gates and garbage outputs is shown. For the design of this flip-flop, a reversible D latch is also needed. Using this, the design of reversible RAM evaluates a significant reduction of number of gates and garbage outputs along with quantum cost. Finally, the reversible architecture of logic elements of

Reversible and DNA Computing, First Edition. Hafiz Md. Hasan Babu.
© 2021 John Wiley & Sons Ltd. Published 2021 by John Wiley & Sons Ltd.

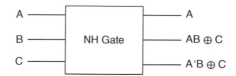

Figure 10.1   Block diagram of 3 × 3 reversible NH gate.

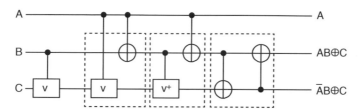

Figure 10.2   Quantum realization of 3 × 3 reversible NH gate.

Plessey FPGA is shown, which will result in less power dissipation and will improve the performance of different applications of FPGAs like DSP, prototyping, etc.

10.1.1   3 × 3 Reversible NH Gate

A 3 × 3 reversible gate is introduced in this subsection, which is known as NH gate. The block diagram of reversible NH gate is shown in Figure 10.1. The quantum equivalent circuit of reversible NH gate has been realized in Figure 10.2.

10.1.2   4 × 4 Reversible BSP Gate

In this subsection, a 4 × 4 reversible gate called "BSP" gate is introduced. The gate is shown in Figure 10.3. The gate can implement all two input Boolean functions. In this chapter, this gate is used to design a reversible MUX to minimize the number of gates and garbage outputs.

10.1.3   4-to-1 Reversible Multiplexer

A 4-to-1 reversible MUX is shown in this subsection. The MUX is designed by the reversible BSP gate without any decoder and is shown in Figure 10.4. The equation of a 4-to-1 MUX can be written as $O = I_0 S'_0 S'_1 + I_1 S_0 S'_1 + I_2 S'_0 S_1 + I_3 S_0 S_1$. Any 4$n$-to-$n$ reversible MUX can be designed using this design approach where n levels are needed to design the 4n-to-n reversible MUX.

Property 10.1.3.1   A 4-to-n MUX can be realized by at least 9 n reversible gates and generates at least 11n garbage outputs.

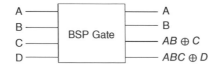

Figure 10.3   Block diagram of 4 × 4 reversible BSP gate.

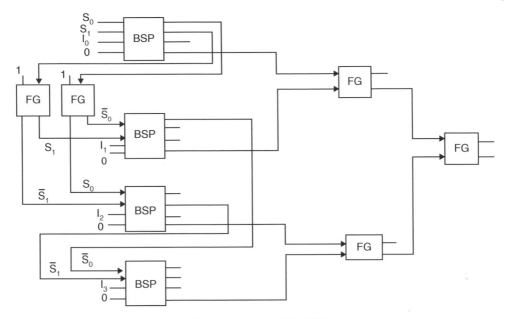

Figure 10.4 4-to-1 reversible MUX.

Proof: According to Figure 10.4, a 4-to-1 ($n = 1$) MUX can be realized by 9(9×1) reversible gates (5 FGs and 4 BSP gates) and generates 11 garbage bits. Similarly, an 8-to-2 ($n = 2$) MUX requires 18 reversible gates and generates 22 garbage bits and so on. Hence, a $4n$-to-n MUX can be realized by at least $9n$ reversible gates which generate at least $11n$ garbage bits.

10.1.4 Reversible D Latch

A design of a D latch is introduced in this subsection with 1 FRG and 1 NH gate, which is shown in Figure 10.5.

Property 10.1.4.1 A reversible D latch can be realized by at least two reversible gates with two garbage outputs.

Proof: The design requires two reversible gates, 1 FRG and 1 NH gate. FRG gate is needed to produce Q and NH gate to produce Q and Q' for the D latch. No further minimization is possible for the design of D latch. In Figure 10.5, only the FRG gate produces two garbage outputs. Hence, a reversible D latch can be realized by at least two reversible gates with two garbage outputs.

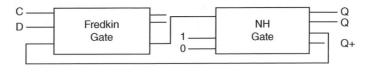

Figure 10.5 Reversible D latch.

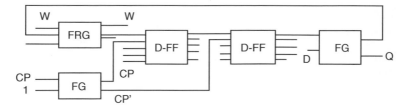

Figure 10.6 Reversible Write-Enabled Master–Slave flip-flop.

10.1.5 Reversible Write-Enabled Master–Slave Flip-Flop

Write enabled master slave flip-flops are needed to design a reversible random access memory (RAM). As data is both read from and written into RAM, each flip-flop should work on two modes: read and write. A design of reversible write enabled master–slave flip-flop is shown in Figure 10.6.

10.1.6 Reversible RAM

In this subsection, a design of a reversible random access memory (RAM) is realized, which is shown in Figure 10.7. Here, the reversible write enabled master–slave D flip-flop of Figure 10.6 is used. From Figure 10.7, it is clear that a reversible RAM produces at least $2^n(8m + 2)$ garbage outputs.

10.1.7 Design of Reversible FPGA

In this subsection, a reversible design of logic element of an SRAM-based FPGA is described. On the highest level, an FPGA consists of programmable logic elements and

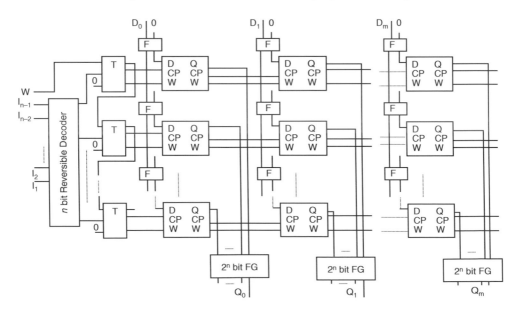

Figure 10.7 Block diagram of a reversible RAM.

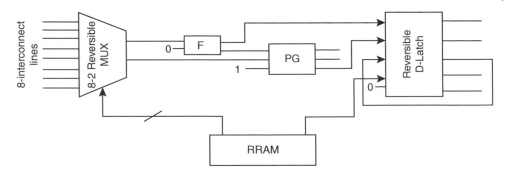

Figure 10.8 A reversible logic element of Plessey FPGA.

programmable routing resources used to interconnect the logic elements. The logic elements implement the combinational and sequential logic functions the user wants to implement in the FPGA, and the routing resources interconnect logic elements to implement the desired function. The reversible design for the logic elements of SRAM-based FPGA is shown. Here the design of Plessey FPGA has been realized using the reversible MUX, reversible D latch and a reversible RAM. The block diagram of the design according to Algorithm 10.1.7.1 is shown in Figure 10.8.

Algorithm 10.1.7.1 Reversible Logic Block Construction Algorithm (A_0 to A_7)

Input: 8-bit interconnection
Output: Q, which will be sent to the next logic block
1: Begin
2: Take selection bits from RRAM to MUX.
3: With the selection bits of Step 2, select two bits out of 8 input bits by the MUX.
4: Copy the first output of Step 3 using a FG.
5: Propagate the second output bit of FG in Step 3 to the first input of Peres gate and propagate the second output bit of Step 3 to the second input of Peres gate. To perform the NAND operation, the third input is 1.
6: Propagate the first output of FG in Step 3 to the first input of the D latch.
7: Propagate the third output of PG to the second input of the D latch.
8: Propagate one control bit to the fourth input of D latch which will make the D latch transparent if the latch is not needed.
9: The fourth output of D latch will return Q.
10: End

Property 10.1.7.1 A reversible logic block of Plessey FPGA can be realized by at least $(g + 22)$ reversible gates, where g is the number of gates, which is required to realize the reversible RAM.

Proof: By Property 10.1.3.1, an 8-to-2 MUX produces at least 18 reversible gates. A D latch produces at least two reversible gates according to Property 10.1.4.1 and the reversible RAM produces at least $2^n(8m + 2)$ reversible gates as shown in Figure 10.7. Let $2^n(8m + 2) = g$. Hence, along with one Feynman gate and one Peres gate, a Plessey FPGA can be realized with at least $g + 18 + 2 + 2 = (g + 22)$ reversible gates.

Property 10.1.7.2   A reversible logic block of Plessey FPGA produces at least $(g + 26)$ garbage outputs, where g is the number of garbage outputs generated from reversible RAM.

Proof: It can be shown, from Property 10.1.3.1, that 8-to-2 MUX produces at least 22 garbage outputs. A D latch produces at least two garbage outputs according to Property 10.1.4.1 and reversible RAM produces at least $(2^n(8m + 2) + n + 1)$ garbage outputs as shown in Figure 10.7. Let $(2^n(8m + 2) + n + 1) = g$. In Figure 10.8, FG generates no garbage outputs, but PG generates two garbage outputs. Hence, as a result the total circuit in Figure 10.8 produces at least $(22 + 2 + 2 + g) = (26 + g)$ garbage outputs.

Property 10.1.7.3   Let gt_f be the number of gates required to realize a reversible logic block of Plessey FPGA, gt_m be the number of gates required to realize a reversible 8-to-2 MUX, gt_d be the number of gates required to realize a reversible D latch, gt_r be the number of gates required to realize a reversible RAM. Then $gt_f >= gt_m + gt_d + gt_r + 2$ where $gt_m >= 18 gt_d >= 2 gt_r >= (8m + 2) \times 2^n$.

Proof: A reversible logic block of Plessey FPGA consists of an 8-to-2 reversible MUX, which requires at least 18 reversible gates according to Property 10.1.3.1. So $gt_m >= 18$. A reversible D latch, which requires at least 2 reversible gates according to Property 10.1.4.1. So, $gt_d >= 2$. A reversible RAM which requires at least $(8m + 2) \times 2^n$ reversible gates according to Figure 10.7. So $gt_r >= (8m + 2) \times 2^n$. Therefore, the total number of gates for a reversible logic block of Plessey FPGA is $gt_f >= gt_m + gt_d + gt_r + 2$, where $gt_{m >= } 18 \, gt_d >= 2$, $gt_r >= (8m + 2) \times 2^n$.

10.2   Reversible PLA

This section describes the idea of reversible PLAs based on ordering of output functions and input variables.

10.2.1   The Design Procedure

An example of the multi-ESOP (exclusive-OR sum-of-product) functions is $F = \{f_1, f_2, f_3, f_4, f_5\}$. Consider the following functions in the reversible PLA:

$$f_1 = ab' \oplus ab'c$$
$$f_2 = ac \oplus a'b'c$$
$$f_3 = ab' \oplus bc' \oplus ab'c$$
$$f_4 = ac$$
$$f_5 = ab' \oplus ac \oplus bc'$$

From now on, the above function F will be confirmed as multiple-output function in this chapter.

Table 10.1 The Products of Functions

Functions	f_1	f_2	f_3	f_4	f_5
Number of Products	2	2	3	1	3

The design is based on the ordering of input variables, which depends on the corresponding order of products. But the order of products will be generated after the optimization of Ex-OR plane. In this subsection, two algorithms are described for the construction of Ex-OR plane followed by the realization of AND plane for generalizing the design. The 3×3 reversible MUX or MG gate is used to design the reversible PLAs, which has minimum quantum cost i.e., 4 as shown in Figure 10.9. MG gate can realize the operation of 2-to-1 multiplexer circuit and able to produce half of minterms generated by two variables. There exist three modes (FG-1, FG-2, and FG-4) of Feynman gate as shown in Figure 10.10 that have been used for the design. One mode of the Toffoli gate TG-3 is also shown in Figure 10.11. Figure 10.12 shows the templates MG-5 and MG-6 of a MUX gate that have been used in the design. In the further discussion, symbol 1, 2, 4 as shown in Figure 10.10 and symbol 5, 6 as shown in Figure 10.12 are used rather than full names FG-1 or MG-5 etc. to represent the particular modes. *DOT* is the cross point in reversible PLA in which no gate is used.

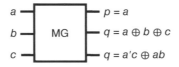

Figure 10.9 3×3 Reversible MUX gate.

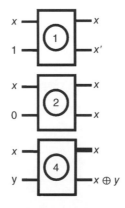

Figure 10.10 Different uses of a Feynman gate.

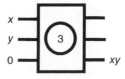

Figure 10.11 One template of toffoli gate.

Property 10.2.1.1 Size of the function $Size(f_i)$: The size of the function is the number of total products, which is expressed as the ESOP.

Example 10.2.1.1 The number of products of f_1 is 2, f_2 is 2, and f_3 is 3. The products of functions are shown in Table 10.1.

Property 10.2.1.2 The product lines are the horizontal lines corresponding to the products of an AND plane. These product lines are used in the Ex-OR plane to generate the output of a particular function consisting of Ex-OR operations. The number of product lines is equal to the number of total products. There are five product lines for five products of the function F.

In the design, the ordering of output functions is related to the $Size(f_i)$. Functions are generated in ascending order based on this criterion. The Ex-OR plane and AND plane are optimized by using MUX and Feynman gates. The realization of Ex-OR plane for the function F is shown in Figure 10.13 by using Algorithm 10.2.1.1.

Property 10.2.1.3 Let n be the number of Ex-OR operations of m output functions and $TDOT$ be the number of cross-points. Then the minimum number of Feynman gates to realize Ex-OR plane is $n + m - TDOT$.

Proof: When there are $TDOT$ cross-points for m functions, the number of additional Feynman gates in Ex-OR plane of reversible PLA is $m - TDOT$. As there are n Ex-OR operations by n Feynman gates, the total number of Feynman gates in the Ex-OR plane of reversible

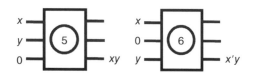

Figure 10.12 Two templates of MUX gate.

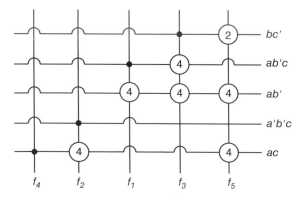

Figure 10.13 Ex-OR plane realization for the function F based on the Algorithm 10.2.1.1.

Algorithm 10.2.1.1 Construction of Ex-OR Plane

1: Begin
2: $TDOT := 0$ [$TDOT$ = Total number of DOT]
3: Sort output functions according to $Size(f_i)$
4: REPEAT Step 5 for each output function
5: **if** Sizeof (f_i) of f_i is one **then**
6:   **if** product p_j exists **then**
7:     use FG-2
8:   **else**
9:     assign a line for product (p_j) and use DOT
10:     $TDOT := TDOT + 1$
11:   **end if**
12: **else**
13:   **if** all product(s) p_j **then**
14:     use FG-2 for the top most line and FG-4 for others
15:   **else**
16:     assign the upper lines for products p_j and use DOT for the top most and FG-4 for existings
17:     $TDOT := TDOT + 1$
18:   **end if**
19: **end if**
20: End

PLA is $n + m - TDOT$. For multi-output function F, the number of outputs (m) is 5 and the number of Ex-OR operations is 6. The number of $TDOTs$ is 4 in Figure 10.13. So, the number of Feynman gates$= n + m - TDOT = 6 + 5 - 4 = 7$.

Property 10.2.1.4 Let p be the number of products and $TDOT$ be the number of cross-points in the Ex-OR plane of reversible PLA. Then the minimum number of garbages to realize Ex-OR plane of reversible PLA is $p - TDOT$.

Proof: As there are $TDOT$ cross-points and p products for m output functions, the total number of garbage outputs in the Ex-OR plane of reversible PLA is $p-TDOT$. Consider Figure 10.13 for multi-output function F. The number of products (p) is 5 and number of cross-points ($TDOT$) is 4. So, the number of garbage outputs is $p - TDOT = 5 - 1 = 4$. The realization of Ex-OR plane generates the order of products as shown in Figure 10.13 and the AND plane is constructed according to this order by using MUX and Feynman gates.

In AND plane, the Feynman gates are used to copy or recover fan-out problem and the MUX gates are used for AND operations. The generation of complementary forms of input literals are unnecessary for the AND plane because MUX and FG are used together to generate all the minterms of two variables without having any dedicated lines of complemented forms of input variables. The realization of the reversible PLA is shown in Figure 10.14 by using Algorithms 10.2.1.1 and 10.2.1.2.

Algorithm 10.2.1.2 Construction of AND Plane

1: Begin
2: $TDOT := 0\ [TDOT = $ Total number of DOT]
3: REPEAT Step 4 for each product (p_j)
4: **if** l_j is the first literal of p_j **then**
5:     **if** l_j is in complemented form **then**
6:         apply FG-1
7:     **else**
8:         **if** l_j is further used **then**
9:             apply FG-2
10:        **else**
11:            use DOT and $TDOT := TDOT + 1$
12:        **end if**
13:    **end if**
14: **else**
15:     **if** l_j in complemented form **then**
16:         MG-6
17:     **else**
18:         apply MG-5
19:     **end if**
20: **end if**
21: End

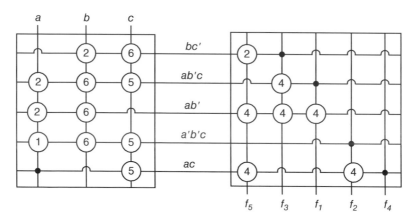

Figure 10.14 Design of reversible PLAs for multi-output function *F*.

Property 10.2.1.5 Let p be the number of products and $TDOT$ be the number of cross-points in the AND plane of a reversible PLA. then the minimum number of Feynman gate to realize AND plane of reversible PLA is $p - TDOT$.

Proof: The cross-points reduce the use of Feynman gates for any particular product line. So, in response to *TDOT* cross-point and p products for m output functions, the total number of Feynman gates in the AND plane of reversible PLA is $p - TDOT$.

Consider Figure 10.14 for multi-output function F. The number of products (p) is 5 and number of cross-points ($TDOT$) is 1. So, the number of Feynman gates is $p - TDOT = 5 - 1 = 4$.

Property 10.2.1.6 Let q be the number of AND operations among products in the AND plane of a reversible PLA. Then the minimum number of MUX gates to realize AND plane of a reversible PLA is q.

Proof: As there are q AND operations for p products of m output functions, the total number of MUX gates to realize the AND plane of a reversible PLA is q.

Property 10.2.1.7 Let l be the number of inputs and q be the number of AND operations among products and $TDOT$ be the number of cross-points in the AND plane of reversible PLA. The minimum number of garbages to realize the AND plane of a reversible PLA is $l + q - TDOT$.

Proof: As there are q AND operations for p products of m output functions F as defined in Subsection 10.2.1 l inputs and $TDOT$ cross-points in the AND plane, the total number of garbages in the AND plane of reversible PLA is $l + q - TDOT$.

10.2.1.1 Delay Calculation of a Reversible PLA

In this subsection, the delay of reversible PLAs is calculated in a greedy approach. The calculation is divided into two phases: (i) AND plane delay (APD (p_i)) and (ii) Ex-OR plane delay (XPD (p_i)) in terms of product lines (horizontal lines). Then both of the delays are merged with respect to both planes. The multi-output function F is used to calculate the delay. First, the delay of AND plane is calculated and then the delay of Ex-OR plane is calculated. Figure 10.15 and Figure 10.16 show the delay calculation of AND plane and Ex-OR plane, respectively. To calculate the delay, the following things are considered:

1. Gate (Via) is represented as circle (DOT).
2. Delay of any gate is 1 and via (DOT) denotes 0.
3. Decimal values show the delay of corresponding circle.

10.2.1.2 Delay Calculation of AND Plane

For AND plane, every gate updates its delay by comparing the delay of neighboring gates at left (L) and top (T) and then, it propagates the updated delay to neighboring gates placed in right (R) and bottom (B) sides, as shown in Figure 10.15. Each circle in Figure 10.15 represents the delay of particular point and arrows show the path of delay propagation.

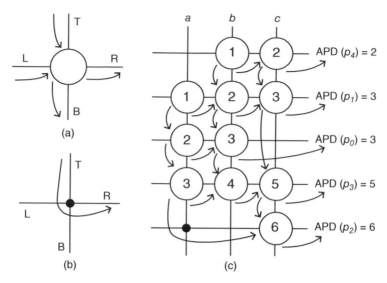

Figure 10.15 Delay calculation of AND plane: (a-b) delay propagation path of a gate and a cross-point and (c) overall delay propagation path for AND plane.

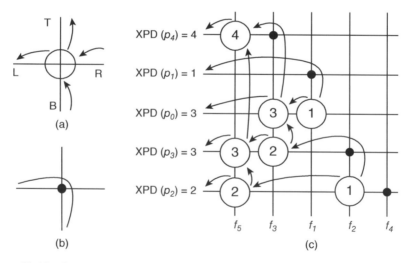

Figure 10.16 Delay calculation of Ex-OR plane: (a-b) delay propagation path of a gate and a cross-point and (c) overall delay propagation path for Ex-OR plane.

10.2.1.3 Delay Calculation of Ex-OR Plane

For Ex-OR plane, every gate updates its delay by comparing the delay of neighboring gates at right (R) and bottom (B) and then, it propagates the updated delay to neighboring gates placed in left (L) and top (T) sides as shown in Figure 10.16.

10.2.1.4 Delay of Overall Design

After calculating the delay of both planes, the delay of product lines having maximum value is the final delay of the overall design of reversible PLAs. Algorithm 10.5 is introduced here for the calculation of delay of reversible PLAs. According to the design, the delay propagation of AND (Ex-OR) plane is top-bottom-right (bottom-top-left).

Algorithm 10.2.1.4.1 Delay Calculation of Reversible PLAs

1: Begin
2: Calculate APD (p_i) (XPD (p_i)) of each product lines of AND (XOR) plane
3: Delay := MAX {APD (p_i) + XPD (p_i)} where $i = 1$ to n ($n =$ total number of product)
4: End

10.3 Summary

This chapter is divided into two parts, namely, reversible field programmable gate array (FPGA) and reversible programmable logic array (PLA). In the first part of this chapter, a design of reversible architecture of the logic element of Plessey FPGA is described, which can result in significant power savings. The reversible architecture is designed using reversible MUX, D latch, NAND, and a reversible random access memory (RAM). A 4×4 reversible gate is introduced for designing an efficient reversible MUX. To design the reversible RAM, a reversible write-enabled master–slave flip-flop is shown. In the second part of this chapter, a design of reversible PLA is described that is able to realize multi-output ESOP (exclusive-OR sum-of-product) functions by using a 3×3 reversible gate, called MG (MUX gate). Also, an algorithm has been introduced to calculate the critical path delay of reversible PLAs. The minimization processes consist of algorithms for ordering of output functions, followed by the ordering of products. Some properties on the numbers of gates, garbages, and quantum costs of reversible PLAs are also shown.

11

Reversible RAM and Programmable ROM

Two memory elements are described in this chapter namely random access memory (RAM) and programmable read only memory (PROM). RAM's internal architecture can be viewed as a two dimensional array of memory cells. Each memory cell stores a single bit of data. In this chapter, a reversible architecture of RAM is described with a reversible gate namely FS gate and some sequential circuits. With the help of properties, the efficiency of reversible logic synthesis of RAM has also been proved. PROM is used to store the value that is unchangeable easily. There is no scope to lose the value if power is off. It also provides security because it is not possible to modify. In this reversible PROM, a reversible decoder is introduced that has low quantum cost and no garbage output. An AND plane and an Ex-OR plane are also described the number of garbage output and quantum cost.

11.1   Reversible RAM

In this section, the design procedure of a reversible RAM is described. In the process of the design, a *3 × 3* reversible gate is introduced namely FS gate which has been described with its quantum realization. Using this gate, an $n \times 2^n$ decoder has been realized in this section. As flip-flop is an essential and core part of RAM for bit storage, it is successfully managed to realize a *D* flip-flop and write-enabled master–slave *D* flip-flop.

11.1.1   3 × 3 Reversible FS Gate

A 3 × 3 reversible gate is introduced in this subsection, which is known as FS gate. The block diagram of reversible FS gate is shown in Figure 11.1. The reversibility of the gate is shown in Table 11.1. The quantum equivalent circuit of reversible FS gate has been realized in Figure 11.2 where template matching technique has been used to calculate the overall cost and each block has been marked from 1 to 6 where block 5 and 6 don't incur any cost.

Property 11.1.1.1   A 3 × 3 reversible FS gate with the following outputs $P = A$, $Q = A'B'$ $\oplus$ C and $R = AB \oplus C$ has the minimum quantum cost i.e., 4.

Proof: In Figure 11.2, the quantum equivalent circuit of a 3 × 3 reversible FS gate has been shown, where each of the blocks marked as 1 to 4 has quantum cost one and blocks 5 and

Reversible and DNA Computing, First Edition. Hafiz Md. Hasan Babu.

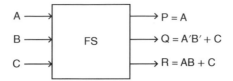

Figure 11.1 Block diagram of 3 × 3 reversible FS gate.

Table 11.1 Truth Table of 3 × 3 Reversible FS Gate.

A	B	C	P	Q	R
0	0	0	0	1	0
0	0	1	0	0	1
0	1	0	0	0	0
0	1	1	0	1	1
1	0	0	1	0	0
1	0	1	1	1	1
1	1	0	1	0	1
1	1	1	1	1	0

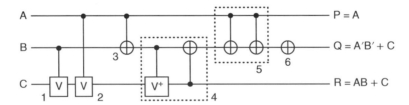

Figure 11.2 Quantum realization of 3 × 3 reversible FS gate.

6 don't have any quantum cost i.e., 0. So the total quantum cost of the FS gate is minimum i.e., 4.

11.1.2 Reversible Decoder

In this subsection, the use of the reversible FS gate is explored to generate a reversible decoder. The 2×2^2 reversible decoder as well as $n \times 2^n$ are described here. An algorithm is also introduced here to construct an n-input decoder in this subsection. Two 3 × 3 reversible FS gates and one Feynman gate have been used to produce a reversible implementation of a 2×2^2 decoder, which is shown in Figure 11.3.

Figure 11.4 shows 3×2^3 reversible decoder whereas the realization of an $n \times 2^n$ reversible decoder has been shown in Figure 11.5 according to the Algorithm 11.1.2.0.1.

Property 11.1.2.1 An $n \times 2^n$ reversible decoder can be realized with at least 2^{n-1} number of reversible gates.

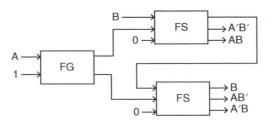

Figure 11.3 2×2^2 Reversible decoder.

Algorithm 11.1.2.0.1 $n \times 2^n$ Reversible Decoder

1: Begin
2: Take one $(n-1) \times 2^{(n-1)}$ reversible decoder. Outputs of this block are considered to be of level L_{i-1}.
3: **for** each level L_i, where $i = 3$ **to** n **do**
4:     **for** $j = 1$ **to** 2^{i-1} **do**
5:         Take one FS gate F_j
6:         **if** $j = 1$ **then**
7:             $F_j[I_1] = n_{th}$ input of the decoder
8:         **else**
9:             $F_j[I_1] = F_{j-1}[O_1]$
10:        **end if**
11:        $F_j[I_2 = j_{th}$ output from L_{i-1}
12:        $F_j[I_3] = 0$
13:    **end for**
14: **end for**
15: End

Proof: According to Figure 11.3, a 2×2^2 reversible decoder can be realized with $2^2 - 1 = 3$ reversible gates using one Feynman gate and two FS gates. In the same way 3×2^3 decoder has $2^3 - 1 = 7$ gates. Thus, it can be said that an $n \times 2^n$ reversible decoder can be realized with at least 2^{n-1} reversible gates.

Property 11.1.2.2   An $n \times 2^n$ reversible decoder generates at least $n - 1$ garbage outputs.

Proof: A 2×2^2 reversible decoder generates $2 - 1 = 1$ garbage output using Algorithm 11.1.2.0.1. In the same way, a 3×2^3 decoder generates $3 - 1 = 2$ garbage outputs. Hence an $n \times 2^n$ decoder can be realized with at least $n - 1$ garbage outputs.

11.1.3   Reversible D Flip-Flop

In this subsection, a reversible D flip-flop is introduced, which can be used to store bits of data in a reversible memory. One FRG and one F2G have been used to design the D flip-flop, which is shown in Figure 11.6.

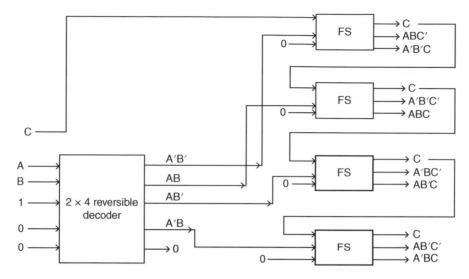

Figure 11.4 3×2^3 Reversible decoder.

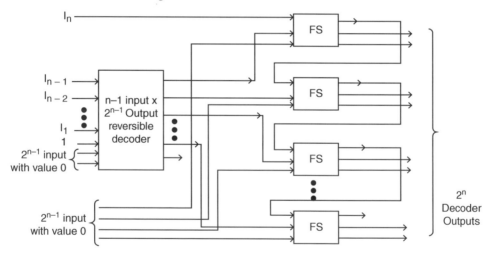

Figure 11.5 $n \times 2^n$ reversible decoder.

11.1.4 Reversible Write-Enabled Master–Slave D Flip-Flop

Since data is both read from and write into RAM, every flip flop should be able to work on two modes: read and write. So in the design of reversible RAM, a write-enabled master–slave D flip-flop is a necessary component. In this section, Figure 11.7 shows the design of reversible write-enabled master–slave D flip-flop, which consists of three Fredkin gates and four Feynman double gates.

11.1.5 Reversible Random Access Memory

The reversible design of RAM is described in this subsection. A RAM is a two-dimensional array of flip-flops. There are 2^n rows where each row contains m flip-flops. Each time only

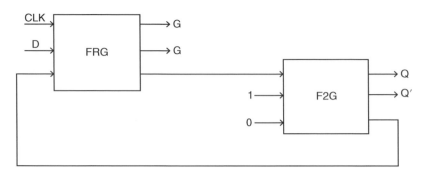

Figure 11.6   Reversible D flip-flop.

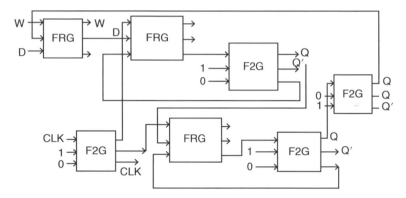

Figure 11.7   Reversible write-enabled master–slave D flip flop.

one of the 2^n output lines of the decoder is active, which selects one row of flip-flops of the RAM. Whether a read or a write operation of the circuit is performed or not depends on the W input. The algorithm of a reversible RAM is given in Algorithm 11.1.5.1. Figure 11.8 shows the implementation of Algorithm 11.1.5.1 to realize the $m \times 2^n$ bit RAM.

Property 11.1.5.1   Let gt_r be the number of gates required to realize an $m \times 2^n$ bit reversible RAM where n be the number of selection bits and m be the number of data bits in the reversible RAM. Then $gt_r >= 2^n(8m + 2) - 1$.

Proof: The decoder part of a reversible RAM requires 2^{n-1} gates. And in the $m \times 2^n$ reversible RAM, there are $m \times 2^n$ write-enabled master–slave D flip-flops, where each flip flop has seven gates. In addition, the RAM requires 2^n Toffoli gates and $m \times 2^n$ Feynman gates for AND and copy operations, respectively. Therefore, $gt_r >= 2^n - 1 + 7 \times m \times 2^n + 2^n + m \times 2^n$. So $gt_r >= 2^n(8m + 2) - 1$.

Property 11.1.5.2   Let n be the number of selection bits and m be the number of data bits in the reversible RAM. Let gar_r be the number of garbage outputs generated from the reversible RAM. Then $gar_r >== 2^n(7m + 3) + n$.

Algorithm 11.1.5.1 $m \times 2^n$ bit Reversible RAM

1: Begin
2: Take one $n \times 2^n$ reversible decoder
3: **for** each row $i = 1$ **to** 2^n in RAM **do**
4:     Take one Toffoli gate T_i
5:     **if** $i = 1$ **then**
6:       $T_i[I_1] =$ Write input (either 0 or 1)
7:     **else**
8:       $T_i[I_1] = T_{i-1}[O_1]$
9:     **end if**
10:     $T_i[I_2] = i_{th}$ output of the decoder
11:     $T_i[I_3] = 0$
12:     **for** each $j = 1$ **to** m **do**
13:       Take one reversible write-enabled master–slave D flip-flop FF_j
14:       $FF_j[D] = D_j$, primary data input of the RAM
15:       **if** $j = 1$ **then**
16:         $FF_j[CP] = T_j[O_2]$
17:         $FF_j[W] = T_j[O_3]$
18:       **else**
19:         $FF_j[CP] = CP$ output FF_{j-1}
20:         $FF_j[W] = W$ output FF_{j-1}
21:       **end if**
22:     **end for**
23: **end for**
24: End

Proof: According to Property 11.1.2.1, a reversible RAM generates $n - 1$ garbage outputs from its decoder. There are m numbers of Feynman gates at the last row in the design of the RAM, where each Feynman gate generates one garbage output. Again, there are 2^n rows in the last column in the design of the RAM, where each row requires a master–slave D flip-flop, which generates three garbage outputs. Moreover the write-enabled master–slave D flip-flop generates six garbage outputs and Toffoli gate at the last row generates one garbage output. Again $m(2^n - 1)$ garbage outputs are generated from the 2^n bit FG shown in Figure 11.8. Therefore, $gar_r >= n - 1 + m + 1 + 3 \times 2^n + 6 \times m \times 2^n 1 + m(2^n - 1)$.

So, $gar_r >= 2^n(7m + 3) + n$.

11.2 Reversible PROM

In this section, the design of 2-to-4 reversible decoder and the design of PROM are described that will help to analyze the reversible PROM.

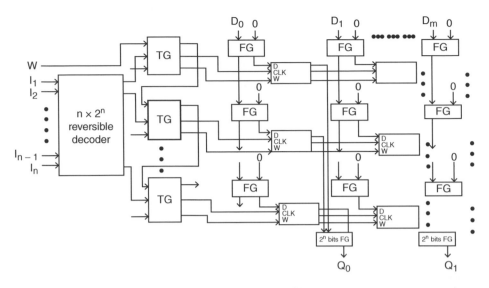

Figure 11.8 Reversible RAM.

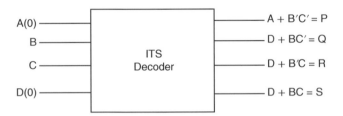

Figure 11.9 Reversible ITS decoder.

11.2.1 Reversible Decoder

A decoder is a device that reverses the operation of encoder and it has multiple inputs and multiple outputs. The inputs and outputs follow the rule of n-to-2^n. Here n is the number of inputs and 2^n is the number of outputs. If input becomes $n = 1, 2, 3 \ldots$ then output will be $2^1 = 2, 2^2 = 4, 2^3 = 8 \ldots$. The binary encoded input number of a decoder is passed to PROM, which is referred to as address. The address line is presented by every signal line. In this subsection, a reversible decoder named ITS decoder is introduced, which is shown in Figure 11.9. This decoder has quantum cost of 4. The quantum representation of the ITS decoder is shown in Figure 11.10. Table 11.2 shows the truth table of reversible ITS decoder. The one-to-one mapping between input vectors and output vectors are got from the Table 11.2.

11.2.2 Design of Reversible PROM

In the design of reversible PROM, the reversible ITS decoder, AND plane, and Ex-OR plane are used. Reversible PROM uses decoder as input; as well as AND plane and Ex-OR plane are used as output, which is shown in Figure 11.11. The output of decoder is selected as

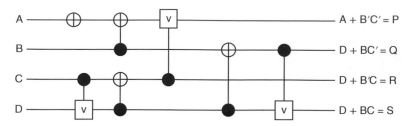

Figure 11.10 Quantum representation of reversible ITS decoder.

Table 11.2 Truth Table of Reversible ITS Decoder

A	B	C	D	P	Q	R	S
0	0	0	0	1	0	0	0
0	0	0	1	1	1	1	1
0	0	1	0	0	0	1	0
0	0	1	1	0	1	0	1
0	1	0	0	0	1	0	0
0	1	0	1	0	0	1	1
0	1	1	0	0	0	0	1
0	1	1	1	0	1	1	0
1	0	0	0	0	0	0	0
1	0	0	1	0	1	1	1
1	0	1	0	1	0	1	0
1	0	1	1	1	1	0	1
1	1	0	0	1	1	0	0
1	1	0	1	1	0	1	1
1	1	1	0	1	0	0	1
1	1	1	1	1	1	1	0

AND array plane's input terms. The output of AND plane is used as Ex-OR plane's input. Final output of Ex-OR plane shows the output of PROM. Every line of input and output is known as address line and every bit is known as word.

PROM consists of an AND plane and EX-OR plane. Where AND plane is fixed and Ex-OR plane is programmable. A large number of equations can have a PROM. To design the reversible PROM, Feynman gate and MUX gate are used. A reversible gate named TI gate is also introduced. Its quantum cost is four. Example of the multi-ESOP (exclusive sum of product) functions is $F = \{f_1, f_2, f_3, f_4, f_5\}$. In the reversible PROM, the function F is as follows:

$$f_1 = ab' \oplus ab'c$$
$$f_2 = ac \oplus a'b'c$$

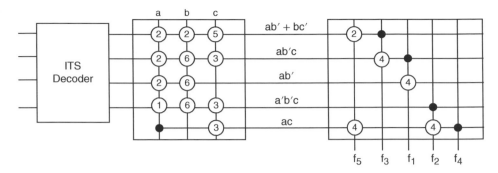

Figure 11.11 Block diagram of reversible PROM.

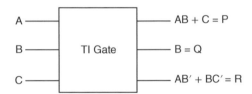

Figure 11.12 Reversible TI gate.

$$f_3 = ab' \oplus bc' \oplus ab'c$$
$$f_4 = ac$$
$$f_5 = ab' \oplus ac \oplus bc'$$

The input variables depend on products. In EX-OR plane, the products of function will be generated. Two algorithms have been established for the reversible PROM: one is for Ex-OR plane and another is for AND plane. The 3 × 3 reversible TI gate has been used to get reversible PROM. It has been used to reduce the number of garbage outputs and quantum cost of AND plane and Ex-OR plane. The quantum cost of TI gate is four and garbage is two. The TI gate is shown in Figure 11.12. Table 11.3 shows the truth table of reversible TI gate. The one-to-one mapping between input vectors and output vectors are shown in Table 11.3.

There exist three modes of Feynman gate such as FG-1, FG-2, and FG-4, which are shown in Figure 11.13 and they have been used in the design of reversible PROM. Besides these modes, one mode of Toffoli gate TG-3 is also shown in Figure 11.14. In addition, two modes of TI-5 and TI-6 gates are shown in Figure 11.15. It is noted that POINT is used in the design where no gate is used.

Property 11.2.2.1　The size of the function is the number of total products that are expressed as the ESOP.

Example 11.2.2.1　The number of products of f_1 is 2, f_2 is 2, and f_3 is 3. The product of functions is shown in Table 11.4.

If anyone wants, the number of products and functions can be increased. In this subsection, a reversible PROM is explained. A reversible PROM consists of an AND plane and

Table 11.3 Truth Table of Reversible ITS Gate.

A	B	C	P	Q	R
0	0	0	0	0	0
0	0	1	1	0	0
0	1	0	0	1	1
0	1	1	1	1	0
1	0	0	0	0	1
1	0	1	1	0	1
1	1	0	1	1	1
1	1	1	0	1	0

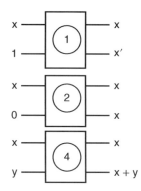

Figure 11.13 Different uses of Feynman gate.

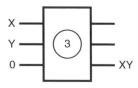

Figure 11.14 Template of Toffoli gate.

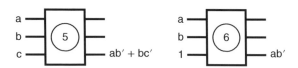

Figure 11.15 Two templates of TI gate.

Ex-OR plane. Particular functions for AND plane and Ex-OR plane are used in the design. The AND plane and Ex-OR plane are shown in Figure 11.16.

In this design, the order of output functions is related to the number of product. Algorithm 11.2.2.1 is used for the order of output functions in the Ex-OR plane of PROM circuit and Algorithm 11.2.2.2 is used for the design of AND plane of PROM circuit.

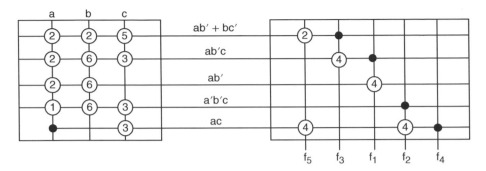

Figure 11.16 The combined design of AND plane and Ex-OR plane.

Algorithm 11.2.2.1 The Algorithm for Ex-OR Plane

1: Begin
2: *TPOINT* := 0 [*TPOINT*= total number of DOT]
3: Arrange the output function according their products.
4: **REDO** Step 5 for every output function of F.
5: **if** a single product or all products of a function exits **then**
6:    use FG-2.
7: **else**
8:    Assign product and use POINT
9:    Here *TPOINT* := *TPOINT* + 1
10: **end if**
11: For Ex-OR operation, use FG-4.
12: End

Property 11.2.2.2   Let x be the number of Ex-OR operations and y be the number of output functions. Let *TPOINT* be the number of cross points. Then the minimum number of Feynman gates in the Ex-OR plane is $x + y$ - *TPOINT*.

Example 11.2.2.2   In Ex-OR plane in Figure 11.16, the number of Ex-OR operations x, is 3, the number of output functions, y is 5, and the number of *TPOINT* is 4. So, the total number of Feynman gates in the Ex-OR plane in Figure 11.16 is $x + y - TPOINT = 4 + 5 - 4 = 5$.

Property 11.2.2.3   If the number of product is p and the total number cross points is *TPOINT*. Then the garbage of Ex-OR plane is p – *TPOINT*.

Table 11.4   The Product of Functions.

Functions	f_1	f_2	f_3	f_4	f_5
Number of Products	2	2	3	1	3

Algorithm 11.2.2.2 The Algorithm for Designing AND Plane

1: Begin
2: *TPOINT:=0* [here *TPOINT* is the total number of *POINT*]
3: **REDO** Step 4 to get every product.
4: **if** it is the first input of **then**
5:    **if** it is in complementary form **then**
6:       use FG-1
7:    **end if**
8: **else**
9:    **if** there are inputs **then**
10:       use FG-2
11:    **end if**
12: **else**
13:    use *POINT* and *TPOINT* := *TPOINT* + 1
14: **end if**
15: **if** it is in complementary form **then**
16:    use TI-6
17: **else**
18:    use TG-3
19: **end if**
20: **if** there are products **then**
21:    use TI-5
22: **end if**
23: End

Example 11.2.2.3 In the Ex-OR plane in Figure 11.16, the number of products, p, is 5 and the number of cross points, *TPOINT*, is 4. So, the number of garbage outputs in the Ex-OR plane is $p - TPOINT = 5 - 4 = 1$.

11.3 Summary

This chapter is divided into two parts, namely reversible RAM and reversible PROM. In the first part of this chapter, the reversible logic synthesis of a RAM is described with a 3×3 reversible gate named as FS. In the way of designing a reversible RAM, an $n \times 2^n$ reversible decoder, reversible D flip-flop, and write-enabled master–slave D flip-flops, are also designed. Moreover, some properties for designing reversible decoder as well as reversible RAM are also described. In the second part of this chapter, design of reversible PROM is discussed. A reversible decoder named ITS and another reversible gate TI are also introduced. In addition, an AND-plane and an Ex-OR plane are described for designing the reversible PROM.

12

Reversible Arithmetic Logic Unit

Recent developments in reversible logic has improved the quantum computer algorithms and its corresponding computer architectures. Reversible logic is extensively being considered as the potential logic design style for implementation in modern nanotechnology and quantum computing with minimal impact on physical entropy. Significant contributions have been made in the literature towards the design of reversible logic gate structures and arithmetic units. However, not many efforts are directed toward the design of a reversible arithmetic logic unit (ALU). The binary logic circuits that were built using traditional irreversible gates certainly lead to energy dissipation, regardless of the technology used to realize the gates. In this chapter, a design of reversible ALU is introduced.

12.1 Design of ALU

An Arithmetic logic unit (ALU) is a data processing unit which is an important part of the central processing unit (CPU). Various types of CPUs are available but every CPU contains an ALU. In the following subsections, conventional and reversible ALUs are described with some of their properties.

12.1.1 Conventional ALU

The DM74LS181 is reference logic of a 4-bit ALU, which can perform 16 logic operations on two variables and a variety of arithmetic operations. This ALU provides 16 arithmetic operations such as add, subtract, comparison, double and 12 other arithmetic operations. It also provides 16 logic operations of two variables such as Ex-OR, comparison, AND, NAND, OR, NOR, and 10 other logic operations. The logic diagram of conventional ALU is shown in Figure 12.1.

12.1.2 The ALU Based on Reversible Logic

The multi-function ALU based on reversible logic gates mainly contains the reversible function generator and the reversible controlled unit. The reversible function generator and the reversible controlled unit are cascaded by some n-Toffoli gates and NOT gates. An arbitrary bit reversible ALU modules can be realized in this way. The output signals are reused to

Reversible and DNA Computing, First Edition. Hafiz Md. Hasan Babu.
© 2021 John Wiley & Sons Ltd. Published 2021 by John Wiley & Sons Ltd.

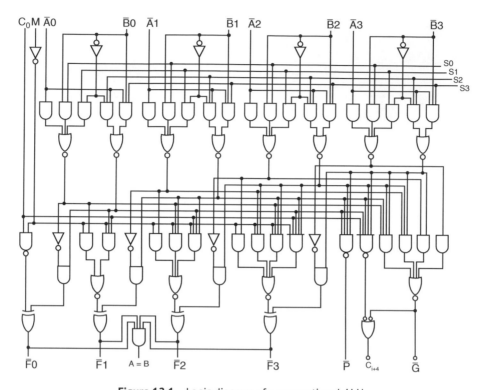

Figure 12.1 Logic diagram of a conventional ALU.

reduce the cost of circuit design as much as possible during the cascading of the reversible function generator and the reversible controlled unit.

12.1.2.1 The Reversible Function Generator

The main feature of a function generator is to process the input information A_i and B_i under the control of the parameters S_0, S_1, S_2, and S_3, and finally it produces X_i and Y_i at the output side, where X_i is the combined function on A_i and B_i controlled by the parameters S_3 and S_2 and Y_i is the combined function on A_i and B_i controlled by the parameters S_1 and S_0. The logic expressions of the function generator are as follows:

$$X_i = \overline{S_3 A_i B_i + S_0 A_i \overline{B_i}}$$ (12.1.2.1.1)

$$Y_i = \overline{A_i + S_0 B_i + S_1 \overline{B_i}}$$ (12.1.2.1.2)

According to the above logical expressions, the reversible function generator is shown in Figure 12.2, and its corresponding block diagram is shown in Figure 12.3.

12.1.2.2 The Reversible Control Unit

The reversible controlled unit is shown in Figure 12.4, which is to complete the summation of three inputs P, Q, and C_i where $P = X_i, Q = Y_i$. It is important to note that X_i and Y_i are derived from the reversible function generator and are added with the carry signal C_i to get the final result F_i.

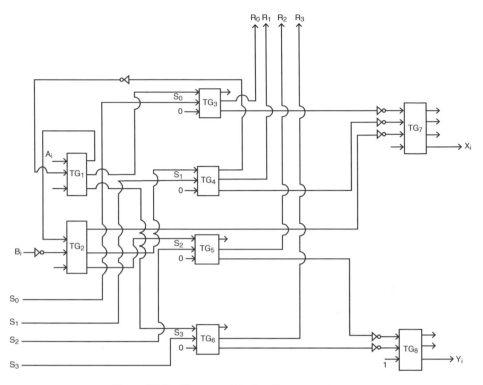

Figure 12.2   The reversible function generator.

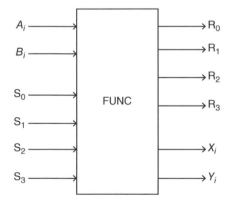

Figure 12.3   Block diagram of reversible function generator.

The logical expression of the reversible controlled unit is as follows:

$$F_i = P \oplus Q \oplus C_i \qquad\qquad (12.1.2.2.1)$$

where signals P and Q come from the signal of the reversible function generator X_i and Y_i and C_i is the carry signal that is added with them to produce F_i as the result. The reversible controlled unit with three-input/output bits uses two 2×2 Toffoli gates and produces the final output with two garbage outputs. The block diagram of the reversible control unit is shown in Figure 12.5.

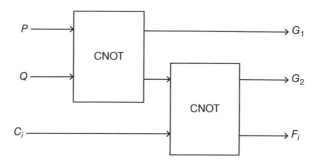

Figure 12.4   The reversible control unit.

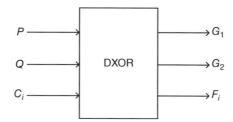

Figure 12.5   Block diagram of the reversible control unit.

12.2   Design of Reversible ALU

To design the reversible ALU with the minimum cost, 3×3 Toffoli gates and NOT gates are used to cascade the reversible function generator and the reversible control unit. Further, any arbitrary bit reversible ALU modules can be realized through the above cascading. The reversible ALU is shown in Figure 12.6, which is cascaded by reversible function generators, reversible control units, and 3×3 Toffoli gates. The reversible ALU performs operations on the binary numbers $A = (A_7, \dots, A_0)$ and $B = (B_7, \dots, B_0)$.

In Figure 12.6, the first operand A, the second operand B, the control signals S_0 to S_3, the carry signal C_i and the control signal M are used as input signals in the reversible ALU, while the result, $F = (F_7, \dots, F_0)$, the carry output signal C_{out} and the garbage outputs are used as the output signals. In addition, the outputs R_0, R_1, R_2, and R_3 of reversible function generator, FUNC_0 are respectively seen as the inputs S_0, S_1, S_2, and S_3 of reversible function generator FUNC_1 and so on.

The value of the first and the second inputs of the reversible control unit DXOR_0 to DXOR_7 are equal to the value of outputs X_i and Y_i of function generator FUNC_0 TO FUNC_7. The value of the third input S c_i of reversible control unit DXOR_0 to DXOR_7 have some connection with the signal M.

Their relationship of the signal can be expressed as follows: When $i = 0$, then

$$C_0 = \overline{C}_{in} + M \qquad\qquad (12.2.1)$$

When $i = 1$, then

$$C_1 = \overline{Y_0 + X_0 C_{in}} + M \qquad\qquad (12.2.2)$$

When $i > 2$, then

$$C_i = \overline{Y_i + X_i C_{i-1}} + M \qquad\qquad (12.2.3)$$

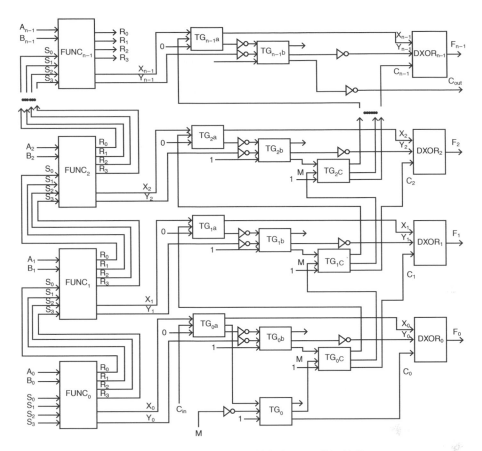

Figure 12.6   The design of 16-bit reversible ALU.

The outputs signals X_i and Y_i are generated after the input signals A_i, B_i, and S_0 to S_3 passed through the reversible function generator X_i and Y_i, which are used as the first two input signals of reversible control unit DXOR$_i$ respectively, and also make operation with control signal M to get the third input signal C_i of reversible control unit DXOR $_i$ based on the Equations 12.2.1 to 12.2.3.

12.3   Summary

This chapter describes the design of an efficient reversible ALU. Using reversible logic gates instead of traditional logic AND/OR gates, a reversible ALU whose purpose is the same as traditional ALU is constructed. Comparing with the number of input bits and the rejected bits of the traditional ALU, the reversible ALU significantly decrease the use and loss of information bits. The reversible 16-bit ALU decreases the information bits use and loss by reusing the logic information bits logically and realizes the aim of lowering power ingestion of logic circuits.

13

Reversible Control Unit

A control unit is a circuit that directs operations within the computer's processor by directing the input and output of a computer system. A control unit consists of two decoders, a sequence counter, and a number of control logic gates. It fetches the instruction from instruction register (IR). Counter and instruction register of a control unit requires flip-flops. If the flip-flop (FF) is designed in an optimized way, the components that consist of flip-flops will also be optimized.

13.1 An Example of Control Unit

The block diagram of a 16-bit control unit is shown in Figure 13.1. Here, the instruction register consists of 16-bits. The operation code (bit 12 to bit 14) is decoded by the 3-to-8 decoder. The outputs of the decoder are $D_0, D_1, \ldots, D_7$. Bit 0 to bit 11 and bit 15 are fed to the control logic gates. The 4-bit sequence counter counts from 0 to 15. The outputs of the counter are decoded into 16 timing signals $T_0, T_1, \ldots, T_{15}$ by the 4-to-16 decoder.

13.2 Different Components of a Control Unit

In this section, the reversible implementation of different components of the control unit is presented. In addition, two reversible gates (namely, HL gate and BJ gate) are introduced that are used to design decoder and $J - K$ flip-flop. The complexities of n-to-2^n decoder are also described in this section. The design of reversible $J - K$ flip-flop, sequence counter, instruction register, and control gates are described here. Finally, construction procedure and the complexities of the control unit are illustrated.

13.2.1 Reversible HL Gate

In this subsection, a 4×4 reversible gate, namely HL gate, is introduced. The gate is shown in Figure 13.2. The truth table of the HL gate is shown in Table 13.1. It can be verified from the truth table that the input pattern corresponding to a particular output pattern can be uniquely determined and vice versa. Figure 13.3 shows that the quantum cost of the HL gate is seven.

Reversible and DNA Computing, First Edition. Hafiz Md. Hasan Babu.
© 2021 John Wiley & Sons Ltd. Published 2021 by John Wiley & Sons Ltd.

Instruction Register

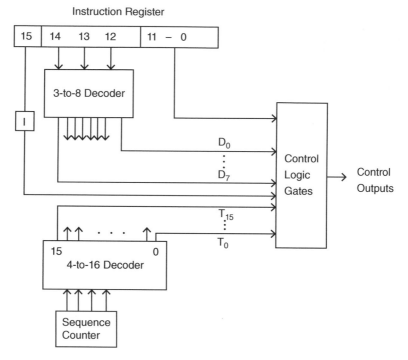

Figure 13.1 Block diagram of a 16-bit control unit.

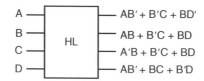

Figure 13.2 Block diagram of reversible HL gate.

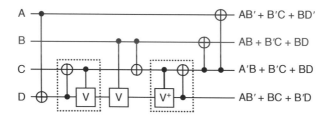

Figure 13.3 Quantum realization of reversible HL gate.

13.2.2 Reversible BJ Gate

In this subsection, a 4 × 4 reversible gate, namely BJ gate, is introduced. The block diagram of the gate is shown in Figure 13.4. The truth table of the BJ gate is shown in Table 13.2. It can be verified from the truth table that the input pattern corresponding to a particular output pattern can be uniquely determined and vice versa. The quantum cost of the BJ gate

Table 13.1 Reversibility of 4 × 4 Reversible HL Gate.

INPUT				OUTPUT			
A	B	C	D	AB'⊕B'C⊕BD'	AB⊕B'C⊕BD	AB'⊕B'C⊕BD	AB'⊕BC⊕B'D
0	0	0	0	0	0	0	0
0	0	0	1	0	0	0	1
0	0	1	0	1	0	1	0
0	0	1	1	1	1	1	1
0	1	0	0	1	1	1	0
0	1	0	1	0	1	0	0
0	1	1	0	1	0	1	1
0	1	1	1	0	1	0	1
1	0	0	0	1	0	0	1
1	0	0	1	1	0	0	0
1	0	1	0	0	1	1	1
1	0	1	1	0	1	1	0
1	1	0	0	1	1	0	0
1	1	0	1	0	0	1	0
1	1	1	0	1	1	0	1
1	1	1	1	0	0	1	1

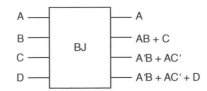

Figure 13.4 Block diagram of reversible BJ gate.

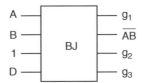

Figure 13.5 NAND implementation of reversible BJ gate.

is 12. The reversible BJ gate can be used as universal gate. The NAND implementation of the BJ gate is shown in Figure 13.5.

13.2.3 Reversible 2-to-4 Decoder

In this subsection, two approaches are introduced to design a reversible 2-to-4 decoder. In the first approach, the reversible 2-to-4 decoder has been designed using one FG gate

Table 13.2   Reversibility of 4 × 4 Reversible BJ Gate.

INPUT				OUTPUT			
A	B	C	D	A	AB⊕C	A'B⊕AC'	A'B⊕AC'⊕D
0	0	0	0	0	0	0	0
0	0	0	1	0	0	0	1
0	0	1	0	0	1	0	0
0	0	1	1	0	1	0	1
0	1	0	0	0	0	1	1
0	1	0	1	0	0	1	0
0	1	1	0	0	1	1	1
0	1	1	1	0	1	1	0
1	0	0	0	1	0	1	1
1	0	0	1	1	0	1	0
1	0	1	0	1	1	0	0
1	0	1	1	1	1	0	1
1	1	0	0	1	1	1	1
1	1	0	1	1	1	1	0
1	1	1	0	1	0	0	0
1	1	1	1	1	0	0	1

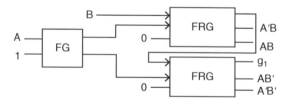

Figure 13.6   2-to-4 Reversible decoder using FG and FRG gate.

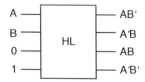

Figure 13.7   2-to-4 Reversible decoder using HL gate.

and two Fredkin gates. This design produces one garbage output and it requires 11 quantum cost. The second approach uses the reversible HL gate to design the reversible 2-to-4 decoder. This circuit does not produce any garbage output and it requires 7 quantum cost. Figure 13.6 and Figure 13.7 show two different approaches of the design of a reversible 2-to-4 decoder.

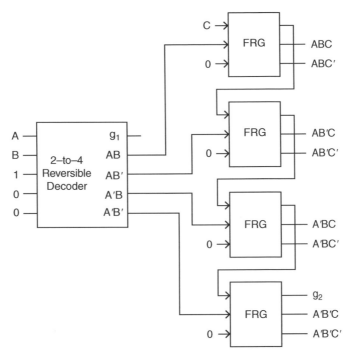

Figure 13.8   Reversible 3-to-8 decoder (Approach 1).

13.2.4   Reversible 3-to-8 Decoder

A reversible 3-to-8 decoder can be designed using one 2-to-4 reversible decoder and four Fredkin gates. Two approaches of the design of the 3-to-8 decoder are shown in Figure 13.8 and Figure 13.9. The first approach of the reversible decoder produces two garbage outputs, whereas the second approach produces one garbage output.

13.2.5   Reversible n-to-2^n Decoder

In this subsection, a generalized reversible n-to-2^n decoder is constructed and the complexities of the circuit are proved. A reversible n-to-2^n decoder is designed with a $(n-1)$-to-$2^{(n-1)}$ reversible decoder and 2^{n-1} Fredkin gates. Figure 13.10 and Figure 13.11 show two approaches of the design of the n-to-2^n decoder.

Property 13.2.5.1   An n-to-2^n reversible decoder (Approach 2) can be realized by at least $2^n - 3$ reversible gates, where n is the number of bits and $n \geq 2$.

Proof: The above statement is proved by mathematical induction.

A 2-to-4 ($n = 2$) decoder (Approach 1) is constructed using one HL gate. So, a 2-to-4 decoder requires at least $1(= 2^2 - 3)$ reversible gates. So, the statement holds for the base case $n = 2$.

Assume that the statement holds for $n = k$. So, a k-to-2^k decoder can be realized by at least $2^k - 3$ reversible gates.

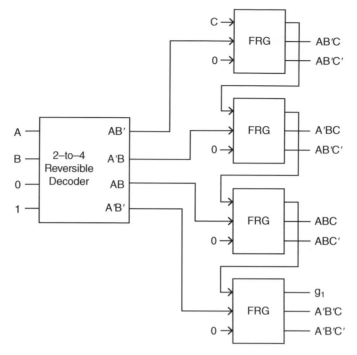

Figure 13.9 Reversible 3-to-8 decoder (Approach 2).

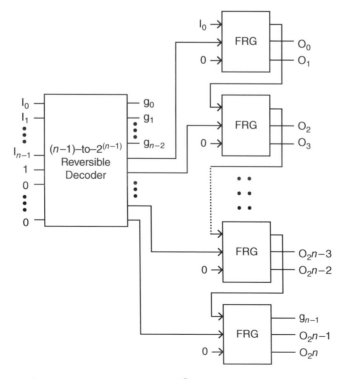

Figure 13.10 Reversible n-to-2^n decoder (Approach 1).

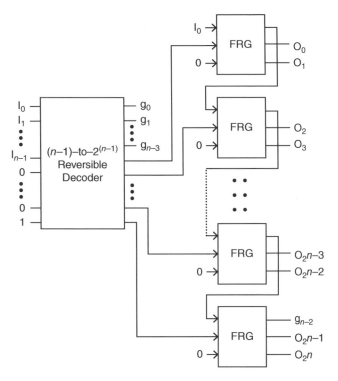

Figure 13.11 Reversible n-to-2^n decoder (Approach 2).

A $(k+1)$-to-$2^{(k+1)}$ decoder is constructed using k-to-2^k decoder and 2^k FRG gates. So, the total number of gates required to construct a $(k+1)$-to-$2^{(k+1)}$ decoder is at least $2^k - 3 + 2^k = 2.2^k - 3 = 2^{(k+1)} - 3$.

So, the statement holds for $n = k + 1$.

Therefore, an n-to-2^n reversible decoder (Approach 2) can be realized by at least $2^n - 3$ reversible gates.

Property 13.2.5.2 An n-to-2^n reversible decoder (Approach 2) generates at least $n - 2$ garbage outputs, where n is the number of bits and $n \geq 1$.

Proof: The above statement is proved by induction.

A 2-to-4 ($n = 2$) decoder (Approach 2) requires only one HL gate. All of the outputs of the HL gates are used to design a 2-to-4 decoder. So, no garbage output is produced by the HL gate as shown in Figure 13.13. So, a 2-to-4 decoder generates at least 0 ($= 2 - 2$) garbage output. So, the statement holds for the base case $n = 2$.

Assume that, the statement holds for $n = k$. So, a k-to-2^k decoder generates at least $k - 2$ garbage outputs.

A $(k+1)$-to-$2^{(k+1)}$ decoder is constructed using k-to-2^k decoder and 2^k FRG gates. A k-to-2^k decoder generates at least $k - 2$ garbage outputs and only the last FRG gate produces one garbage output. So, total number of garbage outputs generated by a $(k+1)$-to-$2^{(k+1)}$ decoder is at least $k - 2 + 1 = (k + 1) - 2$.

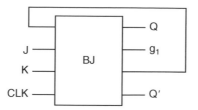

Figure 13.12 Reversible JK flip flop.

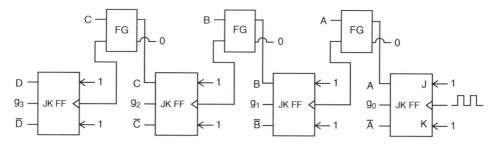

Figure 13.13 4-bit reversible sequence counter.

So, the statement holds for $n = k + 1$.

Therefore, an n-to-2^n reversible decoder (Approach 2) generates at least $n - 2$ garbage outputs, $n \geq 2$.

13.2.6 Reversible JK Flip-Flop

In this subsection, the design of JK flip-flop is shown. The reversible BJ gate is used to construct a JK flip-flop. Figure 13.12 shows the design of a JK flip-flop.

13.2.7 Reversible Sequence Counter

In this subsection, a 4-bit sequence counter is designed that counts from 0 to 15. The sequence counter is designed using four JK flip-flops and four Feynman gates. The JK flip-flops change states with the positive clock edge and the counter counts from 0 to 15. The design of sequence counter produces four garbage outputs, shown in Figure 13.13.

The sequence counter shown in Figure 13.13 can be expanded for any number of bits. An n-bit sequence counter requires n JK flip flops and $n - 1$ Feynman gates (total $2n - 1$ gates), it produces n garbage outputs and it requires $12n + (n - 1) = 13n - 1$ quantum cost.

13.2.8 Reversible Instruction Register

Instruction register is a high-speed circuit that holds an instruction for decoding and execution. A reversible 16-bit instruction register is designed using 16 HNFG gates and 16 JK flip flops. The JK flip-flops take the inputs through HNFG gates and with the change of the clock pulse they produce the normal and complemented outputs. The instruction register produces 17 garbage outputs as shown in Figure 13.14.

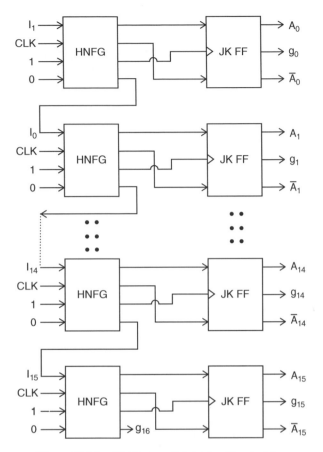

Figure 13.14 16-bit reversible instruction register.

The instruction register shown in Figure 13.14 can be expanded for any number of bits. An n-bit instruction requires n JK flip-flops and n HNFG gates (total $2n$ gates), it produces $n + 1$ garbage outputs, and it requires $12n + 2n = 14n$ quantum cost.

13.2.9 Control of Registers and Memory

The control inputs of the registers are LD (load), INR (increment), and CLR (clear). If it is needed to derive the gate structure associated with the control inputs of address register (AR), then the control functions that change the content of AR are as follows:

$$R\,T_0 : AR \leftarrow PC$$
$$R\,T_2 : AR \leftarrow IR\,(0 - 11)$$
$$D_7IT_3: AR \leftarrow M\,[AR]$$
$$R\,T_0 : AR \leftarrow 0$$
$$D_5\,T_4: AR \leftarrow AR + 1$$

Here, D_5 and D_7 are decoder operations and T_0, T_2, T_3, and T_4 are timing signals. The above control functions can be combined into the following Boolean operations:

$$LD\,(AR) = R\,T_0 + R\,T_2 + D_7\,IT_3$$

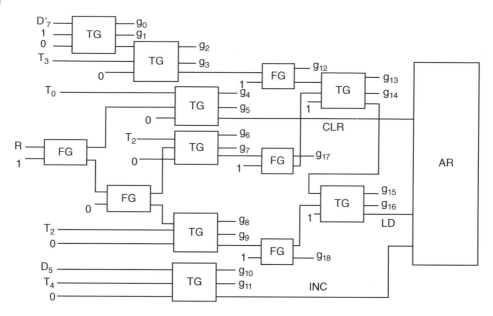

Figure 13.15   Reversible control gates associated with AR.

$$CLR\,(AR) = RT_0$$
$$INC\,(AR) = D_5\,T_4$$

Here, the first statement specifies transfer of information from a register or memory to AR. The second statement clears AR and the last statement increments AR by 1.

The gate structure associated with the above example is shown in Figure 13.15. In a similar fashion, the control gates can be derived for the other register as well as the logic needed to control the read and write inputs of memory.

13.2.10   Construction Procedure and Complexities of the Control Unit

The control unit consists of two decoders, a sequence counter, and a number of control logic gates. It fetches the instruction from the instruction register. The inputs to the control logic gates come from two decoders, the *I* flip-flop and bits 0 through 11 of IR. The outputs of the control logic circuit are:

- Signals to control the inputs of the registers
- Signals to control the read and write inputs of memory
- Signals to set, clear or complement the flip-flops

The steps to construct a 16-bit control unit are described in Algorithm 13.2.10.1.

Property 13.2.10.1   Let ga_{cu} be the number of gates required to realize a reversible *n*-bit control unit (where, *n* is the number of bits) of a processor, ga_{ir} be the number of gates required to realize an *n*-bit instruction register, ga_{d1} be the number of gates required to realize an *m*-to-2^m decoder (where $m < n$), ga_{sc} be the number of gates required to realize a $(log_2 n)$-bit sequence counter, ga_{d2} be the number of gates required to realize a $log_2 n$-to-*n*

Algorithm 13.2.10.1 Construct a Reversible Control Unit

Input: 8-bit interconnection

Output: Q, which will be sent to the next logic block

1: Begin
2: Construct a 3-to-8 decoder.
3: Take the inputs of the 3-to-8 decoder from the instruction register (bit 12 to bit 15).
4: Construct a 4-bit sequence counter.
5: Take the inputs of the 4-to-16 decoder from the outputs of the sequence counter.
6: Construct a control logic gate.
7: Take the inputs of the control logic gate from the instruction register (bit 0 to bit 11 and bit 15), 3-to-8 decoder (D_0 to D_7) and 4-to-16 decoder (T_0 to T_{15}).
8: The outputs of the control register provide necessary signals to carry out the operation of the instruction register.
9: End

---

decoder and ga_{cl} be the number of gates required to realize control logic associated to AR. Then,

$$ga_{cu} >= ga_{ir} + ga_{d1} + ga_{sc} + ga_{d2} + ga_{cl} = 2log_2 n + 2^m + 2n - 9.$$

Proof: According to Subsection 13.2.8, an n-bit instruction set requires at least $2n$ reversible gates. So $ga_{ir} >= n$. An n-bit instruction has m-bit opcode (where $m < n$). An m-to-2^m reversible decoder is required to decode the opcode. According to Property 13.2.5.1, an m-to-2^m decoder requires at least $2^m - 3$ reversible gates. So $ga_{d1} >= 2^m - 3$. A $(log_2 n)$-bit reversible sequence counter counts from 0 to $n - 1$. According to subsection 13.2.7, a $(log_2 n)$-bit sequence counter requires at least $2log_2 n - 1$ reversible gates. Thus, $ga_{sc} >= 2log_2 n - 1$. A $log_2 n$-to-n decoder is required to decode the output of the sequence counter. A $log_2 n$-to-n decoder requires at least $2log_2 n - 3 = n - 3$ reversible gates. So $ga_{d2} >= n - 3$. It is note that the number of gates of the control logic associated to IR depends on the instruction set of a computer. In this case, the number of gates (shown in Figure 13.15) for the given instruction set is 13, where $ga_{d2} >= 15$. Therefore, the total number of gates for reversible n-bit control unit of a processor is:

$$g_{cu} > = ga_{ir} + ga_{d1} + ga_{sc} + ga_{d2} + ga_{cl}$$
$$= n + (2^m - 3) + (2log_2 n - 1) + (n - 3) + 13$$
$$= 2log_2 n + 2^m + 2n - 9.$$

Property 13.2.10.2 Let go_{cu} be the number of garbage outputs generated by a reversible control unit of a processor, ga_{ir} be the number of garbage outputs produced by the n-bit instruction register, go_{d1} be the number of garbage outputs for an m-to-2^m decoder (where $m < n$), go_{sc} be the number of garbage outputs for a $(log_2 n)$-bit sequence counter, go_{d1} be the number of garbage outputs produced by a $log_2 n$-to-n decoder, and go_{cl} be the number of garbage outputs generated by the control logic gates. Then

$$go_{cu} >= go_{ir} + go_{d1} + go_{sc} + go_{d2} + go_{cl} = 2log_2 n + m + n + 16.$$

Proof: An n-bit instruction register produces at least $n + 1$ garbage outputs. So, $go_{ir} >= n + 1$. An n-bit instruction has m-bit opcode (where $m < n$). According to Property 13.2.5.2, an m-to-2^m reversible decoder generates at least $m - 2$ garbage outputs. So, $go_{d1} >= m - 2$. According to subsection 13.2.7, a $(log_2 n)$-bit sequence counter generates at least $log_2 n$ garbage bits. So, $go_{sc} >= log_2 n$. A $log_2 n$-to-n decoder is required to decode the output of the sequence counter. A $log_2 n$-to-n decoder produces at least $log_2 n - 2$ garbage outputs. Thus, $go_{d2} >= log_2 n - 2$. The number of garbage outputs produced by the control logic associated to IR depends on the instruction set of a computer. In this case, the number of garbage outputs for the given instruction set is 19, as shown in Figure 13.15. So $go_{d2} >= 19$. Therefore, the total number of garbage bits generated by a reversible control unit of a processor is

$$go_{cu} > = go_{ir} + go_{d1} + go_{sc} + go_{d2} + go_{cl}$$
$$= (n + 1) + (m - 2)log_2 n + (log_2 n - 2) + 19$$
$$= 2log_2 n + m + n + 16.$$

13.3   Summary

This chapter presented a reversible control unit. Two 4×4 reversible gates, namely HL gate and BJ gate, are introduced to design the reversible decoder and JK flip-flop. An algorithm has been shown to design a reversible control unit. On the way to design the control unit, reversible decoder, sequence counter, instruction register and control logic gates are also constructed. In addition, some properties on the numbers of gates and garbage outputs of the control unit were presented.

Part II

Reversible Fault Tolerance

An Overview About Fault-Tolerance and Testable Circuits

Fault is an error occurring in the systems that forces the system to deviate from its normal behavior. In both traditional and reversible logic circuits the complexity of generating tests for all possible faults in a logic circuit can be minimized through the use of fault models that cover a particular set of fault possibilities. The fault models vary according to the type of description that is being considered, which in turn varies according to the level of abstraction.

Conventional circuits dissipate energy to reload missing information because of overlapped mapping between input and output vectors. Reversible computing is gaining popularity in various fields such as quantum computing, DNA informatics, and CMOS technology etc. Reversibility recovers energy loss and prevents bit error by including a fault-tolerant mechanism. Reversibility recovers bit loss but is not able to detect bit error in circuit. A fault-Tolerant reversible circuit is capable of preventing error at outputs.

Fault tolerance is the property that enables a system to operate accurately in the presence of the failure of one or more of its components. Fault tolerance in reversible circuit reflects robustness of the system. Fault-tolerant systems are capable of detecting and correcting faults. If the logic circuit itself is made of fault-tolerant components, then the detection and correction of faults in circuit become cheaper, easier, and simpler. To achieve fault tolerance, the first step is to identify occurrence of fault. One approach is the parity preserving technique. Any fault that affects only one signal is detectable at the circuit's primary outputs in parity-preserving reversible circuits. Hence, the fault tolerance can be incorporated in reversible computing using parity-preserving reversible logic circuits.

Reversible and DNA Computing, First Edition. Hafiz Md. Hasan Babu.
© 2021 John Wiley & Sons Ltd. Published 2021 by John Wiley & Sons Ltd.

Conservative logic is an alternative for achieving fault tolerance in reversible computing and has capability of multi-bit fault detection, but it is an expensive and complex resultant reversible circuit. A reversible logic gate is called conservative if the number of logical ones of its input lines equals the number of logical ones of their output. Hence, detection of multiple bit errors is possible at the circuit outputs.

Testing of a reversible circuit is another important part of reversible logic design. Testing is necessary to understand whether the reversible circuit performs its operation correctly. Errors or faults frequently occur in a reversible circuit while performing operation. Fault testing in a reversible logic circuit is commonly used for the detection of faults occurred in the circuit.

There are two types of testing of which one is offline (non-concurrent testing) and another is online (concurrent testing), again both can be combined. In offline testing, the circuit will be taken out of normal operations and can be tested by applying a number of test vectors to the circuit for which the correct output values for the circuit are known. Thus, a key element in offline testing approaches for a given fault model is the computation of test sets that are complete for the model under consideration. An input vector that is used for testing a circuit offline for fault detection is known as a test vector. A set of test vectors is known as a test set. A test set is complete if it is capable of detecting all faults in the fault set S, and such a test set is minimal if it contains the fewest possible vectors. Sometimes additional modification in circuitry is required, in which case the approach is referred to as a design-for test (DFT) approach.

In online testing approaches, fault can be identified while the circuit is operating normally. It is performed during the normal operation of the circuit. This may require the addition of circuitry to enable the detection of faults while the circuit is being used in normal operation. Thus, offline testing approach requires extra overhead to detect fault in circuit.

So some differences of applying online testing and offline testing can be summarized as: online approaches for fault detection represent testing of design in their normal operation while offline requires extra overhead to detect faults. Offline testing approaches use test vectors to detect faults in the circuit whereas online testing approaches detect faults in normal circuit operation. Online testing works while the system performs its normal operation, allowing faults to be detected in real time. Offline testing requires the system or a part of the system to be taken out of operation to perform testing, and generally involves the application of a set of test vectors that will detect all possible faults under a given fault model (a complete test set).

This part of the book is divided into two types of contents, namely reversible fault-tolerant circuits and reversible online testable circuits. Some backgrounds and preliminary studies about reversible fault-tolerant logic gates have already been included in different chapters. Some approaches of designing different reversible fault-tolerant adders such as full adder, carry skip adder, carry look-ahead adder, and ripple carry adder are given in Chapter 14. Chapter 15 shows an approach of a reversible fault-tolerant multiplier circuits. Chapter 16 shows the design of a reversible fault-tolerant floating-point division circuits. An approach for designing a reversible fault-tolerant decoders is described in Chapter 17. Similarly, Chapter 18 shows the design approaches of a reversible fault-tolerant shifter and rotator techniques. Chapter 19 describes the design approaches of reversible fault-tolerant programmable logic devices such as programmable logic array, programmable array logic, and field programmable gate array. Finally, Chapter 20 describes the design procedure of a reversible fault-tolerant arithmetic logic unit (ALU) with its QCA implementation. In addition, Chapter 21 describes the realization of online testable reversible circuits using NAND blocks. Moreover, Chapter 22 shows an approach for designing some reversible online testable circuits. Finally, Chapter 23 describes the applications of reversible computing.

14

Reversible Fault-Tolerant Adder Circuits

Fault tolerance is the property that permits a system to continue functioning properly in the event of the failure of some of its constituents. If the system itself is made of fault-tolerant constituents, then the detection and correction of faults become easier and simple. In communication and many other systems, fault tolerance is realized by parity. Therefore, parity-preserving reversible circuits will be the future design tendencies toward the development of fault-tolerant reversible systems in nanotechnology and a gating network will be parity preserving if its distinct gates are parity preserving. Thus, it needs parity-preserving reversible logic gates to construct parity preserving reversible circuits. This chapter presents a 4×4 parity-preserving logic gate named IG. It is parity preserving, that is, the parity of the inputs matches the parity of the outputs. IG is universal in the sense that it can be used to synthesize any arbitrary Boolean function. It is shown that a fault-tolerant reversible full-adder circuit can be realized using only two IGs. The presented design does not produce any unnecessary garbage outputs. The key concern of the reversible logic design is to minimize the number of garbage outputs. The presented fault-tolerant full-adder (FTFA) block can be used to realize other fault-tolerant arithmetic logic circuits in nanotechnology such as ripple carry adder, carry look-ahead adder, carry-skip logic and multiplier/divisor circuits.

14.1 Properties of Fault Tolerance

Reversibility recovers bit loss, but it is not capable of detecting bit error in circuits. Fault-tolerant reversible circuits are capable of preventing error at outputs.

Property 14.1.1 A fault-tolerant (FT) gate or conservative reversible gate is a gate in which the Hamming weight of input and output are equal.

Let the input and output vectors of any fault-tolerant gate be $I_v = \{I_0, I_1, \ldots, I_{n-1}\}$ and $\{O_v = O_0, O_1, \ldots, O_{n-1}\}$, where the following Equation (14.1.1) and Equation (14.1.2) must be preserved:

$$I_v \leftrightarrow O_v \tag{14.1.1}$$

$$I_0 \oplus I_1 \oplus \cdots \oplus I_{n-1} = O_0 \oplus O_1 \oplus \cdot \oplus O_{n-2} \tag{14.1.2}$$

Reversible and DNA Computing, First Edition. Hafiz Md. Hasan Babu.
© 2021 John Wiley & Sons Ltd. Published 2021 by John Wiley & Sons Ltd.

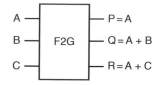

Figure 14.1 Feynman double gate.

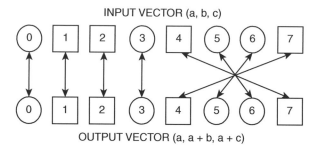

INPUT VECTOR (a, b, c)

OUTPUT VECTOR (a, a + b, a + c)

Figure 14.2 Feynman double gate preserves fault tolerance over input–output unique mapping.

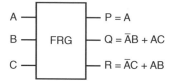

Figure 14.3 Fredkin gate.

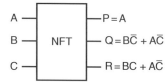

Figure 14.4 New fault-tolerant gate.

14.1.1 Parity-Preserving Reversible Gates

Reversibility recovers bit loss but is not able to detect bit error in circuits. Fault tolerant reversible circuit is capable to prevent error at outputs. Since, reversible gates have the equal number of input and output lines, a sufficient requirement for parity preservation of a reversible gate that each gate be parity preserving. In other words, input and output parity of bits should be equal. Here it needs parity preserving reversible logic gates to construct fault tolerant reversible circuits. A few parity preserving logic gates have been introduced in the literature. Among them, a 3×3 Feynman double gate (F2G) is depicted in Figure 14.1 and its fault tolerance over input–output unique mapping is presented in Figure 14.2. A 3×3 Fredkin gate (FRG) is depicted in Figure 14.3. Also, a 3×3 parity preserving reversible gate, namely new fault tolerant gate (NFT) is depicted in Figure 14.4 and a 4×4 parity-preserving HC gate (PPHCG) is presented in Figure 14.5.

This subsection introduces a 4×4 parity preserving reversible gate named IG gate, depicted in Figure 14.6. The gate is one-through, which means one of the input variables

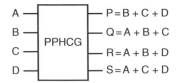

Figure 14.5 Parity-preserving HC gate.

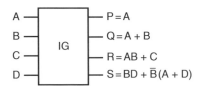

Figure 14.6 Parity-preserving IG gate.

Table 14.1 Truth Table of the Parity-Preserving IG Gate.

A	B	C	D	P	Q	R	S
0	0	0	0	0	0	0	0
0	0	0	1	0	0	0	1
0	0	1	0	0	0	1	0
0	0	1	1	0	0	1	1
0	1	0	0	0	1	0	0
0	1	0	1	0	1	0	1
0	1	1	0	0	1	1	0
0	1	1	1	0	1	1	1
1	0	0	0	1	1	0	1
1	0	0	1	1	1	0	0
1	0	1	0	1	1	1	1
1	0	1	1	1	1	1	0
1	1	0	0	1	0	1	0
1	1	0	1	1	0	1	1
1	1	1	0	1	0	0	0
1	1	1	1	1	0	0	1

is also output. The corresponding truth table of the gate is shown in Table 14.1. It can be verified from the truth table that the input pattern corresponding to particular output pattern can be uniquely determined. The reversible IG gate is parity preserving. This is readily verified by comparing the input parity $A \oplus B \oplus C \oplus D$ to the output parity $P \oplus Q \oplus R \oplus S$. The IG gate is universal in the sense that it can be used for implementing arbitrary Boolean functions as shown in Figure 14.7, Figure 14.8, and Figure 14.9.

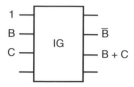

Figure 14.7 Parity-preserving IG gate as a NOT gate.

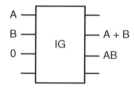

Figure 14.8 Parity-preserving IG gate as AND gate and Ex-OR gate.

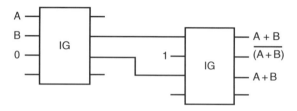

Figure 14.9 Parity-preserving IG gate as Ex-OR gate, Ex-NOR gate and OR gate.

14.2 Reversible Parity-Preserving Adders

In this section, at first, design of reversible fault-tolerant full-adder circuit is described, which is composed of NFT and Feynman double (F2G) gates. Then, the designs of fault-tolerant carry skip adder (CSA), Carry look-ahead adder (CLA) and ripple carry adders are described.

14.2.1 Fault-Tolerant Full Adder

The quantum cost of NFT gate (shown in Figure 14.10) and Fredkin (FRG) gates are the same, i.e., 5. But the quantum cost of Feynman double (F2G) gate is 2. Here, the NFT and F2G gates are used, because of the reusability of the adder circuit for CSA and CLA.

Property 14.2.1.1 Single NFT full adder (SNFA) is a fault-tolerant full adder circuit that consists of one NFT gate and three F2G gates, where the quantum cost of SNFA is 11 and the total number of garbage output of SNFA is 3, as shown in Figure 14.11.

Proof: Let a, b, and c_{in} be the inputs of a full adder circuit, where s and c_{out} are the corresponding outputs. There are three different states at the inputs (a, b, and c_{in}) where the outputs (s and c_{out}) produce same patterns, as shown in Table 14.2. For any parity-preserving reversible circuit, total number of EVEN or ODD parity at input or output

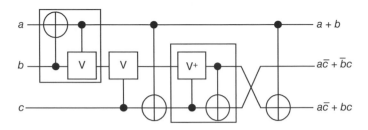

Figure 14.10 Quantum representation of NFT gate.

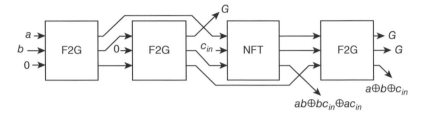

Figure 14.11 Design of single NFT full adder.

Table 14.2 Input-Output Patterns of a Full Adder

Input			Output	
a	b	C_{in}	s	C_{out}
0	1	1	0	1
1	1	0	0	1
1	0	1	0	1

is equal. Table 14.2 shows that the all input patterns are EVEN but the corresponding output patterns are ODD. Turning three ODD patterns at output into EVEN by adding two extra bits is not possible because two bits can represent 2^2 different states where 00 and 11 (01 and 10) are EVEN (ODD) only. So, the reversible full adder circuit requires at least three garbage outputs to make itself reversible fault-tolerant full adder.

14.2.2 Fault-Tolerant Carry Skip Adder

Fast carry emission is the main concern of CSA, and it has two characteristics:

1. The full adder propagates c_{in} to c_{out} if any operand is equal to logical 1; and
2. It generates carry itself (c_{out} is independent on c_{in}).

Property 14.2.2.1 Propagate operation is a simple Ex-OR operation between two operands, which is responsible for only bypassing the carry of previous stage to next stage.

Let $X = (x_0, x_1, x_2, \ldots, x_{n-1})$ and $Y = (y_0, y_1, y_2, \ldots, y_{n-1})$ be two n-bit operands where propagate operation p_i of ith stage can be defined using x_i and y_i as follows:

$$p_i = x_i \oplus y_i$$

Property 14.2.2.2   Generate operation is an AND operation that enables current stage of adder to generate carry for next stage. So the generate operation of ith stage is as follows:

$$g_i = x_i y_i$$

Property 14.2.2.3   A reversible fault-tolerant CSA (RFT-CSA) consists of SNFAs and FRGs to perform summation and propagate carry, respectively, which reduce the delay or bypassing carry due to the recalculation of carry for the next stage. If any input is equal to a logical 1, then it propagates the carry input to the carry output.

Property 14.2.2.4   An n-bit RFT-CSA can be realized by using $(3n + 1)$ F2Gs, n FRGs, and n NFTs.

Proof: n SNFAs are needed to realize n-bit RFT-CSA (each SNFA consists of three F2Gs and one NFT) to generate sum (s_i) and propagate operation (p_i) where $i = 0, 1, 2, \ldots, (n-1)$. And n FRGs are needed for performing AND operation among n propagate operations with c_{in}. So, the calculation of the number of NFT (NFT_{CSA}), the number of FRG (FRG_{CSA}) and the number of F2G $(F2G_{CSA})$ to implement n-bit RFT-CSA is as follows:

$$NFT_{CSA} = n$$
$$FRG_{CSA} = n \; and$$
$$F2G_{CSA} = 3n$$

But RFT-CSA needs another extra F2G to generate final carry, c_{out} by performing Ex-OR operation between c_{n-1} and $(p_{n-1}, p_{n-2} \cdots p_0 c_{in})$. So,

$$F2G_{CSA} = 3n + 1$$

Therefore, an n-bit RFT-CSA can be realized by using $(3n + 1)$ F2Gs, n FRGs and n NFTs.

Property 14.2.2.5   An n-bit RFT-CSA can be realized with $(n + 5)$ critical path delay.

The design of 4-bit RFT-CSA is shown in Figure 14.12, which uses the full adder (SNFA) circuit.

Finally, the total garbage outputs (GB_{CSA}) and quantum cost (QC_{CSA}) of n-bit RFT-CSA can be written as follows:

$$GB_{CSA} = 4n$$
$$QC_{CSA} = 5 \times 2n + 2 \times (3n + 1)$$
$$= 16n + 2$$

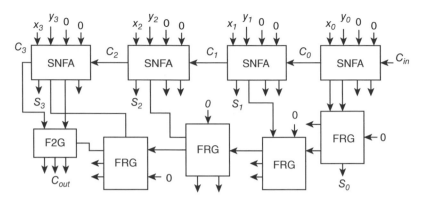

Figure 14.12   Design of fault-tolerant CSA.

14.2.3   Fault-Tolerant Carry Look-Ahead Adder

This subsection introduces the design of reversible fault-tolerant carry look-ahead adder (RFT-CLA) circuit as shown in Figure 14.13. The design of RFT-CLA is based on NFT and F2G gates where the carry is generated before the summation. Some important properties of fault-tolerant carry look-ahead adder are discussed below:

Property 14.2.3.1   A reversible fault tolerant carry look-ahead adder (RFT-CLA) with serially connected n $SNFA_s$ works as a carry generator itself where the carry output of ith stage is produced before the summation s_i ($i = 0, 1, 2, \ldots, (n-1)$).

Property 14.2.3.2   An n-bit RFT-CLA can be realized by using n NFTs and n F2Gs.

Property 14.2.3.3   The delay of n-bit RFT-CLA ($D_{RFT\text{-}CLA}$) can be minimized to (($n+3$).

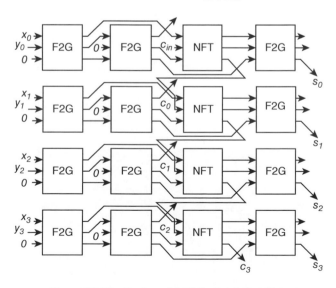

Figure 14.13   Design of 4-bit fault-tolerant CLA.

Proof: According to the definition, the delay of any circuit is the number of maximum gates laying on contiguous path of any input to output. The delay of a parallel adder circuit depends on carry propagation from c_{in} to c_{out}). Any n-bit RFT-CLA needs n SNFAs where the delay of RFT-CLA, $D_{RFT-CLA} = 4n$. Because the carry input (c_i) of i_{th} stage is generated by spending 1 units delay where $i = 0, 1, 2, \ldots, (n-1)$. In first stage, extra two units delay is needed because of the first carry output (c_0) generation, which is related to operands at first stage. On the other hand, the last stage has extra single unit delay because the final sum is generated after one stage of the generation of final carry (c_{out}). So, the delay calculation for n-bits RFT-CLA is as follows:

$$D_{RFT-CLA} = n + 3$$

Therefore, an n-bit RFT-CLA can be realized by using $(n+3)$ unit delay.

Property 14.2.3.4   An n-bit RFT-CLA can be realized with minimum quantum cost $11n$ quantum cost.

Property 14.2.3.5   An n-bit RFT-CLA can be realized with minimum garbage $3n$ garbage outputs.

14.2.4   Fault-Tolerant Ripple Carry Adder

The full adder is the basic building block in a ripple carry adder. To construct a fault-tolerant ripple carry adder, here parity-preserving IG gate is used, depicted in Figure 14.6. The design of the FTFA is shown in Figure 14.14.

The reversible ripple carry adder using the FTFAs is shown in Figure 14.15, which is obtained by cascading the full adders in series.

It can be inferred from Figure 14.14 and Figure 14.15 that the ripple carry adder architecture uses only $2N$ reversible parity-preserving IG gates for N-bit addition and produces only $3N$ garbage outputs.

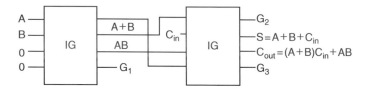

Figure 14.14   FTFA circuit.

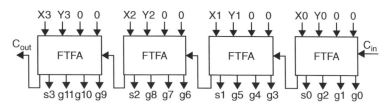

Figure 14.15   Fault-tolerant ripple carry adder.

14.3   Summary

This chapter presents a 4×4 parity-preserving reversible gate called IG gate and demonstrates its universality by realizing all possible Boolean functions. A unique fault-tolerant reversible full adder circuit using the parity-preserving IG gates has also been presented. Reversible logic implementation of the optimized fault-tolerant carry skip adder has also been shown. This chapter also covered the designs of minimum cost fault-tolerant carry skip adder (RFT-CSA) and carry look-ahead adder circuits. Both designs have used the structure of fault-tolerant full adder (RFT-FA or SNFA) circuit. In addition, fault-tolerant ripple carry adder using the fault-tolerant full adder (FTFA) block is described. Several properties have been discussed to make the designs of RFT-CSA and RFT-CLA more generalized for an n-bit fault-tolerant adder.

15

Reversible Fault-Tolerant Multiplier Circuit

One of the most challenging issues in circuit design is power consumption. Reversible logic is one of the ways for power optimization. Addition and multiplication are two heavily used arithmetic operations in many computational units. High-speed multipliers are required for the processors to work competently. In this chapter, an optimal design of a fault-tolerant reversible $n \times n$ multiplier circuit is described, where n is the number of bits of the operands of multiplier. Two algorithms have been presented to construct the partial product generation (PPG) circuit and the multi-operand addition (MOA) circuit of the multiplier. A fault-tolerant reversible gate, namely, LMH gate, is also shown to produce an optimal multiplier. In addition, several theorems on the numbers of gates, garbage outputs, and quantum cost of the fault-tolerant reversible multiplier have been presented to show its optimality.

15.1 Reversible Fault-Tolerant Multipliers

There are several reversible fault tolerant multipliers available in the market. One available multiplier uses Fredkin gate (FRG) for partial product generation and Islam gate (IG) for multi-operand addition. Another available multiplier uses FRG gate for partial product generation. It also uses one modified islam gate (MIG) as half adder and two MIG gates as full adder. In addition, there is a signed fault tolerant reversible multiplier that uses FRG gate, Feynman double gate (F2G), and modified new fault-tolerant (MNFT) gates to generate partial products. This signed multiplier also uses five variable parity-preserving gate (F2PG) and MIG gates to add the partial products.

15.1.1 Reversible Fault-Tolerant $n \times n$ Multiplier

In this subsection, a fault-tolerant reversible multiplier circuit is introduced that is based on parallel multiplication. The multiplication operation is done in two steps:

- Step I: Partial product generation (PPG)
- Step II: Multi-operand addition (MOA)

The operation of a reversible multiplier is shown in Figure 15.1. It consists of 16 partial products of the X and Y inputs to perform 4×4 multiplications. To construct the optimized fault-tolerant reversible $n \times n$ multiplier, a fault-tolerant reversible gate is introduced.

Reversible and DNA Computing, First Edition. Hafiz Md. Hasan Babu.
© 2021 John Wiley & Sons Ltd. Published 2021 by John Wiley & Sons Ltd.

Partial Product Generation				X_3	X_2	X_1	X_0
				Y_3	Y_2	Y_1	Y_0
				P_{03}	P_{02}	P_{01}	P_{00}
Multi Operand Addition			P_{13}	P_{12}	P_{11}	P_{10}	
		P_{23}	P_{22}	P_{21}	P_{20}		
	P_{33}	P_{32}	P_{31}	P_{30}			
Z_7	Z_6	Z_5	Z_4	Z_3	Z_2	Z_1	Z_0

Figure 15.1   Working procedure of a 4×4 multiplier circuit.

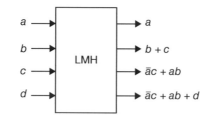

Figure 15.2 4×4 Reversible LMH gate.

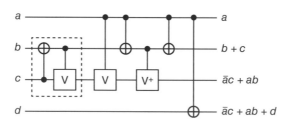

Figure 15.3   Quantum realization of LMH gate.

15.1.2   LMH Gate

In this subsection, a 4×4 reversible gate, namely LMH gate, is shown. The gate is shown in Figure 15.2. The corresponding truth table of the gate is shown in Table 15.1. It can be verified from the truth table that the input pattern corresponding to a particular output pattern can be uniquely determined and vice versa; and it is parity preserving. Quantum realization of the LMH gate is shown in Figure 15.3.

15.1.3   Partial Product Generation

LMH and Fredkin gates are used to generate partial product. Each LMH gate takes two operands as input and produces copy of both operands and product of two operands. Each Fredkin gate takes two operands as input and produces one operand and product of two operands. Figure 15.4 shows the PPG circuit of 4×4 multiplier.

The design can be generalized for $n \times n$ multiplier. Generalized algorithm for $n \times n$ PPG circuit is shown in Algorithm 15.1.3.1. Figure 15.5 shows the generalized design of a fault-tolerant $n \times n$ PPG circuit.

Table 15.1 Truth Table of 4×4 Reversible LMH Gate.

Input				Output				Parity (0 = even, 1 = odd)
a	b	c	d	p	q	r	s	
0	0	0	0	0	0	0	0	0
0	0	0	1	0	0	0	1	1
0	0	1	0	0	1	1	1	1
0	0	1	1	0	1	1	0	0
0	1	0	0	0	1	0	0	1
0	1	0	1	0	1	0	1	0
0	1	1	0	0	0	1	1	0
0	1	1	1	0	0	1	0	1
1	0	0	0	1	0	0	0	1
1	0	0	1	1	0	0	1	0
1	0	1	0	1	1	0	0	0
1	0	1	1	1	1	0	1	1
1	1	0	0	1	1	1	1	0
1	1	0	1	1	1	1	0	1
1	1	1	0	1	0	1	1	1
1	1	1	1	1	0	1	0	0

Algorithm 15.1.3.1 Partial Product Generation

1: Begin
2: **for** $i = 0$ **to** $n - 1$ **do**
3:   **for** $j = 0$ **to** $n - 1$ **do**
4:     **if** $i = n - 1$ or $j = n - 1$ **then**
5:       Input:= $\{Xi, Yj, O\}$
6:     **else**
7:       Input:= $\{Xi, Yj, 0, O\}$ and output:= $\{Xi, Yj, XiYj, XiYj\}$
8:     **end if**
9:     **if** $i = n - 1$ **then**
10:       Output:= $\{Xi, \overline{xi}Yj, XiYj\}$
11:     **else**
12:       Output:= $\{yj, Xi\overline{yj}, XiYj\}$
13:     **end if**
14:   **end for**
15: **end for**
16: End

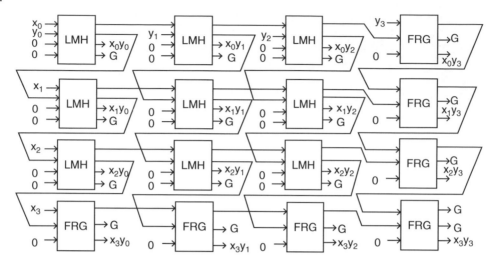

Figure 15.4 4 × 4 Partial product generator circuit.

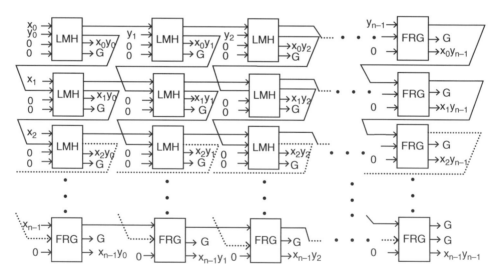

Figure 15.5 Generalized architecture of fault-tolerant PPG.

15.1.4 Multi-Operand Addition

To realize 4×4 multiplier, eight output bits are required. To make the add operation of the partial products, here it needs four half adders and eight full adders. In this circuit, MIG gate is used as half adder and SNFA gate is used as full adder. Figure 15.6 shows the design of multi operand addition circuit.

This design is generalized for $n \times n$ multiplier. Generalized algorithm for $n \times n$ MOA circuit is shown in Algorithm 15.1.4.1. Figure 15.7 shows the generalized design of $n \times n$ MOA circuit.

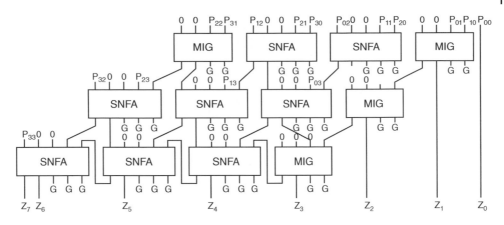

Figure 15.6 4×4 multi-operand addition circuit.

Algorithm 15.1.4.1 Multi-Operand Addition

1: Begin
2: **for** $i = 0$ **to** $n - 2$ **do**
3:   **for** $j = 0$ **to** $n - 1$ **do**
4:     **if** $(j = 0)$ or $(i = 0$ and $j = (n - 1))$ **then**
5:       Apply MIG
6:     **else**
7:       Apply SNFA
8:     **end if**
9:   **end for**
10: **end for**
11: End

Property 15.1.4.1 A fault-tolerant reversible $n \times n$ multiplier requires $n(2n - l)$ gates, where n is the number of bits of the operands of the multiplier.

Proof: For $n \times n$ multiplication, n^2 partial products are generated. To generate each partial product, one gate is required. So, in PPG circuit, n^2 gates are needed. To add the partial products, a multi-operand addition circuit is required. In MOA circuit, n half adders and $n(n - 2)$ full adders are needed. Each half adder and full adder requires one MIG gate and one SNFA gate, respectively. So, the number of required gates is $n + n(n - 2) = n(n - l)$.

Property 15.1.4.2 A fault-tolerant reversible $n \times n$ multiplier produces $4n(n - l) + 1$ garbage outputs, where n is the number of bits of the operands of multiplier.

Proof: From Property 15.1.4.1, it is found that the PPG circuit requires n^2 gates. Among them, all gates except the last FRG produces one garbage output while the last FRG produces two garbage outputs. So, in PPG circuit, the number of garbage outputs is $(n^2 - 1) + 2 = n^2 + 1$.

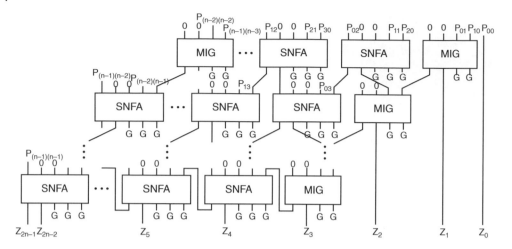

Figure 15.7 Generalized architecture of fault-tolerant MOA.

Property 15.1.4.1 shows that the MOA circuit requires n MIG gates and $n(n-2)$ SNFA gates. Each of MIG and SNFA gate produces two and three garbage outputs, respectively. So, in MOA circuit, the number of garbage outputs is $2n + 3n(n-2) = n(3n-4)$.

Therefore, a fault-tolerant reversible $n \times n$ multiplier circuit produces $(n^2 + l) + n(3n-4) = 4n(n-l) + 1$ garbage outputs.

Property 15.1.4.3 A fault-tolerant reversible $n \times n$ multiplier requires $17n(n-l) + 1$ quantum cost, where n is the number of bits of the operands of multiplier.

Proof: The PPG circuit generates n^2 partial products. LMH and Fredkin gates are used in the PPG circuit. The last column and the last row are constructed with Fredkin gates while others were constructed with LMH gates. The numbers of Fredkin and LMH gates are $(2n-1)$ and $(n-1)^2$, respectively. The quantum cost of each Fredkin and LMH gate is five and six, respectively. So, the quantum cost of PPG circuit is $5(2n-1) + 6(n-1)^2 = 6n^2 - 2n + 1$.

The MOA circuit requires n MIG gate and $n(n-2)$ SNFA gates. The quantum cost of each MIG and SNFA gate is 7 and 11, respectively. So, the quantum cost of MOA circuit is $7n + 11n(n-2) = n(11n-15)$.

Therefore, the quantum cost of a fault tolerant reversible $n \times n$ multiplier circuit is $(6n^2 - 2n + 1) + n(11n - 15) = 17n(n-1) + 1$.

15.2 Summary

This chapter presents an optimal methodology for designing a fault-tolerant reversible $n \times n$ multiplier, where n is the number of bits of the operands of multiplier. An algorithm is also provided to design the multiplier. In addition, a fault-tolerant reversible gate is introduced, namely LMH gate. It has also been shown that the circuit has been constructed with the optimum numbers of gates, garbage outputs, quantum costs, and constant inputs.

16

Reversible Fault-Tolerant Division Circuit

Division is the most difficult operation in computer arithmetic. Nowadays, people use a hardware module divider to implement the division algorithm. Conventionally sequential circuits are used to implement the divider. The division circuit can be used in the arithmetic unit of a processor. In this chapter, a fault-tolerant reversible division circuit is introduced.

16.1   Preliminaries of Division Circuits

In arithmetic and algebraic computation, the basic operations are addition, subtraction, multiplication, and division. It is a fundamental problem to find efficient algorithms for division, as it seems to be the most difficult of these basic operations. The division operation has been regarded as a significant verification challenge. This operation is usually implemented as a sequence of basic operations rather than a dedicated logic circuit. As a result, it exhibits both an exceptionally wide array of intricate corner cases and an immense state space, challenging both simulation and formal methods.

16.1.1   Division Algorithms

Designers of processors with enhanced arithmetic and logic unit (ALU) are always on the lookout for algorithms that perform basic operations, especially division. The algorithm should be such that it involves processes requiring compact hardware while satisfying efficiency constraints. In this subsection, a brief discussion about some division algorithms with their properties is shown.

   The binary floating-point division algorithm deals with the implementation of low latency software for binary floating-point division with correct rounding to nearest. The approach targets a VLIW integer processor of the ST200 family and is based on fast and accurate programs for evaluating some particular bivariate polynomials. With the ST200 compiler, the speed-up observed with this approach is by a factor of almost 1.8.

   Another approach of division, namely floating-point division algorithm, presents the AMD-K7 IEEE 754 and 87 compliant floating-point division process. Highly accurate initial approximations and a high-performance shared floating point multiplier assist in achieving low-division latencies at high operating frequencies. This algorithm also

Reversible and DNA Computing, First Edition. Hafiz Md. Hasan Babu.
© 2021 John Wiley & Sons Ltd. Published 2021 by John Wiley & Sons Ltd.

describes a novel time-sharing technique, which allows independent floating point multiplication operations to proceed while division is in progress.

Floating-point division using a Taylor-series expansion algorithm presents the implementation of a fused floating-point divide unit based on a Taylor-series expansion algorithm. By this algorithm, the resulting arithmetic unit also exhibits high throughput and moderate latency as compared with other floating-point unit (FPU) implementations of leading architectures. Moreover, this algorithm achieves fast computation by using parallel powering units such as squaring and cubing units, which compute the higher-order terms significantly faster than traditional multipliers with a relatively small hardware overhead.

16.2 The Division Method

In this section, a method for binary floating-point division is shown as well as an algorithm for binary floating-point division. Although floating-point (FP) divisions are less frequent in applications than other basic arithmetic operations, reducing their latency is often an issue. Since low latency implementations may typically be obtained by expressing and exploiting instruction parallelism, intrinsically parallel algorithms tend to be favored. In this chapter, an algorithm is shown to get the desired result, which is described below.

In the first step, working with floating-point numbers requires some understanding of the internal representation of data. Researchers must be aware of the finite precision issues. For floating-point values (single, double, or extended precision), the possibility of accuracy loss must be considered. Floating-point accuracy loss can occur due to two causes.

Cause 1 If a floating-point literal (for example: 0.1f) is assigned to a floating-point variable, the floating-point variable might not have sufficient bits to hold the desired value without introducing some errors. The floating-point literal might require a very large number or even an infinite number of bits for infinite precision representation.

Cause 2 If operations are floating points, then each step (operation) can introduce its specific error. This happens because, in the case of some operations, the computed result cannot be stored with its complete precision.

For example, if multiplication of two numbers, S_1 bits with S_2 bits (this is true for integer types and for floating-point types) occurs, then the result requires $S_1 + S_2$ bits for complete precision. Additive operations introduce a relatively low error. Multiplication introduces a relatively high error. It is important to understand that the floating-point accuracy loss (error) is propagated through calculations and it is the role of the programmer to design an algorithm which is correct. A floating-point variable can be regarded as an integer variable with a power of two scales. If the floating-point variable to an extreme value is applied, the scale will automatically be adjusted. A floating-point variable can accumulate a significant error and/or become denormalized.

In the second step, the algorithm is providing methods for accelerating floating-point divisions of the form x/y, when y is known before x, either at compile-time (i.e., y is a constant; in such a case, much pre-computation can be performed) or at runtime. This method arrives at the result more quickly but just as accurately as by using division: here it needs a correctly rounded value, as required by the IEEE 754 Standard for FP arithmetic. Divisions by constants are a clear application of the content of this chapter. There are other applications; for instance, when many divisions by the same y are performed (an example is Gaussian elimination). In this chapter, the focus is on rounding to the nearest value only.

In the third step, the most widespread solutions for multiplicative division are shown. An accurate and fast computation of the division operation has become mandatory. Efficient implementations produce significant improvements in system performance. This step uses an initial low-precision approximation to get the result. The initial approximation is used to perform a number of iterations to obtain the result accurate to a required number of bits. The number of iterations depends on the precision of the seed value and the required precision of the result. The IEEE 754 Standard requires that division results must be rounded, following the strategies described in this standard.

To achieve a high-speed and low-power-consuming divider, a division algorithm is shown. This technique will work for binary floating-point numbers. The division method consists of four steps: First, it considers floating-point data and rounding. Second, it performs correctly rounded division. Third, it performs correct rounding from one-sided approximations. Finally, it calculates the result of the division operation.

16.2.1 Floating-Point Data and Rounding

The floating-point data here are $\pm 0, \pm\infty$, quiet or signaling Not-a-Numbers (qNaN, sNaN), as well as normal binary floating-point numbers:

$$x = (-1)^{s_x}.m_x.2^{e_x}, \tag{16.2.1}$$

with $s_x \in \{0,1\}$, $m_x = (1.m_{x,1} \ldots m_{x,p-1})2$ and $e_x \in \{e_{min}, \ldots, e_{max}\}$. (In particular subnormal numbers will not be considered.) The precision p and external exponents e_{min} and e_{max} are assumed to be integers such that $p \geq 2$ and $e_{min} = 1 - e_{max}$. The rounding attribute chosen here is "to nearest even" (round ties to even) and will be referred to as **RN**. The standard requires that $RN(x/y)$ be returned whenever x and y are as in Equation 16.2.1).

16.2.2 Correctly Rounded Division

That RN (x/y) essentially reduces to a correctly rounded ratio of significant can be recalled as follows. First, using RN $(-x) = -$RN (x) gives $RN(x/y) = (-1)^{s_r}$. $RN(|x/y|)$, where s_r is the XOR of s_x and s_y. Then taking $c = 1$ if $m_x \geq m_y$, and 0 otherwise, one has $|x/y| = l. 2^d$, where

$$l = 2m_x/m_y.2^{-c}, d = e_x - e_y - 1 + c \tag{16.2.2.1}$$

Since both m_x and m_y are in Equations 16.2.1) and 16.2.2.1), l is in Equation 16.2.2.1), and $l . 2^d$ will be called the *normalized representation* of $|x/y|$. Tighter enclosures of l can in fact be given in the following property.

Property 16.2.2.1   If $m_x \geq m_y$ then $l \in [1, 2 - 2^{1-p}]$ else $l \in (1, 2 - 2^{1-p})$.

Proof: If $m_x \geq m_y$ then $c = 1$, and it is deduced from $1 \leq m_x \leq 2 - 2^{1-p}$ and $0 < 1/m_x \leq 1/m_y \leq 1$ that $1 \leq l \leq 2 - 2^{1-p}$. If $m_x < m_y$ then $m_x \leq m_y - 2^{1-p}$. Thus, $l \leq 2 - 2^{2-p}/m_y$.

Hence, using $m_x \geq 1$ and $1/m_y > 1/2$, it is obtained $1 < l < 2 - 2^{1-p}$ as desired. These enclosures of l will be used explicitly when handling underflow. For now, it is simply noted that both of them given $\text{RN}(l) \in [1, 2 - 2^{1-p}]$, so that $\text{RN}(x/y) = \text{RN}(l) . 2^d$. If $e_{min} \leq d \leq e_{max}$ then it will return the normal number $\text{RN}(x/y) = (-1)^{s_r} . m_r . 2^{e_r}$, where

$$s_r = s_x \oplus s_y, m_r = \text{RN}(l), e_r = d \qquad (16.2.2.2)$$

and "." denotes the multiplication sign.

Else, d is either smaller than e_{min} or greater than e_{max}. Since subnormal are not supported, some special values will be returned in either case. Thus, to get $\text{RN}(x/y)$ the main task consists of deducing from m_x and m_y the correctly rounded value $\text{RN}(l)$ in Equation 16.2.2.2), the sign s_r and exponent e_r being computable in parallel and at lower cost.

16.2.3 Correct Rounding from One-Sided Approximations

Among the many methods known for getting $\text{RN}(l)$, the one is focused here is the algorithm uses *one-sided approximations*.This method reduces the computation of RN (l) to that of an approximation v of l such that

$$-2^{-p} < l - v \leq 0 \qquad (16.2.3.1)$$

Here v is representable with, say, k-bits while l has in most cases an infinite binary expansion $(1.l_1 \dots l_{p-1}l_p \dots)2$. Once such a v is known, correct rounding follows easily and all that remains is to deduce from each possible pair (m_x, m_y) a value v that satisfies Equation 16.2.3.1.

16.2.4 The Algorithm for Division Operation

A binary number can be thought of as composed of strings of units, strings of zeros, strings of units including isolated zeros, and strings of zeros including isolated units. For example, the binary number in Figure 16.1 has a string of units from the 2^{-1} to 2^{-4} positions, a string of zeros from 2^{-5} to 2^{-8} positions, a string of units from the 2^{-9} to 2^{-13} positions with an isolated zero at the 2^{-11} position and a string of zeros from the 2^{-14} to 2^{-18} positions with an isolated unit at the 2^{-16} position. The significance of this string-type decomposition of a binary number depends on three elementary observations: (i) a string of units from 2^{-p} to 2^{-q} positions contributes a value of $(2^{-p+1} - 2^{-q})$ to the magnitude of the number, for example, $0.11111 = 1 - 0.00001 = 2^0 - 2^{-5}$; (ii) an isolated unit in the 2^{-p} position in a string of zeros contributes 2^{-p}, for example, $0.00100 = 2^{-3}$; (iii) an isolated zero in the 2^{-q} position decreases the magnitude of the number of 2^{-q}, for example, $0.11011 = 0.11111 - 0.00100 = 2^0 - 2^{-3} - 2^{-5}$. Altogether, then, as shown in Figure 16.1, it can be written as the binary number $0.111100001101100100 = 2^0 - 2^{-4} + 2^{-8} - 2^{-11} - 2^{-13} + 2^{-16}$. A convenient shorthand notation is to place "+" or "−" above those positions of a binary numbers whose values respectively contribute to or decrease the magnitude of the number. Thus, it can be written the above number as follows: The generality of these elementary observations is clear.

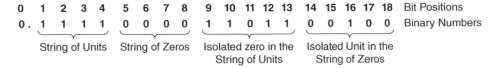

0	1	2	3	4	5	6	7	8	9	10	11	12	13	14	15	16	17	18	Bit Positions
0 .	1	1	1	1	0	0	0	0	1	1	0	1	1	0	0	1	0	0	Binary Numbers

String of Units   String of Zeros   Isolated zero in the String of Units   Isolated Unit in the String of Zeros

Figure 16.1 Illustration of the decomposition of a binary number.

Smith and Weinburger [16] utilized such a decomposition of binary numbers in developing a technique of rapid multiplication, for only those additions to or subtractions from the partial product need to be performed that are indicated by the decomposition of the multiplier. For example, $(0.1101)(0.1101100100) = (0.1101)(2^0 - 2^{-3} - 2^{-5} + 2^{-8})$ requires only four operations. Clearly there can be different decompositions for the same nonzero binary number. Rules for obtaining a "best" decomposition, in the sense of having the least number of terms, are given by Smith and Weinburger (discussed by Ledley), Lehman, Tocher, and Reitwiesner [16-19]. For the rapid multiplication, a best decomposition must be obtained. The method of rapid division reverses a rapid multiplication process.

Algorithm 16.2.4.1 has the advantage of extreme simplicity and in a majority of cases significantly reduces the number of operations required for division. Algorithm 16.2.4.1 achieves maximum efficiency in reducing the number of additions and subtractions to a minimum. Robertson [20] allows alternative decomposition of the quotient to accommodate at most one addition or subtraction for each two binary positions and exhibits sets of comparisons of the leading digits of the divisor and current remainder by which his process is controlled. It is allowed alternative decomposition here, within narrower limits, to yield a "best" decomposition, and here exhibits the precise configurations of leading bits of the divisor and remainder, which must be examined to conform to the limits.

Reitwiesner [19] develops a "canonical" minimum decomposition for the quotient, and demonstrates how to obtain it by comparisons made at each bit of the quotient; but his comparisons are made over the full lengths of the divisor and remainder. Tocher [18] also requires comparison of the full lengths of the divisor and remainder, by virtue of his employment of the factor $2/3$. Other writers using different approaches have been unable to achieve maximum efficiency. The principal factor in the development of the algorithm is the acceptance of alternative "best" decompositions whenever convenient to do so. Thus, it is not necessary to employ the "canonical" form developed in most of the cited references.

In Algorithm 16.2.4.1, the number of positions shifted may be one less than, equal to, or more than that which is required to normalize $D^{(s)}$. This occurs because simple normalization will not necessarily result the best decomposition for Q. If it is known that the best decomposition is of Q, then each successive shift (from left to right) should stop with the binary point at the right of a position requiring an operation.

The procedure is summarized by Algorithm 16.2.4.1; the process ends when it reaches the number of bits desired in the quotient. It is assumed that the binary point is to the far left of each word, that the denominator Y is positive and normalized (its most significant bit a unit), that the numerator D is positive, and that $D<Y$, with D either normalized or with a single zero after the binary point. The first step is to form the first remainder, $D' = D - Y$.

Algorithm 16.2.4.1 Algorithm for the Binary Floating Point Division Operation

---

Division ($D = YQ$, Y is normalized)

1: Begin
2: set $i = 0$
3: set $s = 1$
4: if ($Y > D$), form remainder, $D^S = D^{S-1} - Y$
5: if ($Y < D$), form remainder, $D^S = D^{S-1} + Y$
6: if ($D^S < 0$), set $Q_i = 0$
7: if ($D^S \geq 0$), set $Q_i = 1$
8: normalize D^S
9: calculate correct rounding from one sided approximations of D^S
10: if ($D^S < 0$), set $Q_{0+\alpha} = 0$ and the following bits are units
11: if ($D^S \geq 0$), set $Q_{0+\alpha} = 1$ and the following bits are zeros
12: increment s (for the next step)
13: set $i = i + \alpha$
14: if(i = number of bits), end the process
15: go to Step 4
16: End

Since, with the initial assumptions, D' is negative, the negative loop of Algorithm 16.2.4.1 is followed. If, D' has α zeros to the right of the binary point, then quotient Q has at least $\alpha - 1$ units to the right of the point. (In this initial step, $Q_i = 0$ for $i = 0$ means merely that $Q < 1$). Now, D' is normalized and the second remainder, $D'' = D' + Y$, is formed. If, D'' is negative, then $Q_{0+\alpha} = 0$, and the following bits are units. If, D'' is positive, then $Q_{0+\alpha} = 1$, and the following bits are zeros. The procedure continues in the way, differencing and normalizing each time, and determining Q_i and Q_{i+1} through $Q_{(i+\alpha)+1}$ at each step. For example, consider $D = .1001\ 1111\ 0000\ 1100$ and $Y = .1101$. The division process is shown in Figure 16.2.

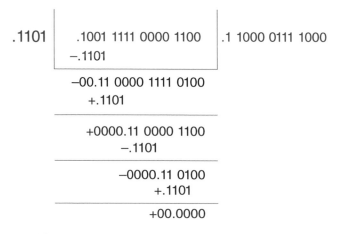

Figure 16.2 Example of a division operation.

Property 16.2.4.1 The time complexity of the binary floating-point division algorithm is $O(log_2 n)$, where n is the number of numerators.

Proof: In the algorithm, there are only $O(log_2 n)$ layers, and each layer has $O(1)$ transmission delay. The delay of making a partial remainder is $O(1)$. The partial remainder then normalized and correctly rounded. Normalizing and calculating correctly rounded one-sided approximations of the partial remainder would need $O(1)$ time. So, $O(log_2 n)$ layers need $O(log_2 n)$ execution time. Therefore, the time complexity of the binary floating-point division algorithm is $O(log_2 n)$.

16.3 Components of a Division Circuit

In order to design reversible fault-tolerant division circuit, reversible multiplexers (MUXs), registers, parallel-in-parallel-out (PIPO) left-shift registers, D latch, rounding and normalization registers and parallel adder are required. A reversible fault-tolerant MUX circuit and D latch are shown in Figure 16.3 and Figure 16.5, respectively, which are used in the division circuit in Section 16.4. The reversible fault-tolerant MUX circuit is composed with a Fredkin gate and the reversible fault-tolerant D latch is composed with RR gate. The block diagram is shown in Figure 16.4 and the truth table is given in Table 16.1. The truth table proves that the RR gate is a reversible fault-tolerant gate.

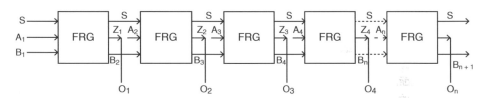

Figure 16.3 2-input n-bit reversible fault-tolerant MUX.

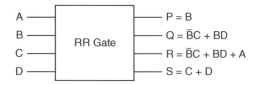

Figure 16.4 Block diagram of RR gate.

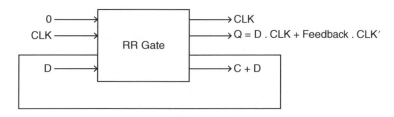

Figure 16.5 Reversible fault-tolerant D latch using RR gate.

Table 16.1 Truth Table of 4 × 4 Reversible RR Gate.

A	B	C	D	B	B'C + BD	B'C + BD ⊕ A	C ⊕ D
0	0	0	0	0	0	0	0
0	0	0	1	0	0	0	1
0	0	1	0	0	1	1	1
0	0	1	1	0	1	1	0
0	1	0	0	1	0	0	0
0	1	0	1	1	1	1	1
0	1	1	0	1	0	0	1
0	1	1	1	1	1	1	0
1	0	0	0	0	0	1	0
1	0	0	1	0	0	1	1
1	0	1	0	0	1	0	1
1	0	1	1	0	1	0	0
1	1	0	0	1	0	1	0
1	1	0	1	1	1	0	1
1	1	1	0	1	0	1	1
1	1	1	1	1	1	0	0

16.3.1 Reversible Fault-Tolerant MUX

Figure 16.3 shows a two-input n-bit reversible fault tolerant MUX using FRG gates, where s is the select input and $A_1A_2A_3 \ldots A_n$ and $B_1B_2B_3 \ldots B_n$ are two inputs. If $s = 0$, then $(Z_1Z_2Z \ldots Z_n) = (A_1A_2A_3 \ldots A_n)$ or if $s = 1$, then $(Z_1Z_2Z \ldots Z_n) = (B_1B_2B_3 \ldots B_n)$. This reversible fault-tolerant MUX requires n FRG gates, generates n garbage outputs and needs $5n$ quantum cost.

16.3.2 Reversible Fault-Tolerant D Latch

Figure 16.5 shows the implementation of reversible fault-tolerant clocked D latch. This D latch is used to implement an n-bit reversible fault-tolerant register. This reversible fault-tolerant D latch requires one RR gate, generates one garbage output, and needs six quantum cost.

16.4   The Design of the Division Circuit

In the design of a reversible fault-tolerant division circuit, some components are introduced that include parallel-in-parallel-out (PIPO) left-shift register, divisor register, rounding register, normalization register, and parallel adder. All these components are fault tolerant. Finally, a reversible fault-tolerant division circuit is also illustrated.

16.4.1   Reversible Fault-Tolerant PIPO Left-Shift Register

Parallel data bits can be loaded into the reversible fault-tolerant PIPO left-shift register. This data bits also can be shifted left and appear in outputs. Control inputs of reversible left-shift register are shown in Table 16.2.

According to Table 16.2, during clock pulse, when SV and E are 0 (low), left-shift is performed. When SV is 0 and E is 1 (high), the parallel load is performed. Therefore, the input bits are loaded into the left-shift register and outputs are available from the Q output. When SV is 1, the reversible-fault tolerant PIPO left-shift register saves its current value; in other words, the reversible PIPO left-shift register do not perform anything (inactive) when SV is 1. Q_i^+ can be obtained as follows:

$$Q_i^+ = SV'.E.I_i + SV'.E'.Q_{i-1} + SV.Q_i \tag{16.4.1.1}$$

Where "." denoted the multiplication sign. Figure 16.6 shows the block diagram of an F2PG gate and Table 16.3 shows the truth table of the F2PG gate. The truth table proves that the F2PG gate is a reversible fault-tolerant gate. Figure 16.7 shows the implementation of Equation 16.4.1.1). The circuit of Figure 16.6 is a fault-tolerant circuit because it is composed

Table 16.2   Control Inputs of a Fault-Tolerant Reversible Left-Shift Register.

SV	E	Output,
0	0	(Left shift)
0	1	(Parallel load)
1	x	(No change)

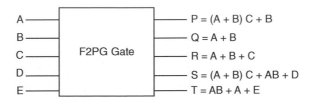

Figure 16.6   Block diagram of F2PG gate.

Table 16.3 Truth Table of the Fault-Tolerant F2PG Gate.

A	B	C	D	E	P	Q	R	S	T
0	0	0	0	0	0	0	0	0	0
0	0	0	0	1	0	0	0	0	1
0	0	0	1	0	0	0	0	1	0
0	0	0	1	1	0	0	0	1	1
0	0	1	0	0	0	0	1	0	0
0	0	1	0	1	0	0	1	0	1
0	0	1	1	0	0	0	1	1	0
0	0	1	1	1	0	0	1	1	1
0	1	0	0	0	1	1	1	0	0
0	1	0	0	1	1	1	1	0	1
0	1	0	1	0	1	1	1	1	0
0	1	0	1	1	1	1	1	1	1
0	1	1	0	0	0	1	0	1	0
0	1	1	0	1	0	1	0	1	1
0	1	1	1	0	0	1	0	0	0
0	1	1	1	1	0	1	0	0	1
1	0	0	0	0	0	1	1	0	1
1	0	0	0	1	0	1	1	0	0
1	0	0	1	0	0	1	1	1	1
1	0	0	1	1	0	1	1	1	0
1	0	1	0	0	1	1	0	1	1
1	0	1	0	1	1	1	0	1	0
1	0	1	1	0	1	1	0	0	1
1	0	1	1	1	1	1	0	0	0
1	1	0	0	0	1	0	0	1	0
1	1	0	0	1	1	0	0	1	1

Table 16.3 (Continued)

A	B	C	D	E	P	Q	R	S	T
1	1	0	1	0	1	0	0	0	0
1	1	0	1	1	1	0	0	0	1
1	1	1	0	0	1	0	1	1	0
1	1	1	0	1	1	0	1	1	1
1	1	1	1	0	1	0	1	0	0
1	1	1	1	1	1	0	1	0	1

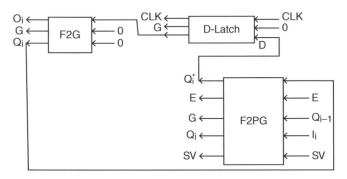

Figure 16.7 Reversible fault-tolerant PIPO left-shift register.

of F2PG and F2G gates, which are fault tolerant. An *n*-bit reversible fault-tolerant PIPO left-shift register can be implemented by 3*n* reversible gates, (3*n* + 1) garbage outputs with 22*n* quantum cost.

16.4.2 Reversible Fault-Tolerant Register

Figure 16.8 shows the implementation of reversible fault-tolerant register with the help of reversible fault-tolerant D latch, which is shown in Figure 16.5. It requires *n* gates, produces *n* garbage outputs and needs 6*n* quantum cost.

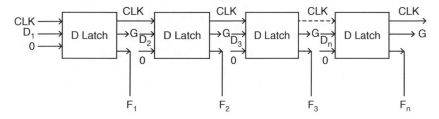

Figure 16.8 Reversible fault-tolerant register.

16.4.3 Reversible Fault-Tolerant Rounding Register

Figure 16.9 shows the implementation of the reversible fault-tolerant rounding register. The register is fault tolerant, as its components D latch and F2G gate are fault tolerant. The rounding register requires two reversible gates, two garbage outputs, and eight quantum cost. It is the first-ever reversible fault-tolerant rounding register.

16.4.4 Reversible Fault-Tolerant Normalization Register

A reversible fault-tolerant normalization register is shown in Figure 16.10, which is composed of one D latch. It is the first-ever reversible fault-tolerant normalization register. This register requires one reversible gate and one garbage output with quantum cost of six.

16.4.5 Reversible Fault-Tolerant Parallel Adder

In Figure 16.11, a new fault-tolerant full-adder gate (NFTFAG) is introduced. If the truth table of NFTFAG is constructed, it can be verified from the truth table that the input pattern corresponding to a particular output pattern can be uniquely determined and vice versa; and it is also parity preserving. The quantum cost of the NFTFAG is nine. The quantum representation of NFTFAG is illustrated in Figure 16.12.

Property 16.4.5.1   The reversible NFTFAG is a fault-tolerant gate.

Proof: Let the input vector and output vector of NFTFAG be $I_v = \{A, B, C, D, E\}$ and $O_v = \{A, A \oplus B, A \oplus B \oplus C, A(B \oplus C) \oplus BC \oplus D, A(B \oplus C) \oplus BC' \oplus E\}$. Ex-ORing all the inputs and outputs, it is: $A \oplus B \oplus C \oplus D \oplus E \oplus A \oplus A \oplus B \oplus A \oplus B \oplus C \oplus A(B \oplus C) \oplus BC \oplus D \oplus A(B \oplus C) \oplus BC' \oplus E$

$$= (A \oplus A \oplus A \oplus A) \oplus (B \oplus B \oplus B) \oplus (C \oplus C) \oplus (D \oplus D)$$

$$\oplus (E \oplus E) \oplus (AB \oplus AB) \oplus (AC \oplus AC) \oplus (BC \oplus BC')$$

$$= B \oplus BC \oplus BC'$$

$$= B \oplus B(C \oplus C')$$

$$= B \oplus B$$

$$= 0$$

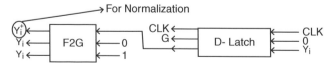

Figure 16.9   Reversible fault-tolerant rounding register.

Figure 16.10   Reversible fault-tolerant normalization register.

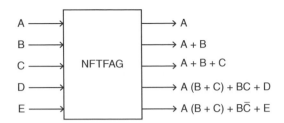

Figure 16.11 Reversible fault-tolerant NFTFAG.

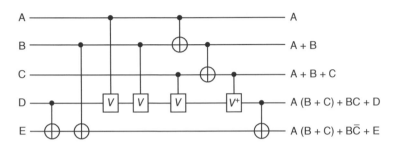

Figure 16.12 Quantum representation of NFTFAG.

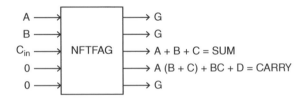

Figure 16.13 NFTFAG as a reversible fault-tolerant full adder.

It is known that the Ex-OR of all inputs and outputs of a reversible fault-tolerant gate is 0. Therefore, the NFTFAG is a fault-tolerant gate.

It is also proved by the truth table shown in Table 16.4 that the NFTFAG is fault tolerant.

The NFTFAG can be used as a fault-tolerant full adder. By asserting zero (0) constant input in the fourth and fifth inputs of NFTFAG, we can arrive at the third output and in the fourth output. Figure 16.13 shows the realization of a reversible fault-tolerant full adder using NFTFAG. First, second and fifth outputs in Figure 16.13 are garbage outputs.

A $(n+1)$-bit reversible fault tolerant parallel adder using NFTFAG is shown in Figure 16.14. As carry-out of the parallel adder is ignored in the fault-tolerant reversible division circuit, implementation of $(n+1)$-bit parallel adder requires n full adders, and the last bit position of the adder requires computing two Ex-OR operations. Thus, it can be implemented using $(n+2)$ reversible gates and $(3n+2)$ garbage outputs with the quantum cost of $9n+4$.

16.4.6 The Reversible Fault-Tolerant Division Circuit

Figure 16.15 shows the design of the reversible fault-tolerant n-bit division circuit. It has two PIPO reversible left-shift registers: one is $(n+1)$-bit named A and the

Table 16.4 Truth Table of the NFTFAG.

A	B	C	D	E	P	Q	R	S	T
0	0	0	0	0	0	0	0	0	0
0	0	0	0	1	0	0	0	0	1
0	0	0	1	0	0	0	0	1	0
0	0	0	1	1	0	0	0	1	1
0	0	1	0	0	0	0	1	0	0
0	0	1	0	1	0	0	1	0	1
0	0	1	1	0	0	0	1	1	0
0	0	1	1	1	0	0	1	1	1
0	1	0	0	0	0	1	1	0	1
0	1	0	0	1	0	1	1	0	0
0	1	0	1	0	0	1	1	1	1
0	1	0	1	1	0	1	1	1	0
0	1	1	0	0	0	1	0	1	0
0	1	1	0	1	0	1	0	1	1
0	1	1	1	0	0	1	0	0	0
0	1	1	1	1	0	1	0	0	1
1	0	0	0	0	1	1	1	0	0
1	0	0	0	1	1	1	1	0	1
1	0	0	1	0	1	1	1	1	0
1	0	0	1	1	1	1	1	1	1
1	0	1	0	0	1	1	0	1	1
1	0	1	0	1	1	1	0	1	0
1	0	1	1	0	1	1	0	0	1
1	0	1	1	1	1	1	0	0	0
1	1	0	0	0	1	0	0	1	0
1	1	0	0	1	1	0	0	1	1
1	1	0	1	0	1	0	0	0	0

Table 16.4 (Continued)

A	B	C	D	E	P	Q	R	S	T
1	1	0	1	1	1	0	0	0	1
1	1	1	0	0	1	0	1	1	0
1	1	1	0	1	1	0	1	1	1
1	1	1	1	0	1	0	1	0	0
1	1	1	1	1	1	0	1	0	1

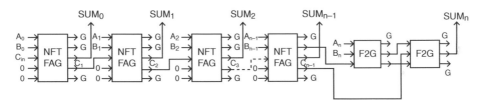

Figure 16.14 $(n+1)$-bit reversible fault-tolerant parallel adder.

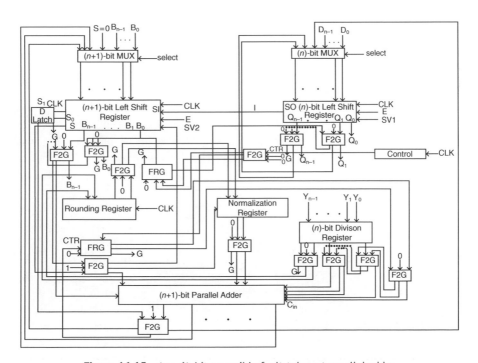

Figure 16.15 $(n+1)$-bit reversible fault-tolerant parallel adder.

Algorithm 16.4.5.1 Algorithm for the Binary Floating Point Divider Circuit

---

1: Begin
2: *select* = 1
3: *count* = 0
4: **while** TRUE **do**
5:    **if** *CLK* is High and $E = 1$ **then**
6:      **if** (*select* = 1), inputs are loaded into left-shift registers **then**
7:        *select* = 0
8:      **else**
9:        outputs from rounding registers are loaded into normalizing register
10:        *count* = *count* + 1
11:      **end if**
12:    $E = 0$
13:    **if** (*CLK* is High), S_1 gets the value of S_0 **then**
14:      *count* = *count* + 1
15:      the values of divisor are the inputs of adder
16:      **if** the MSB of Sum is 0 **then**
17:        Q_0 will be 1 during the next *CLK*
18:      **else**
19:        Q_0 will be 0 during next *CLK*
20:      **end if**
21:    **end if**
22:   **end if**
23: **end while**
24: End

---

other is an n-bit named Q. It contains an n-bit reversible register to store the divisor, one D latch is used to maintain S_1, one FRG is used as a 1-bit, two-input MUX. The design also contains a rounding register to calculate the rounding to the nearest even and a normalization register for normalizing the divisor on each step. Initially, $S = 0$, $B(B_{n-1}B_{n-2}...B_0) = 0$, $D(D_{n-1}D_{n-2}...D_0) =$ dividend, $Y(Y_{n-1}Y_{n-2}...Y_0) =$ divisor, and $CTR = 0$. When the division operation is completed, register $Q(Q_{n-1}Q_{n-2}...Q_0)$ contains the quotient and $B(B_{n-1}B_{n-2}...B_0) = 0$ contains the remainder. If *select* = 1, then two-input $(n + 1)$-bit MUX selects $S = 0$ and $B(B_{n-1}B_{n-2}...B_0) = 0$ and n-bit MUX selects dividend $D(D_{n-1}D_{n-2}...D_0)$. During the clock pulse when $E = 1$ and $SV2 = 0$, the input S_1 and output data from $(n + 1)$-bit MUX are loaded into $(n + 1)$-bit left-shift register. When $SV1 = 0$, outputs from n-bit MUX are loaded into Q in parallel. When $E = 0$, both A and Q act as left-shift registers. Initially, the value is not important, it is important only after the left-shift of $A.Q$ ($A.Q$ means S_0 of register Q is connected to SI of A), thus the value of S is shifted to which is used to select the operation to be performed on A and Y. If it is 1, then $A + Y$ is performed, otherwise $A - Y$ is computed. Addition or subtraction is performed using $(n + 1)$-bit reversible parallel adder. The complement of the most significant bit (MSB) of the sum is loaded into bit position of register Q and $(n + 1)$-bit SUM is loaded into A during next clock pulse when select is 0. It requires $2n + 1$ clock pulses to store the value of quotient into register Q.

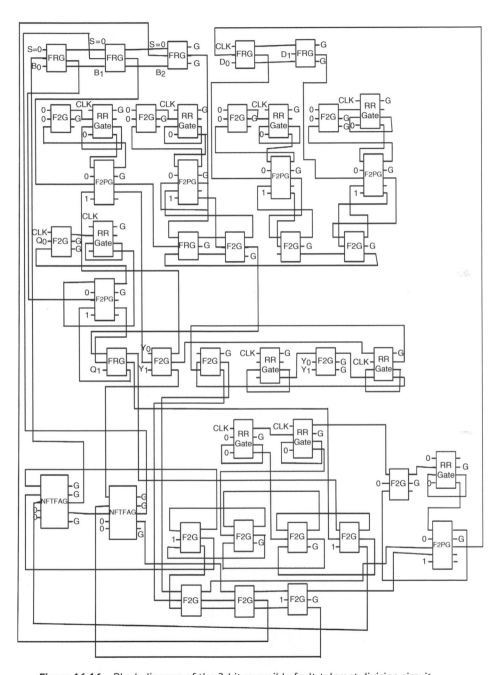

Figure 16.16 Block diagram of the 2-bit reversible fault-tolerant division circuit.

The n-bit divisor is loaded into rounding register through F2G. During next clock pulse, the bits of the divisor $Y(Y_{n-1}Y_{n-2}...Y_0)$ is rounded and loaded into normalization register through another F2G. On each clock pulse, the remainder results from normalization register and loaded into $(n+1)$-bit parallel adder. After $2n+1$ clock pulses, control outputs a high signal. This signal is connected to $SV1$ input of Q register. Thus, Q stores the quotient indefinitely. If S is 0, then remainder restoration is not required, $SV2$ will be high and B will store the remainder indefinitely. If S is 1, then remainder restoration is necessary. During $2n+1$ clock pulse as is 1, restoration is performed by adding Y with A. During the next clock pulse, the correct value of remainder is loaded into A when E is *1*. After remainder restoration, S must be 0. These results $SV2$ to be high and B will store the remainder indefinitely. The working principle of the fault-tolerant reversible design is described in Algorithms 16.2.4.1 and 16.4.6.1.

The construction of a 2-bit reversible fault-tolerant division circuit using Algorithm 16.4.6.1 is shown in Figure 16.16.

16.5 Summary

In this chapter, an n-bit reversible fault-tolerant binary division circuit is described, where n is the number of bits of dividend and divisor. An algorithm is presented for division operation with the optimum time complexity in the design of dividers. The division method consists of four steps: First, it considers floating-point data and rounding. Second, it performs correctly rounded division. Third, it performs correct rounding from one-sided approximations. Finally, it calculates the result of the division operation. The design of the divider circuit shows that it is composed of reversible fault-tolerant multiplexers, parallel-in, parallel-out (PIPO) left-shift registers, D latch, rounding and normalization registers, and parallel adder. The divisor register and the parallel adder have the minimum quantum cost. A Fredkin gate and Feynman double gate are also used to form the divider circuit. Finally, an algorithm is also presented to construct a compact n-bit reversible fault tolerant binary division circuit. In this chapter, an algorithm has also been introduced to reduce the number of steps required for performing division operation.

17

Reversible Fault-Tolerant Decoder Circuit

Decoders are the collection of logic gates fixed up in a specific way such that, for an input combination, all outputs terms are low except one. These *terms* are the *minterms*. Thus, when an input combination changes, two outputs will change. Let, there be n inputs. Thus, the number of outputs will be 2^n. Hardware of digital communication systems relies heavily on decoders as it retrieves information from the coded output. Decoders have also been used in the memory and I/O of microprocessors. This chapter describes the design methodology of reversible fault-tolerant decoder circuit.

17.1 Transistor Realization of Some Popular Reversible Gates

In this section, two very popular reversible gates are described namely, Feynman double gate (F2G) and Fredkin gate (FRG). The block diagrams, quantum realization, transistor realization along with the truth tables are shown in the Subsection 17.1.1 and Subsection 17.1.2.

17.1.1 Feynman Double Gate

The input vector (I_v) and output vector (O_v) for a 3×3 reversible Feynman double gate (F2G) are defined as follows: $I_v = (a, b, c)$ and $O_v = (a, a \oplus b, a \oplus c)$. A block diagram of F2G is shown in Figure 17.1. Figure 17.2 represents the quantum equivalent realization of F2G. From Figure 17.2, we see that it is realized with two 2×2 Ex-OR gate. Thus, its quantum cost is two. According to the design procedure, 12 transistors are required to realize F2G reversibly, as shown in Figure 17.3.

17.1.2 Fredkin Gate

The input and output vectors for a 3×3 Fredkin gate (FRG) are defined as follows: $I_v = (a, b, c)$ and $O_v = (a, a b \oplus ac, a c \oplus ab)$. A block diagram of FRG is shown in Figure 17.4. Figure 17.5 represents the quantum realization of FRG. In Figure 17.5, each rectangle is equivalent to a 2×2 quantum primitive, therefore, its quantum cost is considered as one. Thus, total quantum cost of FRG is five. To realize the FRG, four transistors are needed, as shown in Figure 17.6.

Reversible and DNA Computing, First Edition. Hafiz Md. Hasan Babu.

Figure 17.1 Block diagram of F2G.

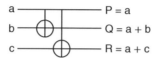

Figure 17.2 Quantum equivalent realization of F2G.

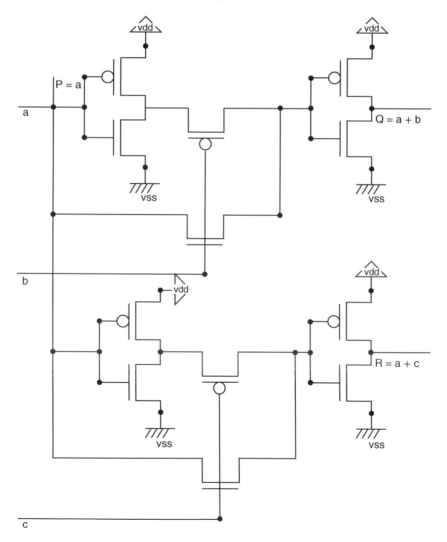

Figure 17.3 Transistor realization of F2G.

Figure 17.4 Block diagram of FRG.

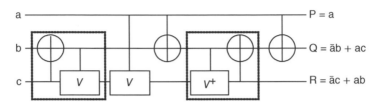

Figure 17.5 Quantum equivalent realization of FRG.

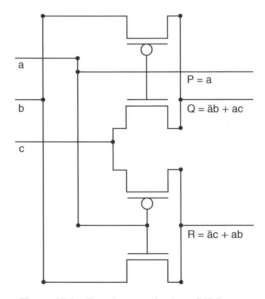

Figure 17.6 Transistor realization of FRG.

Reversible Fredkin and Feynman double gates obey the rules of the fault tolerant (parity preserving) properties. The truth table of Fredkin and Feynman double gates is shown in Table 17.1.

17.2 Reversible Fault-Tolerant Decoder

Considering the simplest case, $n = 1$, here it gets a 1-to-2 decoder. Only a F2G can work as 1-to-2 reversible fault-tolerant decoder (RFD), as shown in Figure 17.7. From now on, a RFD will be denoted as RFD.

Figure 17.8 and Figure 17.9 represent the architecture of 2-to-4 and 3-to-8 RFD respectively. From Figure 17.9, it is found that 3-to-8 RFD is designed using 2-to-4 RFD, thus a schema of Figure 17.8 is created, which is shown in Figure 17.10. Algorithm 17.2.1 presents the design procedure of the n-to-2^n RFD. Primary inputs to the algorithm are n control bits.

Table 17.1 Truth Table of F2G and FRG Gate

Input			Output of F2G			Output of FRG			Parity
A	B	C	P	Q	R	P	Q	R	
0	0	0	0	0	0	0	0	0	Even
0	0	1	0	0	1	0	0	1	Odd
0	1	0	0	1	0	0	1	0	Odd
0	1	1	0	1	1	0	1	1	Even
1	0	0	1	1	1	1	0	0	Odd
1	0	1	1	1	0	1	1	0	Even
1	1	0	1	0	1	1	0	1	Even
1	1	1	1	0	0	1	1	1	Odd

Figure 17.7 1-to-2 Reversible fault-tolerant decoder.

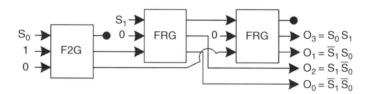

Figure 17.8 Block diagram of the 2-to-4 RFD.

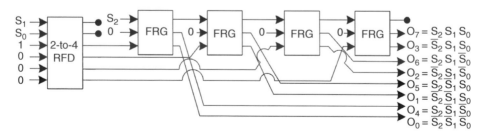

Figure 17.9 Block diagram of the 3-to-8 RFD.

Line 6 of the algorithm assigns the input to the Feynman double gate for the first control bit (S_0), whereas line 9 assigns first two inputs to the Fredkin gates for all the remaining control bits. Lines 10-12 assign third input to the Fredkin gate for $n = 2$.

While line 12 assigns third input to the Fredkin gate through a recursive call to previous RFD for $n > 2$, line 18 returns outputs. The complexity of this algorithm is $O(n)$. According

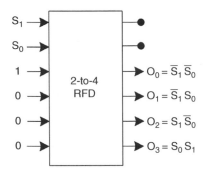

Figure 17.10 Schematic diagram of the 2-to-4 RFD.

Algorithm 17.2.1 Algorithm for the Reversible Fault–Tolerant n-to-2^n Decoder, RFD (S, F2G, FRG)

Input: Data input set $S(S_0, S_1, \ldots;, S_{n-1})$ Feynman double gate (F2G) and Fredkin gate (FRG)

Output: n-to-2^n reversible fault-tolerant decoder circuit

1: Begin
2: i=input
3: o=output
4: **for** $j = 0$ **to** $n - 1$ **do**
5: **if** j=0 **then**
6:   $S_j \rightarrow first.i.$F2G, $1 \rightarrow second.i.$G2G, $0 \rightarrow third.i.$F2G
7: **else**
8:   $S \rightarrow first.i.$FRG, $0 \rightarrow second.i.$FRG
9:   **if** $j = 2$ **then**
10:    $third.o.$F2G $\rightarrow third.i.$FRG
11:   **else**
12:    call RFD($j - 1$), RFD.$o.j \rightarrow third.i.$FRG$_j$
13:   **end if**
14: **end if**
15: **end for**
16: **return** FRG.$o.3$ and FRG.$o.2 \rightarrow$ desired output, remaining F2G.o and FRG.$o \rightarrow$ garbage outputs.
17: End

---

to the algorithm, architecture of n-to-2^n RFD is shown in Figure 17.11. The transistor representations of FRG and F2G using MOS transistors are presented. These representations are finally used to get the MOS circuit of the decoder.

Property 17.2.1 An n-to-2^n reversible fault-tolerant decoder can be realized with at least n garbage outputs and 2^n constant inputs, where n is the number of data bits.

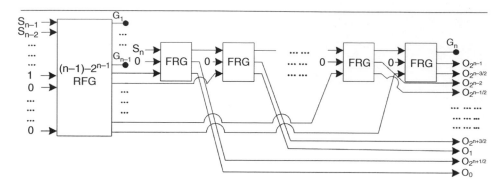

Figure 17.11   Block diagram of the n-to-2^n decoder.

Table 17.2   Truth Table of 1-to-2 Decoder with One Constant Input

Inputs		Outputs		Parity
CI	S_0	O_0	O_1	
0	0	1	0	$I_p=E, O_p=O$
0	1	0	1	$I_p=O_p$
1	0	1	0	$I_p=O_p$
1	1	0	1	$I_p=E, O_p=O$

Proof: An n-to-2^n decoder has n inputs and 2^n outputs. Thus, to maintain the property of reversibility, there should be at least $(2^n - n)$ constant inputs. However, this $(2^n - n)$ constant inputs don't preserve the parity. To preserve the parity, at least n more constant inputs are needed. So, there should be at least n garbage outputs.

Example 17.2.1   Let the value of n be 1. Then, the 1-to-2 reversible fault tolerant decoder can be obtained. As shown in earlier, any reversible circuit should have the equal number of inputs and outputs. In the 1-to-2 decoder, there are two primary outputs (O_0, O_1) and one input (S_0). Hence according to the property of reversibility, a 1-to-2 reversible decoder should have at least one constant input. The value of this constant input can be either 0 or 1. Table 17.2 shows that whatever the value of this constant input, it will never be able to preserve the parity between input and output vectors, which is the prime requirement of the reversible fault tolerant logic circuit. Therefore, to preserve the parity for the 1-to-2 RFD it needs at least one more constant input, i.e., at least two constant inputs are required for the 1-to-2 RFD.

Next, it must prove the existence of combinational circuit which can realize the reversible fault tolerant 1-to-2 decoder by two constant inputs. This can easily be accomplished by the circuit shown in Figure 17.7. By using the truth table, it can be verified that the decoder in Figure 17.7 is reversible and fault tolerant.

Now, in 1-to-2 RFD, there are at least two constant inputs and one primary input, i.e., total of three inputs. Thus, 1-to-2 RFD should have at least three outputs; otherwise it will never comply with the properties of reversible parity preserving circuit. Among these three outputs, only two are primary outputs. So, the remaining output is the garbage output, which holds the Property 17.2.1 for $n=1$.

Property 17.2.2   A 2-to-4 reversible fault-tolerant decoder can be realized with at least 12 quantum cost.

Proof: A 2-to-4 decoder has four different 2×2 logical AND operations. A reversible fault-tolerant AND operation requires at least three quantum cost. So, 2-to-4 reversible fault-tolerant decoder is realized with at least 12 quantum cost.

Example 17.2.2   Figure 17.8 is the proof for the existence of 2-to-4 reversible decoder with 12 quantum cost. Next, it can be proved that it is not possible to realize a reversible fault-tolerant 2-to-4 decoder fewer than 12 quantum cost. In the 2-to-4 decoder, there are four different 2×2 logical AND operations, e.g., $S_1' S_0', S_1' S_0, S_1 S_0', S_1 S_0$. It will be enough if it is proved that it is not possible to realize a reversible fault-tolerant 2×2 logical AND with fewer than three quantum cost. Consider the following:

1. If one quantum cost is used to design the AND, which is not possible according to the above discussion.
2. If two quantum cost are used to design AND, then it needs to use two 1×1 or 2×2 gates. Apparently, two 1×1 gates can not generate the AND. Aiming at two 2×2 gates; there are two combinations which are shown in Figure 17.12 (a) and Figure 17.12(b). In Figure 17.12, the output must be (a, ab) if the inputs are (a, b). The corresponding truth table is shown in Table 17.3.

From Table 17.3, it is found that outputs are not at all unique to their corresponding input combinations (first and second rows have identical outputs for different input

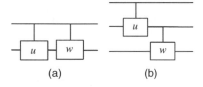

(a)   (b)

Figure 17.12   Combinations of the two 2×2 quantum primitive gates.

Table 17.3   Truth Table of Figure 17.12 (a)

a	b	a	a b
0	0	0	0
0	1	0	0
1	0	1	0
1	1	1	1

Table 17.4 Truth Table of Figure 17.12 (b)

a	b	c	a	ab
0	0	0	0	0
0	0	1	0	0
0	1	0	0	0
0	1	1	0	0
1	0	0	1	0
1	0	1	1	0
1	1	0	1	1
1	1	1	1	1

combinations). So it can't achieve the reversible AND. For Figure 17.12 (b), if inputs are (a, b, c), then the outputs of the lower level will be offered to the next level as a controlled input. This means that the second output of Figure 17.12 (a) has to be ab; otherwise it will never be able to get output ab, since the third output of Figure 17.12 (b) is controlled by the second output. Thereby, according to Table 17.4, it can be asserted that the second combination is impossible to realize the AND no matter how it is set. For the third output of Figure 17.12 (b) (third column of Table 17.4), the input vectors will never be one-to-one correspondent with the output vectors. Therefore, it can be concluded that a combinational circuit for reversible fault-tolerant 2×2 logical AND operation cannot be realized with less than three quantum cost.

The above example clarifies the lower bound in terms of quantum cost of the 2-to-4 RFD. Similarly, it can be proved that the $n - to - 2^n$ RFD can be realized with $5(2^n - 8/5)$ quantum cost when $n \geq 1$, and by assigning different values to n, the validity of this equation can be proved.

Property 17.2.3 An n-to-2^n RFD can be realized with $(2^n - 1)$ reversible fault-tolerant gates, where n is the number of data bits.

Proof: According to the design procedure, an n-to-2^n RFD requires an $(n - 1)$-to-2^{n-1} RFD plus n number of Fredkin gates, which requires an $(n - 2)$-to-2^{n-2} RFD plus $(n - 1)$ Frdekin gates and so on until it reaches to 1-to-2 RFD. 1-to-2 RFD requires a reversible fault-tolerant Feynman double gate only. Thus, the total number of gates required for an n-to-2^n RFD is

$$1 + 2 + 4 + \cdots + n^{th} \ term$$
$$= \frac{2^0(2^n - 1)}{2 - 1} = 2^n - 1$$

Example 17.2.3 From Figure 17.4, it can be found that the 3-to-8 RFD requires a total of seven reversible fault-tolerant gates. If it replaces n with three in Property 17.2.3, the value is seven as well.

Property 17.2.4   Let α, β, and γ be the hardware complexity for a two-input Ex-OR, AND, and NOT operation, respectively. Then an n-to-2^n RFD can be realized with $(2^{n+1} - 2)\alpha + (2^{n+2} - 8)\beta + (2^{n+1} - 4)\gamma$ hardware complexity, where n is the number of data bits.

Proof: In Property 17.2.3, it was proved that an n-to-2^n RFD is realized with a F2G and $(2^n - 2)$ FRG. The hardware complexity of a FRG and a F2G are $2\alpha + 4\beta + 2\gamma$ and 2α, respectively. Hence, the hardware complexity for $n - to - 2^n$ RFD is

$$(2^n - 2)(2\alpha + 4\beta + 2\gamma) + 2\alpha$$
$$= (2^{n+1} - 2)\alpha + (2^{n+2} - 8)\beta + (2n + 1 - 4)\gamma$$

Example 17.2.4   Figure 17.9 showed that the 3-to-8 reversible fault-tolerant decoder requires six Fredkin gates and one Feynman double gate. According to the previous discussion, hardware complexity of a Feynman double gate is 2α, whereas, the hardware complexity of a Fredkin gate is $2\alpha + 4\beta + 2\gamma$. Thus, the hardware complexity of Figure 17.11 is $6(2\alpha + 4\beta + 2\gamma) + 2\alpha = 14\alpha + 24\beta + 12\gamma$. In Property 17.2.4, if value of n is put like $n= 3$, it is exactly $14\alpha + 24\beta + 12\gamma$ as well.

17.3 Summary

This chapter demonstrates the reversible logic synthesis for the n-to-2^n decoder, where n is the number of data bits. The circuits are designed using only reversible fault-tolerant Fredkin and Feynman double gates. Thus, the entire scheme inherently becomes fault tolerant. The algorithm for designing the generalized decoder has been presented. In addition, several lower bounds on the number of constant inputs, garbage outputs, and quantum cost of the reversible fault-tolerant decoder have been described. Transistor simulations of the decoder are shown using standard p-MOS 901 and n-MOS 902 models with delay of 0.030 ns and 0.12 μm channel length, which proved the functional correctness of the decoder circuits.

18

Reversible Fault-Tolerant Barrel Shifter

Reversible fault-tolerant circuits detect faulty signal in its primary outputs through parity checking. In addition, data shift and rotate operations are widely used in arithmetic operations, bit-indexing, and overall in fast encoding. Barrel shifter shifts and rotates multiple bits in a cycle and hence, it attains great importance in designing processors, especially in digital signal processors and low-density parity-check (LDPC) decoders. Among the various barrel shifters, logarithmic one has the simplest structure and is more area efficient as it doesn't require any underneath decoder circuit. In these consequences, this chapter presents the generalized design methodologies of reversible fault-tolerant unidirectional logarithmic barrel shifters.

18.1 Properties of Barrel Shifters

A barrel shifter has n-input and n-output lines for data transmission and k control inputs, where $k = \log_2 n$. Adaptive structure of the basic (n, k) unidirectional logarithmic barrel shifter is shown in Figure 18.1. It has k $(k = \log_2 n)$ stages, which are controlled by k control bits. Control bit S_j ($j = 0$ to $k-1$) of a stage determines whether to shift (or rotate) or not to shift (or rotate) the input data for that stage. If S_j is set to high, then j^{th} stage will shift or rotate the input 2^j times; otherwise, input will remain unchanged.

Property 18.1.1 A reversible unidirectional barrel shifter can be realized with at least k garbage outputs and no constant input, where n is the number of data bits and $k = \log_2 n$.

Proof: The unidirectional barrel shifter with n data bits has k control inputs, where $k = \log_2 n$, i.e., total of $n + k$ inputs. Thus, according to the property of reversibility, it should have at least $(n + k)$ outputs, among which n bits are the primary output. So, the number of garbage output is k, which indicates that no constant input is required.

Reversible and DNA Computing, First Edition. Hafiz Md. Hasan Babu.
© 2021 John Wiley & Sons Ltd. Published 2021 by John Wiley & Sons Ltd.

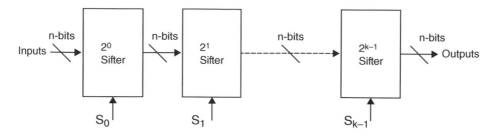

Figure 18.1 Adaptive structure of (n, k) logarithmic barrel shifter.

18.2 Reversible Fault-Tolerant Unidirectional Logarithmic Rotators

Figure 18.2 and Figure 18.3 show the architecture of the (4, 2) and (8, 3) logarithmic rotators, respectively. From these figures, it is found that the rotators are designed using only reversible fault tolerant Fredkin gates. Thus, the rotators preserve parity at all stages.

The unidirectional rotators follow the adaptive structure of logarithmic barrel shifter, *e.g.*, first stage rotates 2^0 bit, second stage rotates 2^1 bits, third stage rotates 2^2 bits, and so on. Rotation occurs only when control signal is set to HIGH. If a control signal from any stage is set to LOW, then instead of rotating, that stage just passes the input to the next stage. Design procedure of the reversible fault tolerant rotator is as follows: Let the data inputs for the (n, k) rotator is $I_0, I_1, I_2, \ldots, I_{n-3}, I_{n-2}, I_{n-1}, I_{n-1}$ and the control inputs are $S_0, S_1, \ldots, S_{k-1}$; where, n represents the number of inputs and k is number of stages which equals to $\log_2 n$. Each of these stages require a chain of $(n - 2^j)$ Fredkin gates (where $j = 0$ to $k-1$). The inputs and outputs for each Fredkin gate in a stage j can be rewritten as $A(i, j), B(i, j), C(i, j)$ and $P(i, j), Q(i, j), R(i, j)$, respectively. Here j represents the jth stage, where, $n - 1 \geq i \geq 0$ and $k - 1 \geq j \geq 0$.

The working procedure of the reversible fault tolerant rotator is as follows: Let $n = 4$. Then, (4, 2) shifter takes $I_0 I_1 I_2 I_3$ as data inputs and S_0 and S_1 as control inputs. If both the control inputs set to HIGH ($S_0 = 1, S_1 = 1$), then data inputs will be rotated $2^0 + 2^1$ times to the right. Sequence of the rotate operation will be $I_1 I_2 I_3 I_0$ for the first stage and $I_3 I_0 I_1 I_2$ for

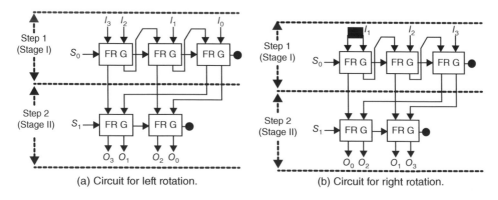

(a) Circuit for left rotation.　　　　(b) Circuit for right rotation.

Figure 18.2 (4, 2) Reversible fault-tolerant unidirectional logarithmic barrel shifter.

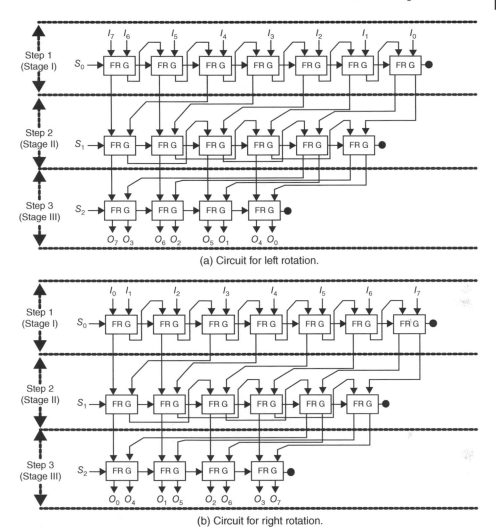

(a) Circuit for left rotation.

(b) Circuit for right rotation.

Figure 18.3 (8, 3) Reversible fault-tolerant unidirectional logarithmic barrel shifter.

the next. On the other hand, if both control inputs are set to LOW, then input sequence will remain unchanged. For example, when control input S_0 is set to LOW, the first Fredkin gate (FRG) of first row will select input I_0, the second will select I_1 and third one will select I_2 also I_3 will be generated from another output of the third FRG. Finally selected four outputs will be used as inputs of the first and second FRG at the second row, which are controlled by the control input S_1. Besides, if the first control input is set to HIGH, then the selection sequence will be I_1, I_2, I_3 and I_0 which are the first, second, and third FRG at the first row, respectively.

Property 18.2.1 Let the value of n be 4. Figure 18.2 shows that the (4, 2) barrel shifter (circuits for rotation) has two two garbage outputs and no constant inputs.

Property 18.2.2   Let FR be the required number of gates for (n, k) reversible fault-tolerant unidirectional logarithmic rotator, where n and k ($k = \log_2 n$) are the number of data bits and control inputs respectively. Then $FR = nk - \sum_{j=0}^{k-1} 2^j$.

Property 18.2.3   A (n, k) reversible fault-tolerant unidirectional logarithmic rotator has k stages and n input bits. According to the design procedure, each jth stage requires $(n\text{-}2^j)$ Fredkin gates, where $j = 0$ to $(k - 1)$. So, the required number of gates for a (n, k) reversible fault tolerant rotator is $(n - 2^0) + (n - 2^1) + (n - 2^2) + \cdots + (n - 2^{k-2}) + (n - 2^{k-1}) = nk - \sum_{j=0}^{k-1} 2^j$

Property 18.2.4   Let QC be the total quantum cost for (n, k) reversible fault tolerant unidirectional logarithmic rotator, where n is the number of data inputs and k is number of control inputs ($k = \log_2 n$). Then, $QC = 5nk - 5 \sum_{j=0}^{k-1} 2^j$

Proof: It is known that the reversible fault-tolerant (n, k) rotator can be realized with reversible fault-tolerant Fredkin gates. According to the properties of the reversible logic synthesis, the quantum cost of each Fredkin gate is 5, which is described in Chapter 17. Therefore, the total quantum cost of the (n, k) reversible fault-tolerant unidirectional logarithmic rotator is $5 \times (nk - \sum_{j=0}^{k-1} 2^j) = 5nk - 5 \sum_{j=0}^{k-1} 2^j$.

Property 18.2.5   Let α, β, and γ be the hardware complexity for two-input EX-OR, AND, NOT calculations, respectively. Let HC be the total hardware complexity for (n, k) reversible fault-tolerant unidirectional logarithmic rotator, where n is the number of data inputs and k is the number of control inputs, which equals to $\log_2 n$. Then $(2nk - 5 \sum_{j=0}^{k-1} 2^{j+1})\alpha + (nk - \sum_{j=0}^{k-1} 2^j)(4\beta + \gamma)$.

Proof: It is known that the reversible fault-tolerant (n, k) unidirectional logarithmic rotator can be realized with Fredkin gates. Hardware complexity of a Fredkin gate is $(2\alpha + 4\beta + \gamma)$. Thus, hardware complexity of (n, k) reversible fault-tolerant unidirectional rotator is $=(2nk - 5 \sum_{j=0}^{k-1} 2^{j+1})\alpha + (4nk - \sum_{k-1}^{j=0} 2^{j+2})\beta + (nk - \sum_{j=0}^{k-1} 2^j)\gamma = (2nk - 5 \sum_{j=0}^{k-1})\alpha + (nk - \sum_{k-1}^{j=0} 2^j)(4\beta + \gamma)$.

Architecture of (n, k) reversible fault-tolerant right rotator is shown in Figure 18.4. From Figure 18.2 and Figure 18.3, it is found that the $(4, 2)$ and $(8, 3)$ reversible fault-tolerant left and right rotators are very similar to each other except the positions of input–output vectors. Thus, the reversible fault tolerant right rotator is shown for only the architecture of (n, k).

18.3   Fault-Tolerant Unidirectional Logarithmic Logical Shifters

Design procedure of the (n, k) reversible fault-tolerant unidirectional logarithmic logical shifters is similar to the rotators but in reverse direction, *i.e.*, S_{k-1} acts as the control signal

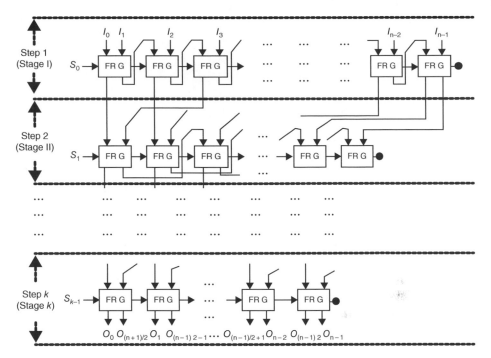

Figure 18.4 (n, k) Reversible fault-tolerant unidirectional logarithmic right rotator.

for the first stage, S_{k-2} for the second stage, and so on. Thus, the k stages of the (n, k) logical shifter is responsible for shifting input data by 2^{k-1} to 2^0 bits. The architecture of the $(4, 2)$ and $(8, 3)$ reversible fault tolerant unidirectional logical shifters are shown in Figure 18.5 and Figure 18.6, respectively.

Inputs to the Fredkin gates for the first stage of the logical shifter are as follows:

$$A(i, k-1) = S_{k-1}, \forall i = (n-1)\ldots0$$
$$B(i, k-1) = i_i, \forall i = (n-1)\ldots0$$
$$C(i, k-1) = 0, \forall i = (n-1)\ldots((n-1) - 2^{k-1}+1)$$
$$C(i, k-1) = R_m, \forall i = ((n-1)-2^{k-1})\ldots0; R \text{ is the third output of the } m \text{ Fredkin gate, where}$$
$$m = (n-1)\ldots((n-1)-2^{k-1} + 1).$$

The controlling input for all Fredkin gates in the first stage of the logical shifter is set to $A(i, k-1) = S_{k-1}$. When $S_{k-1} = 1$, the following values are got at the outputs:

$$P(i, k-1) = S_{k-1}, \forall i = (n-1)\ldots0$$
$$R(i, k-1) = i_i, \forall i = (n-1)\ldots0$$
$$Q(i, k-1) = 0, \forall i = (n-1)\ldots(n-1) - 2^{k-1}+1$$
$$Q(i, k-1) = i_m, \forall i = (n-1)-2^{k-1} \ldots0 \text{ and } m = (n-1) \ldots 2^{k-1}$$

When $S_{k-1} = 0$, the following values are got at the outputs:

$$P(i, k-1) = S_{k-1}, \forall i = (n-1)\ldots0$$
$$Q(i, k-1) = i_i, \forall i = (n-1)\ldots0$$
$$R(i, k-1) = 0, \forall i = (n-1)\ldots0$$

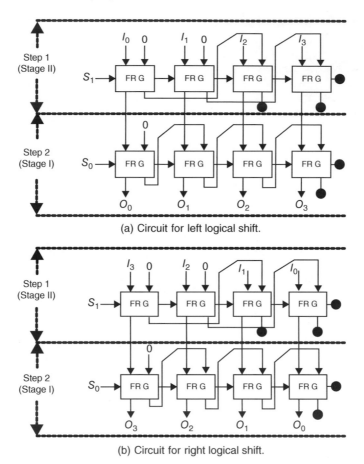

(a) Circuit for left logical shift.

(b) Circuit for right logical shift.

Figure 18.5 (4, 2) Reversible fault-tolerant unidirectional logarithmic logical shifter.

The second stage of (n, k) reversible fault tolerant logical shifters uses inputs from the first stage. From the outputs $Q(i, k-1)$ and $R(i, k-1)$ of the Fredkin gates in the first stage, all $Q(n, k-1)$ and some $R(n, k-1)$ outputs (first $n/2$ outputs) will be the useful outputs, while rest $R(n, k-1)$ outputs (last $n/2$ outputs) will work as the garbage outputs. Design methodology used for the second stage is as follows:

$$A(i, k-2) = S_{k-2}, \forall i = (n-1)...0$$
$$B(i, k-2) = Q(i, k-1), \forall i = (n-1)...0$$
$$C(i, k-2) = 0, \forall i = (n-1)...((n-1)-2^{k-2}+1)$$
$$C(i, k-2) = O_m, \forall i = ((n-1)-2^{k-1})...0; O \text{ is third output of the } m \text{ Fredkin gates, where}$$
$$m = (n-1)...((n-1) - 2^{k-1} + 1).$$

Since S_{k-2} works as the input A of the Fredkin gates, the following values are obtained at the outputs when $S_{k-2} = 1$:

$$P(i, k-2) = S_{k-2}, \forall i = (n-1)...0$$
$$R(i, k-2) = Q(i, k-1), \forall i = (n-1)...0$$
$$Q(i, k-2) = 0, \forall i = (n-1)...(n-1)-2^{k-2}+1$$
$$Q(i, k-1) = Q(m, k-1), \forall i = (n-1)-2^{k-2}...0 \text{ and } m = (n-1)...(n-1) - 2^{k-2} + 1.$$

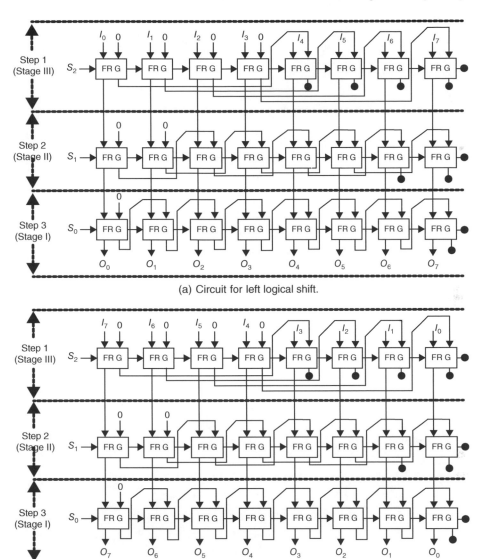

(a) Circuit for left logical shift.

(b) Circuit for right logical shift.

Figure 18.6 (8,3) Reversible fault-tolerant unidirectional logarithmic logical shifter.

When $S_{k-2} = 0$, the following values are obtained at the outputs.

$$P(i, k-2) = S_{k-2}, \forall i = (n-1)\ldots0$$
$$Q(i, k-2) = Q(i, k-1), \forall i = (n-1)\ldots0$$
$$R(i, k-2) = Q(i, k-1), \forall i = (n-1)\ldots0$$

Similar design strategies are used for the remaining stages. Architecture of the (n, k) reversible fault tolerant right logical shifter is shown in Figure 18.7. Since, there is almost no difference between the logical left shifters and the right shifters, except for the input–output

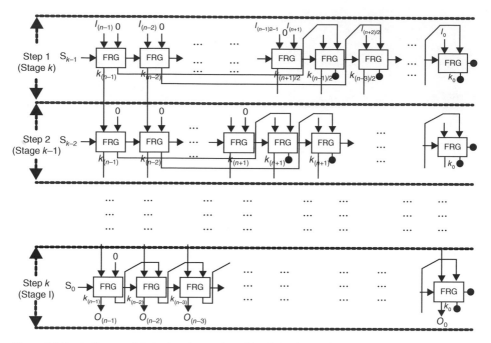

Figure 18.7 *(n,k)* reversible fault-tolerant logarithmic logical shifter (circuit for right logical shift).

combination hence the architecture of (n, k) reversible fault-tolerant left logical shifter is not shown here. As shown in Figure 18.5 and Figure 18.6, only reversible fault-tolerant Fredkin gate is used here. Thus, the designs of the logical shifters also preserve parity.

Property 18.3.1 Let F be the number of gates for a (n, k) reversible fault-tolerant logical shifter, where n is the number of data inputs and $k = \log_2 n$. Then, $F = nk$.

Proof: An (n, k) logical shifter has k stages (0 to $k - 1$) and n input bits, where $k = \log_2 n$. According to the design procedure, n Fredkin gates are required for each stage. Therefore, the total number of Fredkin gates required for reversible fault tolerant (n,k) logical shifter is $\sum_0^{k-1} n = nk$.

Property 18.3.2 An (n, k) reversible fault-tolerant unidirectional logical shifter can be realized with $5nk$ quantum cost and $(2nk\alpha + 4nk\beta + nk\gamma)$ hardware complexity, where n is the number of data bits, $k = \log_2 n$; α, β, and γ are the hardware complexity for two-input Ex-OR, AND, NOT operations, respectively.

Proof: In Property 18.3.1, it is proved that the reversible fault-tolerant (n, k) logical shifter can be realized with nk reversible fault-tolerant Fredkin gates. Similarly, like Property 18.2.4 and 18.2.5, it can be proved that the reversible fault tolerant unidirectional logical shifter can be realized with $5nk$ quantum cost and $(2nk\alpha + 4nk\beta + nk\gamma)$ hardware complexity.

18.4   Summary

This chapter presented the design methodologies of the (n, k) reversible fault-tolerant unidirectional barrel shifters, where n is the number of data bits and $k = \log_2 n$. Several properties on the numbers of garbage outputs and constant inputs were described. It was also proved that the rotator circuits are constructed with the optimum garbage outputs and constant inputs. By articulating several theoretical explanations, the efficiency and supremacy of all the designs are also proved. Most interestingly, all the designs have the capability of detecting errors at circuit's primary outputs.

19

Reversible Fault-Tolerant Programmable Logic Devices

This chapter focuses on the designs of reversible fault-tolerant programmable logic devices. There are different types of programmable logic devices. Among them, three of the programmable logic devices are discussed here, namely programmable logic array (PLA), programmable array logic (PAL), and field programmable gate array (FPGA). The design procedures are briefly described in the following sections.

19.1   Reversible Fault-Tolerant Programmable Logic Array

In this section, a regular structure of reversible fault-tolerant programmable logic array (RFTPLA) has been introduced and corresponding algorithms for construction of RFT-PLAs are described. The algorithms can realize ESOP (exclusive sum-of-products) operations in terms of multi-output functions by using minimum numbers of gates, garbage bits and quantum costs. In the design, the unique properties of FRG (Fredkin gate) and F2G (Feynman double gate) are used for the realization of the designs.

PLA allows sum-of-products (SOP) for implementing Boolean functions. The typical implementation consists of input buffers, the programmable AND-matrix followed by the OR-matrix, and output buffers.

Property 19.1.1   PLA consists of two planes, the first one is AND plane and the second one is OR plane known as AND-OR PLA. When the second plane works as XOR, it is called AND-XOR PLA.

Property 19.1.2   AND plane generates product by combining two or more literals. In the design, the AND plane generates a product that uses two types of gates: F2G, which is used to copy or recover fan-out problem and Fredkin gate, which is used for AND operation.

Property 19.1.3   Ex-OR plane uses products that come from AND plane and produces the final output functions by Ex-ORing those products having zero. Here the Ex-OR plane is designed by using only Feynman double gate.

Example 19.1.1   Figure 19.1 shows the design of AND-ExOR PLA consisting of AND plane followed by Ex-OR plane. The AND plane takes inputs whereas the Ex-OR plane

Reversible and DNA Computing, First Edition. Hafiz Md. Hasan Babu.
© 2021 John Wiley & Sons Ltd. Published 2021 by John Wiley & Sons Ltd.

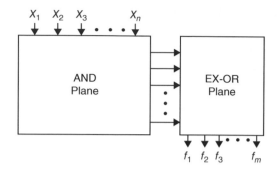

Figure 19.1 AND Ex-OR programmable logic array.

generates output functions and the products are passed from AND plane to Ex-OR plane by using the intermediate lines.

19.1.1 The Design of RFTPLA

At first, let F be a multi-output ESOP functions, where $F = \{f_1, f_2, f_3, f_4, f_5\}$ is defined as follows:

$$f_1 = Ac \oplus \overline{ABC}$$
$$f_2 = A\overline{B} \oplus A\overline{B}C$$
$$f_3 = A\overline{B} \oplus A\overline{B}C \oplus B\overline{C} \qquad (19.1.1.1)$$
$$f_4 = AC$$
$$f_5 = A\overline{B} \oplus AC \oplus B\overline{C}$$

The design is based on the orientation of input and output vectors of the F2G and Fredkin gate (FRG). The input and output vectors of any reversible circuit or the gate containing a unique mapping and the orientation of the input and the output vectors also preserves the uniqueness. For example, Figure 19.2 (a) shows two orientations (F2G-1 and F2G-2) of Feynman double gate and Figure 19.2 (b) shows two orientations (FRG-3 and FRG-4) of Fredkin gate. Input a and c (output p and r) of F2G swap their position to make a different pattern of Feynman gate, as shown in Figure 19.2 (a). On the other hand, output q and r of the FRG swap their position and create another pattern of Fredkin, as shown in Figure 19.2 (b).

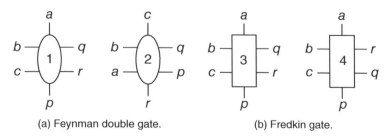

(a) Feynman double gate. (b) Fredkin gate.

Figure 19.2 Four different orientations.

Table 19.1 Frequency Matrix Based on Multi-Output Function, F.

Products	Function					Frequency of Product
	f_1	f_2	f_3	f_4	f_5	
$\bar{a}b$		X	X		X	3
$\bar{a}bc$		X	X			2
ac	X			X	X	3
$b\bar{c}$			X		X	2
$\bar{a}\,\bar{b}\,\bar{c}$	X					1

The order of products is generated after the optimization of Ex-OR plane. In this subsection, two algorithms are described for the minimizations and construction of Ex-OR plane followed by the realization of AND plane for generalizing the design.

Property 19.1.1.1   The frequency of a product is the total number of output functions used by this product.

Proof: Table 19.1 shows the frequency of all products of the multi-output function $F = \{f_1, f_2, f_3, f_4, f_5\}$. For instance, the frequency of ac is the highest i.e., 3 and $a'b'c'$ is the lowest i.e., 1.

Property 19.1.1.2   The cross point in RPLA is used, where no gate is required. This is termed as *DOT*.

Example 19.1.1   In Figure 19.3, each of functions f_1, f_2, f_3, and f_4 uses *DOT* at the top most horizontal line.

Property 19.1.1.3   The products of functions are required to realize the function of ESOP form.

Example 19.1.2   Table 19.2 shows the calculation of the number of products of functions.

In the design, the order of output functions is related to the number of products. Functions are generated in ascending order based on the above criterion and Algorithm 19.1.1.1 realization of Ex-OR plane is defined as follows:

Table 19.2 Calculation of Number of Product of Functions.

Functions	f_1	f_2	f_3	f_4	f_5
Number of Product	2	2	3	1	3

Algorithm 19.1.1.1 Construction of Ex-OR Plane

1: Begin
2: $TDOT := 0 = $ Total number of DOT
3: Sort output functions according to their number of product
4: **for** each output function f_i of F **do**
5:   **if** N_O_P of f_i is one **then**
6:     **if** Product P_j **then**
7:       F2G-2
8:     **else**
9:       Generate P_j in above line and use**DOT** as $TDOT := TDOT + 1$
10:     **end if**
11:   **else**
12:     **if** all products P_j exist **then**
13:       use F2G-2 for all
14:     **else**
15:       Generate product(s) (P_j) on upper line(s) according to frequency of products and use DOT for the top most of F2G-2, where the newly generated products are $TDOT := TDOT + 1$
16:     **end if**
17:   **end if**
18: **end for**
19: End

Proof: The minimum number of Feynman double gates to realize Ex-OR plane is $(n + m - TDOT)$ where:

- $m = $ Number of output functions
- $n = $ Number of Ex-OR operations in m
- $TDOT = $ Number of cross points (DOT)

When there are $TDOT$ cross points for m functions, the number of additional Feynman double gates in Ex-OR plane of RPLA is m-$TDOT$. As there are n Ex-OR operations by n Feynman double gates, the total number of Feynman double gates in the Ex-OR plane of RPLA is $f = n + m - TDOT$.

Example 19.1.3 For multi-output function F of Equation 19.1.1.1, the number of outputs (m) is 5 and the number of Ex-OR operations are 6. Consider Figure 19.2, the number of $TDOT$ is 4. So the number of Feynman double gates is $n + m - TDOT = 6 + 5 - 4 = 7$.

By using Algorithm 19.1.1.1, the realization of Ex-OR plane for function F is shown in Figure 19.3.

Property 19.1.1.4 Let P be the number of products, $TDOT$ be the number of cross points, and x be the number of horizontal series in the Ex-OR plane of RPLA. The minimum number of garbage outputs to realize Ex-OR plane of RPLA is $(P - TDOT + x)$.

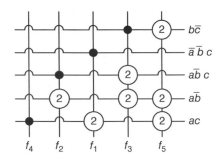

Figure 19.3 Realization of multi-output function (F) based on Algorithm 19.1.1.1.

Proof: As there are *TDOT* cross points and *P* products from output functions, the total number of garbage outputs in the Ex-OR plane of RPLA is $P - TDOT + x$, where x is the number of horizontal series in the Ex-OR plane of RPLA.

Example 19.1.4 Consider Figure 19.3 for function F in Equation 19.1.1.1. In Figure 19.3, the number of products (P) is 5 and the number of cross points (TDOT) is 4. The horizontal series of F2G propagates a garbage bit until the end, where the number of horizontal series is 4. So the number of garbage outputs is $P - TDOT + x = 5 - 1 + 4 = 8$.

The realization of Ex-OR plane generates the order of products as shown in Figure 19.3. The AND plane is constructed according to the order of products. Algorithm 19.1.1.2 describes the construction of AND plane by using Fredkin and Feynman double gates.

In the AND plane, the Feynman double gates are used to recover the fan-out problem and the Fredkin gates are used for AND operations. The generation of complementary form of input literals in AND plane is unnecessary in the AND plane because the Fredkin gate can generate products with or without complemented form of literal, which is not possible by Toffoli gate. By using Algorithms 19.1.1.1 and 19.1.1.2, the realization of the reversible PLA is shown in Figure 19.4.

In this subsection, it is shown that the realization of Ex-OR plane requires the minimum number of gates and garbage outputs, which is based on Algorithm 19.1.1.1. Similarly, Algorithm 19.1.1.2 describes the construction of AND plane.

19.2 Reversible Fault-Tolerant Programmable Array Logic

In this section, the design of a reversible fault-tolerant programmable array logic is shown. The total number of gates, garbage outputs, and quantum cost have been minimized in the design of AND plane of a reversible fault-tolerant programmable array logic (RFTPAL). When the full architecture of reversible PAL remains fault tolerant, the Ex-OR plane is also fault tolerant, and in this case, the reversible PAL requires more number of gates, garbage outputs and quantum cost. The reversible fault-tolerant PAL is useful for designing the digital circuits easily. For example, large functions that have several variables can easily be implemented by using reversible fault-tolerant PAL.

This section combines several ideas to present the construction of a compact reversible fault-tolerant PAL for multi-output functions by introducing the structure based on F2G

Algorithm 19.1.1.2 Construction_Of_AND_Plane

1: Begin
2: $TDOT := 0[TDOT = $ Total number of $DOT]$
3: Sort the output functions according to their number of products
4: **for** each ordered product (P_i) **do**
5:   **if** I_j is the first literal **then**
6:     **if** l_j is in complemented form **then**
7:       apply F2G-1
8:     **else**
9:       **if** l_j is further used **then**
10:         apply F2G-1
11:       **else**
12:         use DOT
13:         $TDOT := TDOT + 1$
14:       **end if**
15:     **end if**
16:   **else**
17:     use FRG-4 and FRG-3 based on ANDing pattern of any product (P_i)
18:   **end if**
19: **end for**
20: End

---

and FRG, heuristic algorithm for the reversible fault-tolerant PAL, ordering of functions as well as ordering of products, optimization of Ex-OR plane and AND plane minimization in terms of products. In the design, the target is to make the reversible PAL fault tolerant. Since the AND planes of PLA and PAL are programmable, the algorithm will minimize the number of gates, garbage outputs, and quantum cost. In the next subsections, the designs of AND plane, OR plane, and minimizations of AND and OR planes are described.

19.2.1 The Design of AND Plane of RFTPAL

As mentioned earlier, a reversible PAL has two planes: AND plane and Ex-OR plane. In the design, two fault-tolerant gates are used, namely F2G and FRG for designing fault-tolerant PAL. F2G is used for copying and complementing the inputs and FRG is used for ANDing the inputs, as required by the outputs. In this design process, the minterms are sorted by the number of frequencies in decreased order. As a result, when the larger number of functions is copied or complemented, the fewer number of functions is automatically generated.

When the AND operation is performed using FRG, the inputs are taken as two bits, e.g., for input A and B, there may be one of four conditions: $AB, A\overline{B}, \overline{A}B$, and $\overline{A}\overline{B}$. If there exist another input along these terms, then it must have only one of the two forms: (*The previous form*) .C or (*The previous form*) .C', if another input is C. The other inputs follow the same rule. This kind of operation will decrease the AND operation, and thus it will decrease the total number of gates, garbage outputs, and quantum cost. Algorithm 19.2.1.1 is addressed here for designing the AND plane of the reversible fault-tolerant PAL. In the

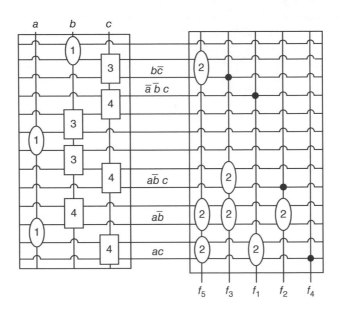

Figure 19.4 Realization of multi-output function F based on Algorithm 19.1.1.1 and Algorithm 19.1.1.2.

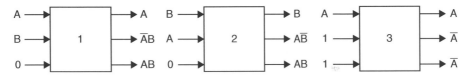

Figure 19.5 Different representations of Fredkin gate.

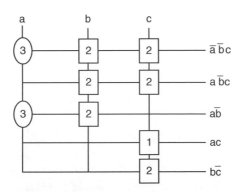

Figure 19.6 The design of AND plane of reversible fault-tolerant PAL.

design, F2G and FRG are used for different combinations of inputs. Moreover, Gates 1, 2 and 3 are used in the design that are shown in Figure 19.5. The design of AND plane of reversible fault-tolerant PAL is given in Figure 19.6.

Algorithm 19.2.1.1 An Algorithm for Designing the AND Plane of the Reversible Fault Tolerant PAL

1: Begin
2: Sort Min-terms
3: $TG := 0$ [TG = Total number of gates]
4: Put all literals into the stack
5: **for** each ordered product **do**
6:   **if** product is the single literal **then**
7:     Apply F2G Gate
8:     Increment the number of gates
9:     Update stack
10: **else**
11:    **if** two literals are complemented **then**
12:      Apply F2G and FRG gates
13:      Increment the number of gates
14:      Update stack
15:    **end if**
16:    **if** Two literals are in normal form **then**
17:      Apply F2G and FRG gates
18:      Increment the number of gates
19:      Update stack
20:    **end if**
21:    **if** one literal is complemented and another is in normal form **then**
22:      Apply F2G and FRG gates
23:      Increment the number of gates
24:      Update stack
25:    **end if**
26: **end if**
27: **end for**
28: End

19.2.2 The Design of Ex-OR Plane of RFTPAL

In the design of RFTPAL, Feynman double gate (F2G) is used for Ex-OR plane. If any other gate is used in the place of F2G in Ex-OR plane, the total number of gates, garbage outputs, and quantum cost is minimized. This section is also considered a new reversible gate, which is called a Feynman extension gate (FEG) here. With the help of this gate, the Ex-OR of any number of inputs is obtained. The inputs and outputs of a FEG gate with n number of Ex-OR literals are shown in Figure 19.7.

If there exist more input literals of Ex-OR plane, the use of FEG is more flexible. When there are N inputs in the function, Feynman extension gate with N is used. For example, if there are two inputs in the function, Feynman extension gate with two inputs will be used. Algorithm 19.2.2.1 presents the design procedure of the Ex-OR plane of a reversible fault tolerant PAL.

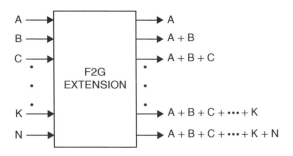

Figure 19.7 Feynman extension gate (FEG).

Algorithm 19.2.2.1 Designing the AND Plane of the Reversible Fault-Tolerant PAL

1: Begin
2: $TG := 0$ [TG = Total number of gates]
3: Calculate the frequency of Min-terms
4: Copy corresponding Min-terms
5: (Suitable Feynman extension gate)
6: Increment the number of gates
7: Apply Ex-OR operation using Feynman extension gates
8: (Suitable Feynman extension gate)
9: Increment the number of gates
10: End

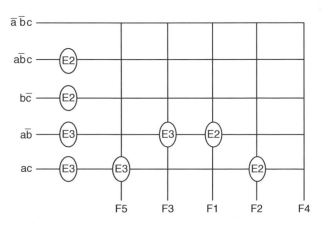

Figure 19.8 The design of the Ex-OR plane of a reversible fault-tolerant PAL using Feynman extension gates.

The design of the minimized Ex-OR plane of a reversible PAL is shown in Figure 19.8. In Figure 19.8, the input literals of the output functions F1, F2, F3, F4, and F5, of the AND plane are Ex-ORed using only two Feynman extension gates with two inputs and two Feynman extension gates with three inputs are used.

19.3 Reversible Fault-Tolerant LUT-Based FPGA

This section presents the design of a reversible fault-tolerant architecture of logic elements of LUT (look-up table) based field programmable gate array (FPGA). The introduced logic elements are master–slave flip-flop, D latch, and multiplexer. A 4×4 and a 6×6 fault-tolerant reversible gates are described for designing efficient reversible fault-tolerant D latch, master–slave flip-flop, and multiplexer.

Usually, FPGA consists of an array of configurable logic block, interconnects, and I/O blocks. FPGA can be configured as needed for each application. Most popular logic blocks are LUT and Plessy logic block. With more inputs, a LUT can implement more logic functions using fewer logic blocks. Thus, it helps to use less routing area. The following subsections illustrate the design layout, algorithms, theoretical properties, and working principles of the logic elements of the LUT-based reversible fault-tolerant FPGA.

19.3.1 Reversible Fault-Tolerant Gates

In this section, two reversible fault-tolerant gates are introduced along with their block diagram, quantum realization, and truth tables.

1. MSH (Mubin-Sworna-Hasan) gate: Figure 19.9 and Figure 19.10 represent the MSH gate (input vector is [A, B, C, D] and output vector is [P, Q, R, S]) with its quantum realization, respectively.
2. MSB (Mubin-Sworna-Babu) gate: Block diagram and quantum realization of the MSB gate (input vector is [A,B,C, D, E, F] and output vector [P, Q, R, S, T, U]) are represented in Figure 19.11 and Figure 19.12, respectively.

19.3.2 Proof of Fault-Tolerance Properties of the MSH and MSB Gates

Table 19.3 demonstrates the unique one-to-one correspondence and parity preservation between inputs (A, B, C, and D) and outputs (P, Q, R, and S) of the MSH gate. For any

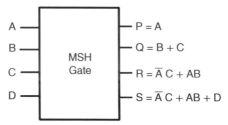

Figure 19.9 The block diagram of reversible fault-tolerant MSH gate.

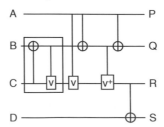

Figure 19.10 The quantum realization of reversible fault-tolerant MSH gate with quantum cost = 6.

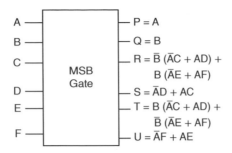

Figure 19.11 The block diagram of reversible fault-tolerant MSB gate.

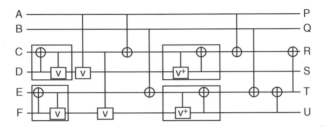

Figure 19.12 The quantum realization of reversible fault-tolerant MSB gate with quantum cost = 12.

input combination, the corresponding output combination shows the same parity (that is $A \oplus B \oplus C \oplus D = P \oplus Q \oplus R \oplus S$) as well as the reversibility of the MSH gate. Similarly, Table 19.4 clarifies the fault tolerance property and the unique mapping between input (A, B, C, D, E, and F) and output (P, Q, R, S, T, and U) of the MSB gate.

19.3.3 Physical Implementation of the Gates

Transistor realizations of the MSH gate and MSB gate are demonstrated in Figure 19.13 and Figure 19.14, which require 10 and 12 transistors, respectively. Now, for MSH gate, suppose,

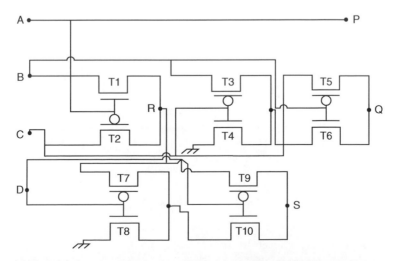

Figure 19.13 The transistor realization of reversible fault-tolerant MSH gate.

Table 19.3 Truth Table of Reversible Fault Tolerant MSH Gate.

Input				Output				Parity
								1 = Odd, 0 = Even
A	B	C	D	P	Q	R	S	
0	0	0	0	0	0	0	0	0
0	0	0	1	0	0	0	1	1
0	0	1	0	0	1	1	1	1
0	0	1	1	0	1	1	0	0
0	1	0	0	0	1	0	0	1
0	1	0	1	0	1	0	1	0
0	1	1	0	0	0	1	1	0
0	1	1	1	0	0	1	0	1
1	0	0	0	1	0	0	0	1
1	0	0	1	1	0	0	1	0
1	0	1	0	1	1	0	0	0
1	0	1	1	1	1	0	1	1
1	1	0	0	1	1	1	1	0
1	1	0	1	1	1	1	0	1
1	1	1	0	1	0	1	1	1
1	1	1	1	1	0	1	0	0

when inputs of $A = 0, B = 1, C = 1$, and $D = 0$ then output P is 1, transistor T_1 is off, and transistor T_2 is on. Thus, C is propagated to R. Since C is 1, the output at R is also 1. Again, both of B and C are 1, transistors T_3 and T_5 are off and T_4 and T_6 are on, and they pass the GND (ground) voltage or 0 to Q. Similarly, $R = 1$ and $D = 0$. Thus, transistors T_7 and T_{10}0 are on and transistors T_8 and T_9 are off. So the output found at S is 1. Finally, the output combination found ($P = 0, Q = 0, R = 1$ and $S = 1$) is obtained from the input combination. ($A = 0, b = 1, C = 1$, and $D = 0$) as tabulated in Table 19.3, which proves the correctness of the physical implementation of the MSH gate. On the other hand, consider the MSB gate for input combination ($A = 0, B = 1, C = 0, D = 1, E = 0$, and $F = 1$), where transistors $T_2, T_4, T_5, T_8, T_{10}0$ and $T_{11}1$ are off and T_1, T_3, T_6, T_7, T_9 and $T_{12}2$ are on. Thus input of D and F are propagated to the output of S and U, respectively, where $S = 1$ and $U = 1$ as both D and F are 1. Again, transistors T_3 and T_9 propagate the input of C and E to the transistors T_6 and $T_{12}2$, respectively, and the outputs obtained at R and T are 0, since both of C and E is 0. Finally, the generated output is ($P = 0, Q = 1, R = 0, S = 1, T = 0$, and $U = 1$) for input combination ($A = 0, B = 1, C = 0, D = 1, E = 0$, and $F = 1$). It can also be shown for any other input combination, which also prove the correctness of the physical implementation of the MSB gate.

19.3.4 Reversible Fault-Tolerant D Latch, Master–Slave Flip-Flop and 4 × 1 Multiplexer

Figure 19.15 presents the architecture of the reversible fault tolerant D latch using one MSH gate to produce the desired output, $S = Clk.feedback \oplus Clk.Data$ with only two garbage

Table 19.4   Truth Table of Reversible Fault-Tolerant MSB Gate

Input						Output						Parity
												1 = Odd, 0 = Even
A	B	C	D	E	F	P	Q	R	S	T	U	
0	0	0	0	0	0	0	0	0	0	0	0	0
0	0	0	0	0	1	0	0	0	0	0	1	1
0	0	0	1	0	1	0	0	0	1	0	1	0
0	0	0	1	1	0	0	0	0	1	1	0	0
0	0	0	1	1	1	0	0	0	1	1	1	1
0	0	1	0	0	0	0	0	1	0	0	0	1
0	0	1	0	0	1	0	0	1	0	0	1	0
0	0	1	0	1	0	0	0	1	0	1	0	0
0	0	1	0	1	1	0	0	1	0	1	1	1
0	0	1	1	0	0	0	0	1	1	0	0	0
0	0	1	1	0	1	0	0	1	1	0	1	1
⋮	⋮	⋮	⋮	⋮	⋮	⋮	⋮	⋮	⋮	⋮	⋮	⋮
1	1	1	0	0	0	1	1	0	1	0	0	1
1	1	1	0	0	1	1	1	1	1	0	0	0
1	1	1	0	1	0	1	1	0	1	0	1	0
1	1	1	0	1	1	1	1	1	1	0	1	1
1	1	1	1	0	0	1	1	0	1	1	0	0
1	1	1	1	0	1	1	1	1	1	1	0	1
1	1	1	1	1	0	1	1	0	1	1	1	1
1	1	1	1	1	1	1	1	1	1	1	1	0

outputs G_1 and G_2 and one constant input. Figure 19.16 illustrates the architecture of the reversible fault-tolerant write-enabled master-slave flip-flop using one FRG, two F2G and two reversible fault-tolerant D latch (described earlier) to produce the desired outputs, Q and Q^+. Figure 19.17 presents the MSB gate, which is used to form a reversible fault-tolerant MUX using Equation 19.3.4.1. The RFT multiplexer is obtained with the least garbage outputs, which is clarified in Property 19.3.4.1.

$$O_{MUX} = \overline{A}\,\overline{B}\,C + \overline{A}\,BE + AB\overline{D} + ABF \qquad (19.3.4.1)$$

Property 19.3.4.1   A $4n - to - n$ RFT multiplexer requires at least $(4n + 1)$ garbage outputs and no constant inputs where n is the number of data bits.

Property 19.3.4.2   The n-input reversible fault-tolerant D latch for an FPGA requires at least n reversible fault-tolerant gates, $2n$ garbage, and $6n$ quantum cost.

Property 19.3.4.3   The n-input RFT write-enabled master-slave flip-flop for an FPGA requires at least $5n$ RFT gates, $7n$ garbage outputs, and $21n$ quantum cost.

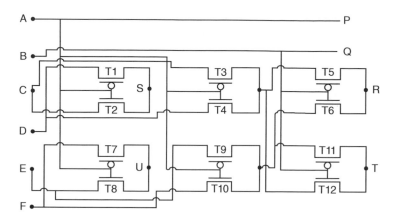

Figure 19.14 The transistor realization of reversible fault-tolerant MSB gate.

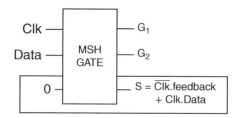

Figure 19.15 The block diagram of reversible fault-tolerant D latch.

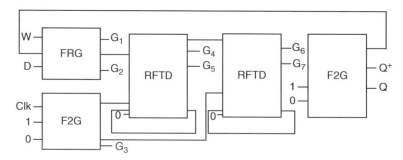

Figure 19.16 The block diagram of reversible fault-tolerant master–slave flip-flop.

19.3.5 Reversible Fault-Tolerant *n*-Input Look-Up Table

An 8×1 RFT multiplexer is introduced using two MSB and one FRG gates and an 16×1 RFT multiplexer is discussed using five MSB gates, which are actually a three-input and four-input LUT as illustrated in Figure 19.18 and Figure 19.19 respectively. In Figure 19.18 and Figure 19.19, 1/0 has been stored using the D latches. Similarly, an *n*-input LUT can be designed with this architecture.

19.3.6 Reversible Fault-Tolerant CLB of FPGA

The components of a 4-LUT CLB (configurable logic block) of an FPGA is designed using reversible fault-tolerant logic. By accumulating the reversible components together,

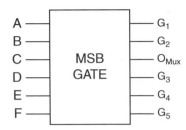

Figure 19.17 The block diagram of a reversible fault–tolerant 4 × 1 multiplexer.

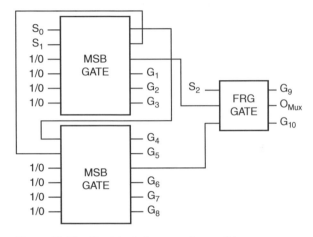

Figure 19.18 The block diagram of reversible fault-tolerant three-input LUT.

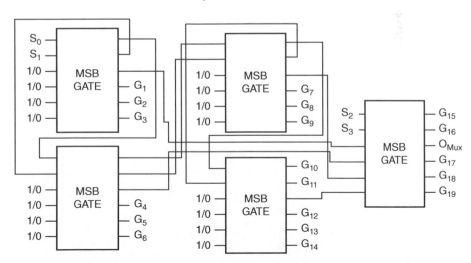

Figure 19.19 The block diagram of reversible fault-tolerant four-input LUT.

Algorithm 19.3.6.1 An *n*-input Reversible Fault-Tolerant (RFT) LUT-Based Configurable Logic Block

---

Input: Input from I/O block

Output: O_{MUX}, which will carry to next logic block

1: Begin
2: **while** I/O block is not empty **do**
3:   $[S_0, S_1, ..., S_n] :=$ selection bits from I/O block;
4:   Apply 4-input RFT LUT where Input:$=\{[S_0, S_1, ..., S_n]\}$; and Output:$=\{Q_0\}$;
5:   Apply RFT Flip Flop where Input:$=\{Q_0$, clock signal$\}$; and output $= \{Q$ or $Q^+\}$;
6:   Apply RFT 2×1 MUX where input $= \{Q_0, Q$ or $Q+\}$; and output $= \{O_{MUX}\}$;
7:   *Selection Bit$_{RFTMUX}$*:$=\{Control_{bit}\}$;
8: **end while**
9: End

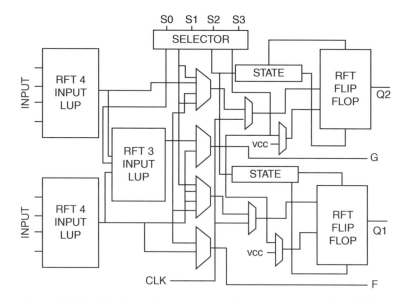

Figure 19.20   The block diagram of reversible fault-tolerant CLB of FPGA.

the CLB of reversible fault-tolerant FPGA is designed as shown in Figure 19.20. Algorithm 19.3.6.1 presents the design procedure of a reversible fault-tolerant CLB of FPGA.

19.4 Summary

This chapter has presented the design methodologies of reversible fault-tolerant programmable logic devices, namely, PLA, PAL, and LUT-based FPGA. A regular structure of a reversible fault-tolerant programmable logic arrays (RFTPLAs) is presented by using different vector orientations of inputs and outputs of Feynman double gate and Fredkin gate. In addition, minimization techniques for both AND and Ex-OR planes are presented. The garbage outputs are used as operational outputs that reduce the number of AND operations in RFTPLAs. RFTPLAs are useful in embedded circuits and other technologies for power consumption and fault tolerance. Later on a reversible fault tolerant programmable array logic (RFTPAL) is described. Since the AND plane is programmable as PLA, the total number of gates, garbage outputs, and quantum cost have been minimized in the design, which remains fault-tolerant as well. The reversible fault-tolerant PAL is useful for designing the digital circuits easily. For example, large functions that have huge numbers of variables can easily be implemented by using RFTPAL. Finally, LUT (look-up table) based configurable logic block (CLB) of FPGA (field programmable gate array) is described. An algorithm for designing the RFT LUT-based logic block for an FPGA has been introduced. In this design, two reversible fault-tolerant gates, namely MSH (Mubin-Sworna-Hasan) and MSB (Mubin-Sworna-Babu) gate, are used.

20

Reversible Fault-Tolerant Arithmetic Logic Unit

In the digital circuit design, the primary factors are low power and a high packing density. In this chapter, a high-performance fault-tolerant arithmetic logic unit (ALU) is showed. ALU is the core of the central processing unit (CPU) of a computer. It performs arithmetic and logical operations. In the following subsections, a 4×4 parity preserving gate named as universal parity preserving gate (UPPG) gate is introduced to optimize the design of the ALU. Here, the *UPPG* gate is used in synthesis of parity-preserving reversible Ex-OR, NAND, NOR, OR, AND, inverter and signal duplication operations. The circuit design of fault-tolerant (FT) reversible ALU is showed dealing with the UPPG gate. This prime design focuses on optimizing the gate counts, constant inputs, garbage outputs, and quantum cost. Some lemmas and algorithm are presented for framing an FT ALU.

20.1 Design of *n*-bit ALU

In this section, the design procedure of FT ALU is described. To design a FT ALU, a 4×4 gate is introduced, namely UPPG gate to optimize the circuit of the ALU. In order to obtain less architecture complexity, group-based cells are designed named as Group-1 PP Cell, Group-2 PP Cell, and Group-3 PP Cell. These group's cells are then extensively used to design 1-bit, 2-bit, and *n*-bit FT ALU.

20.1.1 A 4×4 Parity-Preserving Reversible Gate

In this subsection, a 4×4 parity-preserving reversible gate, namely UPPG, is shown. The block diagram of UPPG is in Figure 20.1. In this gate, one of the inputs (D) is also used as output $(S = D)$ and three other outputs are $P = A \oplus B \oplus D$, $Q = A \oplus CD$, and $R = A \oplus (C + D)$. The truth table given in Table 20.1 shows that every distinct input pattern yields a distinct output pattern, which is bijective and meets the condition of reversibility. The UPPG gate is parity preserving because the hamming weight of input $A \oplus B \oplus C \oplus D$ matches the hamming weight of output $P \oplus Q \oplus R \oplus S$. This parity-preserving gate is used in synthesizing Boolean functions arbitrarily. The utility is shown by Property 20.1.1.1, Property 20.1.1.2, Property 20.1.1.3, and Property 20.1.1.4. Figure 20.1 presents the quantum equivalent circuit of UPPG gate. The quantum cost of UPPG is 7.

Reversible and DNA Computing, First Edition. Hafiz Md. Hasan Babu.

```
A ───┤        ├─── P = A + B + D
B ───┤ UPPG   ├─── Q = A + CD
C ───┤        ├─── R = A + (C + D)
D ───┤        ├─── S = D
```

Figure 20.1 4 × 4 Reversible fault-tolerant UPPG gate.

Table 20.1 Truth Table of 4 × 4 Reversible Fault-Tolerant UPPG Gate

Input				Output			
A	B	C	D	P	Q	R	S
0	0	0	0	0	0	0	0
0	0	0	1	1	0	1	1
0	0	1	0	0	0	1	0
0	0	1	1	1	1	1	1
0	1	0	0	1	0	0	0
0	1	0	1	0	0	1	1
0	1	1	0	1	0	1	0
0	1	1	1	0	1	1	1
1	0	0	0	1	1	1	0
1	0	0	1	0	1	0	1
1	0	1	0	1	1	0	0
1	0	1	1	0	0	0	1
1	1	0	0	0	1	1	0
1	1	0	1	1	1	0	1
1	1	1	0	0	1	0	0
1	1	1	1	1	0	0	1

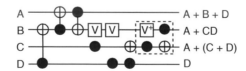

Figure 20.2 Quantum realization of LMH gate.

Property 20.1.1.1 An UPPG gate simultaneously implements Ex-OR, AND, OR, and signal duplication operation.

Proof: If input bit (A) is set as $A = 0$; and three other inputs for the three bits are to be added, the UPPG gate simultaneously implements $P = B \oplus D$, $Q = CD$, $R = C + D$ and $S = D$.

Property 20.1.1.2 A UPPG gate simultaneously implements EX-NOR, NAND, NOR, and Signal duplication operations.

Proof: If input bit (A) is set as $A = 1$, and other three inputs are to be added, the UPPG gate simultaneously implements $P = B \oplus D, Q = (CD)', R = (C + D)'$, and $S = D$.

Property 20.1.1.3 An UPPG gate simultaneously implements NAND, NOR, Inverter, and Signal duplication operation.

Proof: If input bit (A) is set $A = 1$ and another input bit (B) as $B = 0$; and other two inputs for the two bits are to be added, the UPPG gate simultaneously implements $P = D, Q = (CD)', R = (C + D)'$, and $S = D$.

Property 20.1.1.4 The UPPG gate is a fault-tolerant reversible gate.

Proof: The UPPG gate is 4×4 type reversible gate. The UPPG can be represented as: $I_v = (A, B, C, D), O_v = (P = (A \oplus B \oplus D), Q = (A \oplus CD), R = A \oplus (C + D), S = D)$.   It   is known that the hamming weight of input and the hamming weight of output have to be the same in any parity preserving or fault tolerant reversible gate. The hamming weight of input is $A \oplus B \oplus C \oplus D)$ and output of UPPG gate is $(A \oplus B \oplus D)(A \oplus CD) \oplus A \oplus (C + D) \oplus D = A \oplus B \oplus D \oplus CD \oplus (C + D) \oplus D = A \oplus B \oplus CD \oplus (C + D) \oplus D = A \oplus B \oplus CD \oplus (C + D)D + (C + D)D = A \oplus B \oplus CD \oplus CD = A \oplus B \oplus C \oplus D$. Thus, the hamming weight of outputs is equal to the hamming weight of inputs. Hence UPPG gate is a fault-tolerant reversible gate.

20.1.2   1-Bit ALU

In this subsection, a design of 1-bit ALU consisting of the four cells (Group-1 PP cell, Group-2 PP cell, Group-2 PP (Cell)$_{--1}$ and Group-3 PP cell) is described. Each cell is designed by using parity-preserving reversible gates.

20.1.2.1   Group-1 PP Cell
In this subsection, Group-1 PP cell is implemented with the minimum gate counts, constant inputs, garbage outputs and quantum cost which consists of two sub-cell named as an U_3N_F PP cell and U_3F PP cell. These cells are collectively named as Group-1 PP cell. The U_3N_F PP design cell consists of five gates with three types (three NFT gates, one F2G gate, and one UPPG gate) and U_3F PP cell consists of four gate with three types (two F2G gate, one FRG gate, and one UPPG gate), which are depicted in Figure 20.3. The quantum cost of F2G, FRG, UPPG, and NFT are 2, 5, 7, and 5, respectively. Thus, the quantum cost (QC) of this Group-1 PP cell is 40, because

$$QC(Group - 1PPCell) = 3QC(F2G) + 1QC(FRG) + 2QC(UPPG) + 3QC(NFT)$$

$$= 3 \times 2 + 1 \times 5 + 2 \times 7 + 3 \times 5$$

$$= 40$$

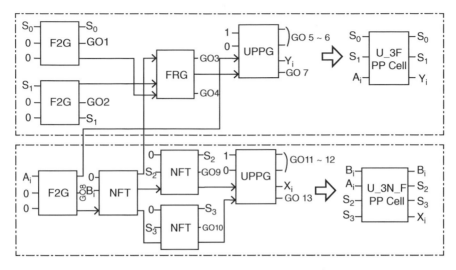

Figure 20.3 Circuit structure of Group-1 PP cell.

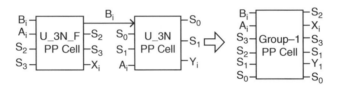

Figure 20.4 Compressed block diagram of Group-1 PP cell.

The Group-1 PP cell takes two input operands (A_i, B_i) along with the four selector bits $(S_0$ to $S_3)$ as together they work to generate Group-1 PP cell output (X_i, Y_i) as implement by using Equations 20.1.2.1.1 and 20.1.2.1.2. The Group-1 cell also produces $(S_0$ to $S_3)$ as output for fan-outs to succeeding cells means a 1-bit ALU can easily design an n-bit ALU. Figure 20.3 and Figure 20.4 depict the circuit structure and compressed block diagram of Group-1 PP cell. The Group-1 PP cell is executed the input operands (A_i, B_i) along with the selector bits $(S_0$ to $S_3)$ by following a logical combination as shown in Table 20.2.

$$X_i = (S_2 A_i B_i' + S_3 A_i B_i)' \tag{20.1.2.1.1}$$

$$Y_i = (A_i + (S_1 B_i' + S_0 B_i))' \tag{20.1.2.1.2}$$

20.1.2.2 Group-2 PP Cell

The output of Group-1 PP cell $(X_i, Y_i, S_0, S_1, S_2, S_3)$ and two more inputs (C_{i-1}, M) are applied to Group-2 PP cell, where C_{i-1} is the input carry of $(i-1)^{th}$ stage and M is mode control bits for the ith stage of the ALU. The Group-2 PP cell generates four outputs $(X_i, C_{i-1}, Y_i, C_i, M)$. The block diagram of Group-2 cell is depicted in Figure 20.5. Group-2 PP cell synthesizes Equations 20.1.2.2.1 and 20.1.2.2.2. The equation for carry-out (C_i) consists of the mode control bit (M), carry-in (C_{in}), and Group-1 PP cell's outputs (X_i, Y_i). In the starting phase the carry-in (C_{in}) and the mode bit (M) are processed to compute the output carry (C_0) using

Table 20.2   Truth Table of Group-1 PP Cell

S_0	S_1	Associated Output Y_i	S_2	S_3	Associated Output X_i
0	0	A_i'	0	0	1
0	1	$(A_i + B_i')'$	0	1	$(A_i B_i)'$
1	0	$(A_i + B_i)'$	1	0	$(A_i B_i')'$
1	1	0	1	1	A_i'

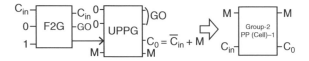

Figure 20.5   Block diagram of Group-2 PP (Cell)$_{-1}$.

Equation (20.1.2.2.3). The computing of C_0 is performed by Group-2 (Cell)$_{--1}$ as shown in Figure 20.5. If set $M = 0$ and $C_{in} = 1$, it means arithmetic operations, but when and $M = 1$ it means logical operations. The list of arithmetic and logical operations executed by the design of the ALU is depicted in Table 20.3.

$$C_i = M + (X_0 C_{in} + Y_0)', \; Ture \; for \; i = 1 \qquad (20.1.2.2.1)$$

$$C_i = M + (X_i C_{i-1} + Y_i)', \; Ture \; for \; i >= 2 \qquad (20.1.2.2.2)$$

$$C_0 = C_{in}' + M \qquad (20.1.2.2.3)$$

In the design of Group-2 PP (Cell)$_{-1}$, one F2G and one UPPG gate are used, which is depicted in Figure 20.5. The quantum cost of Group-2 PP (Cell)$_{-1}$ is 9, because

$$QC \; (Group\text{-}2 \; PP(Cell)_{-1}) = 1QC(F2G) + 1QC(UPPG)$$

$$= 1 \times 2 + 1 \times 7$$

$$= 9$$

In the circuit structure, a Group-2 PP cell consists of three gates of two types ($1 \times FRG + 2 \times UPPG$), as shown in Figure 20.6. The quantum costs of FRG and UPPG gates are 5 and 7, respectively. Thus, the quantum cost of Group-2 PP cell is 19, because

$$QC \; (Group\text{-}2 \; PP \; cell) = 1QC \; (FRG) + 2QC \; (UPPG)$$

$$= 1 \times 5 + 2 \times 7$$

$$= 19$$

20.1.2.3   Group-3 PP Cell

The Group-3 PP cell uses the input of the previous cells. It generates the output function $Z_i = X_i \oplus Y_i \oplus C_i$. The design structure of a Group-3 PP cell is depicted in Figure 20.7. A Group-3 PP cell uses one gate of one type ($1 \times$ UPPG gate). The quantum cost of a UPPG

Table 20.3 Different Functions of Fault-Tolerant ALU

Opcode results	ALU Output		ALU Input			
	Logical function $(M = 1)$	Arithmetic functions $(M = 0, C_{in} = 1)$	S_3	S_2	S_1	S_0
Inverting A	$\overline{A}$	A	0	0	0	0
NOR operation	$(\overline{A}\ \overline{B})$	$A + B$	0	0	0	1
A inhibits B operation	$(\overline{A + \overline{B}})$	$A + \overline{B}$	0	0	1	0
Null operation	0	-1	0	0	1	1
NAND operation	$(\overline{A} + \overline{B})$	$A + A\overline{B}$	0	1	0	0
Inverting B	$\overline{B}$	$A + B + A\overline{B}$	0	1	0	1
XOR Operation	$(\overline{A \oplus B})$	$A - B - 1$	0	1	1	0
B inhibits A operation	$(\overline{A} + B)$	$A\overline{B} - 1$	0	1	1	1
A implicates B operation	$(\overline{A\overline{B}})$	$A + B$	1	0	0	0
XNOR operation	$(\overline{A} \oplus B)$	A+B	1	0	0	1
Signal duplication of B	B	$A + \overline{B} + AB$	1	0	1	0
AND	(AB)	AB-1	1	0	1	1
Unity Operation	1	A+A	1	1	0	0
B implicates A operation	$(\overline{AB})$	$A + B + A$	1	1	0	1
OR operation	(A+B)	$(A + \overline{B} + A)$	1	1	1	0
Signal duplication of A	A	A-1	1	1	1	1

Figure 20.6 Block diagram of Group-2 PP cell.

gate is 7. Therefore, the quantum cost of Group-3 PP cell is:

$$QC \text{ (Group-3 PP cell)} = 1QC \text{ (UPPG)}$$
$$= 1 \times 7$$
$$= 7$$

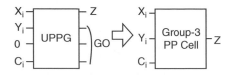

Figure 20.7 Block diagram of Group-3 PP cell.

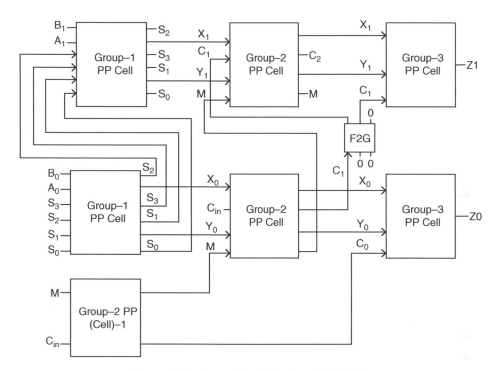

Figure 20.8 Reversible fault-tolerant 2-bit ALU.

20.1.2.4 *n*-bit ALU

In this section, a 2-bit and *n*-bit fault tolerant ALU are depicted in Figure 20.8 and Figure 20.9, which consists of four cells (*n*× Group-1 PP cell, *n*× Group-2 PP cell, 1× Group-2 PP (cell)$_{--1}$ and *n*× Group-3 PP cell). The design procedure of *n*-bit fault tolerant ALU is given in Algorithm 20.1.2.1.

One of the important factors of the compact design of an ALU is its total logical calculation (TLC). Hardware complexity is premeditated by TLC. A compact ALU is constructed with the optimum number of reversible gates (F2G, NFT, FRG and UPPG gates), garbage outputs and quantum cost obtained by Property 20.1.2.1. In Property 20.1.2.2 and Property 20.1.2.3. Let α be te two-input EXOR gate calculation complexity, β be the two-input AND gate calculation complexity and δ be the NOT gate calculation complexity. Let P, Q and R be the respective expressions of the outputs of the gate which are treated as P Expression, Q Expression and R Expression, respectively. Now, the calculation

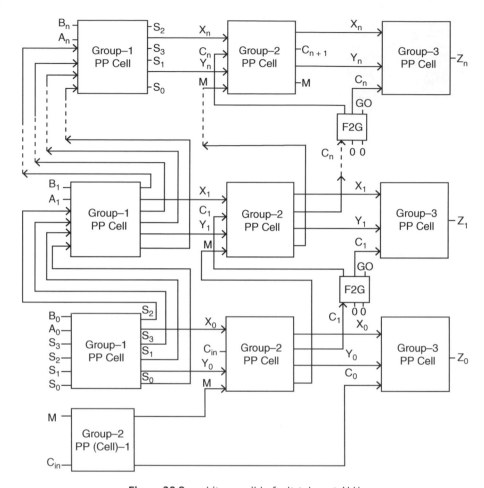

Figure 20.9 *n*-bit reversible fault-tolerant ALU.

complexities of the individual gates F2G, NFT, FRG, and UPPG are as follows:

$T(\text{F2G}) = 2\alpha$

$T_{(\text{NFT})} = (1\alpha)\ (\text{for } P \text{ Expression}) + (1\alpha + 2\beta + 2\delta)\ (\text{for } Q \text{ Expression})$

$+ (1\alpha + 2\beta + 1\delta)\ (\text{for } R \text{ Expression})$

$= 3\alpha + 4\beta + 3\delta$

$T_{(\text{FRG})} = (1\alpha + 2\beta + 1\delta)\ (\text{for } Q \text{ Expression}) + (1\alpha + 2\beta + 1\delta)\ (\text{for } R \text{ Expression})$

$= 2\alpha + 4\beta + 2\delta$

$T_{(\text{UPPG})} = 2\alpha\ (\text{for } P \text{ Expression}) + (1\alpha + 1\beta)\ (\text{for } Q \text{ Expression}) + (1\alpha)\ (\text{for } R \text{ Expression})$

$= 4\alpha + 1\beta$

In the design of ALU, F2G is used for making fan-out. So, the calculation complexities of F2F, NFT, FRG, and UPPG gates for the structure of ALU are obtained as mentioned below.

Algorithm 20.1.2.1 Algorithm for Designing an n–bit Reversible Fault-Tolerant n-bit ALU

Input: n: The number of bits used in the operands A and B.
Select Group-1 PP Cell and take an operand of inputs $(A_n, A_{n-1}, \ldots \ldots A_2, A_1)$, (B_n, B_{n-1}, B_2, B_1), and four selector bits (S_0, S_1, S_2, S_3).
Output: $(Z_n, Z_{n-2} \ldots \ldots, Z_1)$: The outputs depends on the various combinations of selector bits.

1: Begin
2: $TDOT := 0 =$ Total number of DOT
3: Sort output functions according to their number of product
4: **for** each $A_i \in A$ and $B_i \in B$ **do**
5:   Selection of input operands (A_i, B_i) and selector bits (S_0, S_1, S_2, S_3) from Group-1 PP Cell and generate outputs (X_i, S_i, Y_i), where $X_i = (S_2 A_i B_i' + S_3 A_i B_i)'$ and $Y_i = (A_i + (S_1 B_i' + S_0 B_i))'$. It associates to Group-2 PP Cell.
6:   Select Group-2 PP (Cell)$_{-1}$ and acquire inputs (M, C_{in}) and generate outputs M and $C_O = C_{in}' + M$. Where the first output M is applied to the Group-2 PP Cell and C_0 to Group-3 PP Cell.
7:   Select Group-3 PP Cell and acquire input from previous Cells (Group-1 PP Cell, Group-2 PP Cell, Group-2 PP (Cell)$_{-1}$). It finally generate desired output $Z_i = X_i \oplus Y_i \oplus C_i$
8:   Record each Z_i.
9: **end for**
10: End

TLC(T) for Group-1 PP cell is $3 \times (2\alpha)$ (for F2G) $+ 3 \times (3\alpha + 4\beta + 3\delta)$ (for NFT)

$+ 1 \times (2\alpha + 4\beta + 2\delta)$ (for FRG) $+ 2 \times (4\alpha + 1\beta)$ (for UPPG) (as shown in Figure 20.4)

$$= 25\alpha + 18\beta + 11\delta$$

TLC(T) for Group-2 PP cell is $1 \times (2\alpha + 4\beta + 2\delta)$ (for FRG) $+ 2 \times (4\alpha + 1\beta)$ (for UPPG)

$$= 10\alpha + 6\beta + 2\delta$$

TLC(T) for Group-2 PP(cell)_1 is $1 \times (2\alpha)$ (for F2G gate) $+1 \times (4\alpha + 1\beta)$ (for UPPG)

$$= 6\alpha + 2\beta$$

TLC(T) for Group-3 PP cell is $1 \times (4\alpha + 1\beta)$ (for UPPG) $= 4\alpha + 1\beta$

Therefore, TLC $(T)_{\text{2-bit ALU}}$ for the fault-tolerant 2-bit ALU the TLC is:

$$\text{TLC}(T)_{\text{2-bit ALU}} = 2\times \text{(Group-1 PP cell)} + 2\times \text{(Group-2 PP cell)} + 1\times$$

$$\text{(Group-2 PP (cell)}_{-1}) + 2\times \text{(Group-3 PP cell)} + 1\times \text{F2G} = 2 \times (25\alpha + 18\beta + 11\delta +$$
$$2 \times (10\alpha + 6\beta + 2\delta) + 1 \times (6\alpha + 2\beta) + 2 \times (4\alpha + 1\beta) + 2\alpha = 86\alpha + 52\beta + 26\delta$$
(as shown in Figure 20.8)

where

α two input XOR gate calculation
β two input AND gate calculation
δ NOT gate calculation

Property 20.1.2.1 A reversible n-bit fault-tolerant ALU can be designed with $(14n + 2)$ reversible gates and $(21n + 4)$ constant inputs.

Proof: A Group-1 cell has nine gates (3× F2G, 1× FRG, 2× UPPG, and 3× NFT), Group-2 includes three gates (1× FRG and 2× UPPG), Group-2 (Cell)$_{-1}$ includes 2 gates (1× UPPG and 1× F2G) and Group-3 includes one gate (1× UPPG). A 1-bit ALU is constructed with (1× Group-1 cell, 1× Group-2 cell, 1× Group-3 cell, 1× Group-2 (Cell)$_{-1}$ and 1× F2G). Hence, the total number of gates for designing a 1, 2, 3, and n-bit ALU is

$$NOG\ (1\text{-bit}) = NOG\ (\text{Group-1 PP cell}) + NOG\ (\text{Group-2 PP cell}) +$$
$$NOG\ (\text{Group-3 PP cell}) + NOG\ (\text{Group-2 PP(cell)}_{-1} + NOG\ (\text{F2G gate})$$
$$NOG\ (1\text{-bit}) = (9 + 3 + 1) + 2 + 1 = 16 = (13 \times 1) + 2 + (1).$$

Hence, the statement true for the base case $n = 1$.

$$NOG\ (2\text{-bit}) = 2 \times (9 + 3 + 1) + 2 + 2 = 30 = (13 \times 2) + 2 + (2).$$
$$NOG\ (3\text{-bit}) = 3 \times (9 + 3 + 1) + 2 + 3 = 44 = (13 \times 3) + 2 + (3).$$

Assume that the statement is true for $n = m$. Thus, an m-bit fault-tolerant ALU can be designed with $(14m + 2)$ gates.

Again, the Group-1 cell produces 13 constant inputs, Group-2 cell produces 5 constant inputs, Group-2 (Cell)$_{-1}$ produces 4 constant inputs, and Group-3 produces only 1 constant input. Let CI be the total constant inputs. Now, the total constant inputs produce by a 1, 2, and 3 bit ALU is as follows.

$$CI\ (1\text{-bit}) = CI\ (\text{Group-1 PP cell}) + CI\ (\text{Group-2 PP cell}) +$$
$$CI\ (\text{Group-3 PP cell}) + CI\ (\text{Group-2 PP (Cell)}_{-1} + CI\ (\text{F2G gate})$$
$$CI\ (1\text{-bit}) = (13 + 5 + 1) + 4 + 2 = 25 = 19 \times 1 + 4 + 2$$

Hence, the statement true for the base case $n = 1$.

$$CI\ (2\text{-bit}) = 2 \times (13 + 5 + 1) + 4 + 4 = 46 = 19 \times 2 + 4 + 4.$$
$$CI\ (3\text{-bit}) = 3 \times (11 + 5 + 1) + 4 + 6 = 67 = 19 \times 3 + 4 + 6$$

Assume that the statement is true for $n = m$. Hence, an m-bit ALU produces at least $19m + 4 + 2m = (21m + 4)$ constant inputs.

Property 20.1.2.2 A reversible n-bit fault-tolerant ALU can be utilized with $(22n + 3)$ garbage outputs.

Proof: The n-bit ALU has four Group cells such as $n\times$ Group-1 PP cell, 1× Group-2 PP(Cell)$_{-1}n\times$ Group-2 PP cell and $n\times$ Group-3 PP cell. The Group-1 cell produces 13 garbage outputs, Group-2 (Cell)$_{-1}$ produces 3 garbage outputs, Group-2 cell produces 5 garbage outputs, Group-3 cell produces 3 garbage outputs, and the F2G produces 1 garbage output. Let GO be the total garbage outputs. Now, the total garbage outputs can be calculated as follows:

$$GO\ (1\text{-bit}) = GO\ (\text{Group-1 PP cell}) + GO\ (\text{Group-2 PP cell}) + GO\ (\text{Group-3 PP cell})$$
$$GO\ (\text{Group-2 PP (cell)}_{-1} + GO\ (\text{F2G})$$
$$GO(1\text{-bit}) = (13 + 5 + 3) + 3 + 1 = 25 = 21 \times 1 + 3 + 1$$

Hence, the statement true for the base case $n = 1$.

GO (2-bit) $= 2 \times (13 + 5 + 3) + 3 + 2 = 47 = (21 \times 2) + 3 + 2$
GO(3-bit) $= 3 \times (13 + 5 + 3) + 3 + 3 = 69 = (21 \times 3) + 3 + 3$

Hence, the statement is true for $n = m$. An m-bit fault-tolerant ALU uses $(22m + 3)$ garbage outputs.

Property 20.1.2.3 A reversible n-bit fault-tolerant ALU can be realized with $(66n + 9)$ quantum cost.

Proof: An n-bit ALU is constructed with four groups of cells, such as $n \times$ Group-1 PP cell, $1 \times$ Group-2 PP(Cell)$_{-1}$, $n \times$ Group-2 PP Cell, and $n \times$ Group-3 PP cell. The quantum cost of F2G, FRG, NFT, and UPPG gates are 2, 5, 5, and 8, respectively. The Group-1 cell produces 40 quantum cost, Group-2 (Cell)$_{-1}$ produces 9 quantum cost, Group-2 produces 19 quantum cost, Group-3 produces 7 quantum cost, and the F2G produces 2 quantum cost. Let QC be the quantum cost of a reversible circuit. Hence, the total quantum cost is calculated as follows.

QC (1-bit) $=$ QC (Group-1 PP cell) $+$ QC (Group-2 PP cell) $+$ QC (Group-3 PP cell) $+$ QCQC (Group-2 PP (cell)$_{-1}$ $+$ QC (F2G gate)

$$QC(1\text{-bit}) = (40 + 19 + 7) + 9 + 2 = 83 = 66 \times 1 + 9 + 2$$

Hence, the statement true for the base case $n = 1$.

$$QC(2\text{-bit}) = 2 \times (40 + 19 + 7) + 9 + 4 = 145 = (66 \times 2) + 9 + 4$$
$$QC (3\text{-bit}) = 3 \times (40 + 19 + 7) + 9 + 6 = 213 = (66 \times 3) + 9 + 6$$

Assume that the statement true for $n = m$. Hence, an m-bit fault-tolerant ALU uses $(66m + 9 + 2m) = 68m + 9$ quantum cost.

20.2 Summary

This chapter focuses on the design of a reversible fault-tolerant ALU and introduces a parity preserving gate. It has been shown that the reversible gate namely, UPPG, is used to optimize the ALU circuits. An algorithm and some properties are presented in designing the ALU. The ALU generates a number of arithmetic and logical functions with less architectural complexity in terms of gate count, garbage outputs, and quantum cost.

21

Online Testable Reversible Circuit Using NAND Blocks

In recent years, reversible computing system design has attracted a lot of attention. Reversible computing is established on two concepts: logical reversibility and physical reversibility. A computational operation is said to be logically reversible if the logical state of the computational device before the operation of the device can be determined by its state after the process i.e., the input of the system can be regained from the output. Irreversible erasure of a bit in a system leads to generation of energy in the form of heat. An operation is said to be physically reversible if it transforms no energy to heat and produces no entropy. Landauer has shown that for every bit of information lost in logic computations that are not reversible, $kTln2$ joules of heat energy is produced, where k is Boltzmann's constant and T is the absolute temperature at which computation is achieved. The amount of energy dissipation in a system rises in direct proportion to the number of bits that are erased during computation. Bennett showed that $kTln2$ energy dissipation would not occur, if a computation was carried out in a reversible way. Reversible computation in a system can be accomplished if the system is composed of reversible gates.

21.1 Testable Reversible Gates

At present, the available reversible gates can be used to implement random logic functions. However, the testing of reversible circuits as far as researchers are aware have not yet been addressed. The testing of reversible logic gates can be a problem because the levels of logic can be meaningfully higher than in standard logic circuits. This chapter introduces three reversible logic gates: R1 and R2 that can be used in pairs to design testable reversible logic circuits, and R3 that is used to construct two pair two rail checkers. The first gate R1 is used for realizing arbitrary functions while the second gate R2 is employed to integrate online testability features into the circuit. Gates R1 and R2 are shown in Figure 21.1 and Figure 21.2, respectively, the corresponding truth tables of the gates are shown in the Table 21.1 and Table 21.2, respectively. It can be verified from the truth tables that the input pattern conforming to a particular output pattern can be uniquely determined.

The reversible gate R3 is shown in Figure 21.3 and its truth table is given in Table 21.3. Gate R3 differs from gates R1 and R2 in gate width. The gate width of R1 and R2 is 4, while the gate width of R3 is 3. In other words, R3 is a 3×3 reversible gate, while R1 and R2

Reversible and DNA Computing, First Edition. Hafiz Md. Hasan Babu.
© 2021 John Wiley & Sons Ltd. Published 2021 by John Wiley & Sons Ltd.

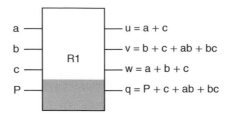

a —
b —
  R1
c —
P —

— u = a + c
— v = b + c + ab + bc
— w = a + b + c
— q = P + c + ab + bc

Figure 21.1   Reversible logic gate R1.

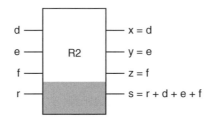

d —
e —   R2
f —
r —

— x = d
— y = e
— z = f
— s = r + d + e + f

Figure 21.2   Reversible logic gate R2.

Table 21.1   Truth Table of Reversible Logic Gate R1

p	a	b	c	u	v	w	q
0	0	0	0	0	0	0	0
0	0	0	1	1	1	1	1
0	0	1	0	0	1	1	0
0	0	1	1	1	1	0	0
0	1	0	0	1	0	1	0
0	1	0	1	0	1	0	1
0	1	1	0	1	0	0	1
0	1	1	1	0	0	1	1
1	0	0	0	0	0	0	1
1	0	0	1	1	1	1	0
1	0	1	0	0	1	1	1
1	0	1	1	1	1	0	1
1	1	0	0	1	0	1	1
1	1	0	1	0	1	0	0
1	1	1	0	1	0	0	0
1	1	1	1	0	0	1	0

are 4×4 reversible gates. The testability feature is not incorporated in gate R3, as it will be used as the basic block for implementing the two-pair two-rail checkers. Gate R1 can implement all Boolean functions. During normal operation input p of R1 gate is set to 0. The OR and the EX-OR functions can be simultaneously implemented on R1, which are shown in Figure 21.4.

Table 21.2 Truth Table of Reversible Logic Gate R2

r	d	e	f	x	y	z	s
0	0	0	0	0	0	0	0
0	0	0	1	0	0	1	1
0	0	1	0	0	1	0	1
0	0	1	1	0	1	1	0
0	1	0	0	1	0	0	1
0	1	0	1	1	0	1	0
0	1	1	0	1	1	0	0
0	1	1	1	1	1	1	1
1	0	0	0	0	0	0	1
1	0	0	1	0	0	1	0
1	0	1	0	0	1	0	0
1	0	1	1	0	1	1	1
1	1	0	0	1	0	0	0
1	1	0	1	1	0	1	1
1	1	1	0	1	1	0	1
1	1	1	1	1	1	1	0

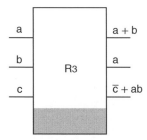

Figure 21.3 Reversible logic gate R3.

Table 21.3 Truth Table of Reversible Logic Gate R3

a	b	c	x	y	z
0	0	0	0	0	1
0	0	1	0	0	0
0	1	0	1	0	1
0	1	1	1	0	0
1	0	0	1	1	1
1	0	1	1	1	0
1	1	0	0	1	0
1	1	1	0	1	1

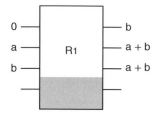

Figure 21.4 Realizations of OR and Ex-OR Gates Using R1 Gate.

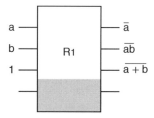

Figure 21.5 Realizations of Ex-NOR and NAND gates using R1 gate.

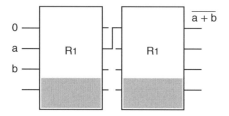

Figure 21.6 Realization of NOR gate using R1 gate.

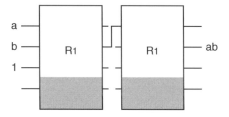

Figure 21.7 Realization of AND gate using R1 gate.

The Ex-NOR function and the NAND functions are acquired by setting input c to 1, which is shown in Figure 21.5. The NOR function can be obtained by cascading two R1 gates, which is implemented in Figure 21.6. An AND gate also requires cascading of two gates, which is shown in Figure 21.7. R1 can transfer a signal at input a to output u by setting input c to 0. The gate R3 is used to implement the two-pair two-rail checkers. Gate R2 is used to transfer the input values at d, e, f to outputs x, y, z. It also produces the parity of the input pattern at output s. The output s of the gate is the complement of the output q when all other gates remain unchanged and p and r are set to 0 and 1, respectively. For example,

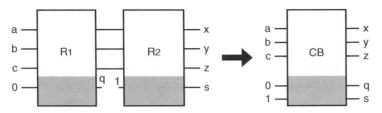

Figure 21.8 The testable logic block using R1 and R2 gates.

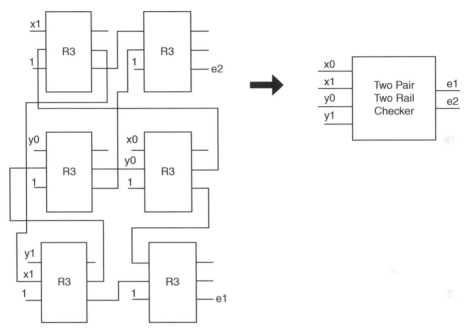

Figure 21.9 Two-pair rail checker.

if input $d\ e\ f\ r = 1000$ is changed to 1001, the output of the gate will change from 1001 to 1000.

A testable logic block can be made by cascading R1 and R2, as shown in Figure 21.8. In this conformation, gate R2 is used to check online whether there is a fault in R1 or in itself. If R1 is fault free, its parity output q and the parity output s of R2 should be complementary; otherwise, the existence of a fault is presumed. Thus, the existence of a fault in the logic block can be identified during the normal operation.

21.2 Two-Pair Rail Checker

A two-pair two-rail checker is constructed using gate R3, which is shown in Figure 21.9. It is composed of eight R3 gates. The error-checking functions of the Two-pair rail checker

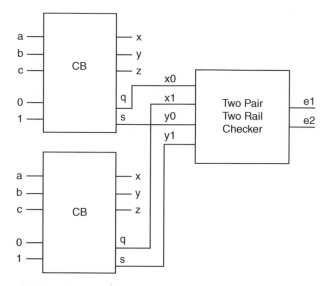

Figure 21.10 Testable block embedded with two-pair rail checker.

are as follows:

$$e_1 = x_0 y_1 + y_0 x_1$$
$$e_2 = x_0 x_1 + y_0 y_1$$

where, x_0/y_0 and x_1/y_1 are complementary.

The fault-free checker will produce complementary output at e_1 and e_2 if the inputs are complementary, otherwise they will be indistinguishable. The block diagram of the testable block along with the two-pair rail checker is shown in Figure 21.10. The outputs q and s of one testable block forms the input x_0 and y_0 for the two-pair rail checker and the outputs of another testable block forms the input x_1 and y_1. Thus, the testable blocks constructed using the gates R1 and R2 can be tested using the two-pair rail checker.

21.3 Synthesis of Reversible Logic Circuits

The implementation of Boolean functions is done using testable NAND blocks. A sum-of-products (SOP) expression can be transferred into reversible logic by converting the SOP expression to NAND-NAND form. Each testable NAND block is realized by cascading the gates R1 and R2. Figure 21.11 shows the implementation of $(ab)'$. If a variable appears more than once in an expression, then a signal duplication gate will be required. Note that, fan-outs are not allowed in reversible logic design.

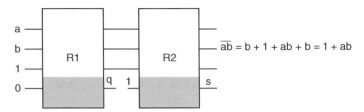

Figure 21.11   Realization of NAND gate using R1 and R2 gates.

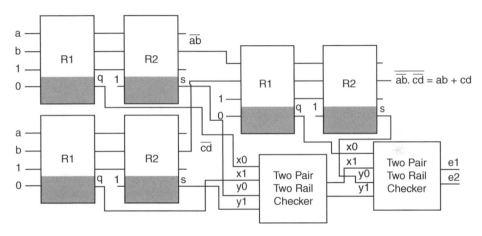

Figure 21.12   Reversible NAND block implementation for the function $ab + cd$.

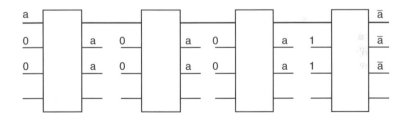

Figure 21.13   Implementation of signal duplication.

The NAND block-based implementation of function $F = ab + cd$ is given below:

1) $ab = ((ab)')' = (1 \oplus ab)', cd = (cd)')' = (1 \oplus ab)'$
2) $ab + cd = ((ab).(cd))' = ((1 \oplus ab).(1 \oplus cd))'$

The implementation of the function is shown in Figure 21.12. The number of signal duplication blocks used instead of fan-outs depends on the number of times the variable appears in the function. For example, the number of blocks required to implement fan-out of variable that appears six times as "a", and three times as its complement "a'" is shown in Figure 21.13.

21.4   Summary

In this chapter, three reversible logic gates have been introduced. To make it easily testable, two gates are cascaded together as a pair. The pair of gates is used to implement any arbitrary function. Thus, the resulting circuits can be tested easily online. In addition, a fault in any of these gates that produces a single-bit output error can be identified by the built-in parity. The two-pair two-rail checker added to the design makes the circuit online testable.

22

Reversible Online Testable Circuits

Testing is an important part of reversible logic, as it is necessary that the resulting reversible circuit must be fault free. A few studies have been done on testable reversible circuits. In 2004 the concurrent testing or online testing for reversible logic circuits was started. In reversible logic, testing can be one of the major problems, as the levels of logic are significantly higher than the standard logic. There exists many testing approaches. In this chapter, a technique to construct online testable circuit is shown. In this manner a universal fault gate is also constructed. Some essential methodologies are also presented to understand the features of the online testable circuit.

22.1 Online Testability

There are two types of testing – one is online (concurrent testing) and another is offline (nonconcurrent testing). While testing a circuit, both approaches can be combined. Online testing is working while the system is performing its normal operation, allowing faults to be detected in real time. Offline testing requires the system or a part of the system to be taken out of operation to perform testing, and generally involves the application of the set of test vectors that will detect all possible faults under a given fault model (a complete test set). In the following subsections, some previous approaches for constructing online testable reversible circuits are described. Some design issues and performance metrics with benefits and limitations of these approaches are also analyzed.

22.1.1 Online Testable Approach Using R1, R2, and R Gates

In this section, an approach of online testability for constructing reversible circuit has been discussed. This approach considers three reversible logic gates R1, R2, and R as shown in Figure 22.1, Figure 22.2, and Figure 22.3. R1 and R2 can be used to construct online testable reversible logic circuits. NAND, OR, EX-OR, and EX-NOR can be constructed using R1 by placing different values on inputs. The last output (q) of R1 is used as a parity output. The first three outputs of a R2 gate are the same as inputs and the last output is parity outputs. A testable block (TB) is constructed by connecting the first three outputs of R1 to the first three inputs of R2 as shown in Figure 22.4 and Figure 22.5. TB generates two parity outputs. Thus a two-pair two-rail checker circuit is required to test the parities of TBs.

Reversible and DNA Computing, First Edition. Hafiz Md. Hasan Babu.
© 2021 John Wiley & Sons Ltd. Published 2021 by John Wiley & Sons Ltd.

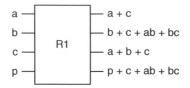

Figure 22.1 Block diagram of R1 gate.

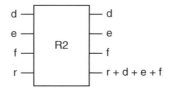

Figure 22.2 Block diagram of R2 gate.

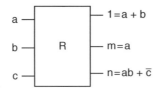

Figure 22.3 Block diagram of R gate.

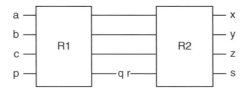

Figure 22.4 Construction of a testable block (TB).

22.1.2 Online Testable Approach Using Testable Reversible Cells (TRCs)

In this section, an approach is discussed to construct an online testable circuit from a reversible circuit. This approach consists of two steps. In the first step, it is needed to transform each $n \times n$ gate G used in the circuit into an $(n + 1) \times (n + 1)$ deduced reversible gate of G, denoted by DRG (G) without modifying its original functionality. An extra input P_{iG} and the corresponding output P_{oG} are added to construct an $(n + 1) \times (n + 1)$ reversible DRG (G). All the outputs of DRG (G) are EXORed with P_{oG} and it is referred as parity output. In the second step, it is needed to construct a testable reversible circuit of G denoted by TRC (G). For TRC (G), X is considered as an $n \times n$ reversible gate, which has the same input and output vectors and the inputs of X pass through to the outputs without any change. Then $n \times n$ reversible X gate is converted to an $(n + 1) \times (n + 1)$ reversible gate, namely DRG (X). P_{iX} is its extra input, and all the outputs of DRG (X) are EXORed with P_{ox} and are referred as parity output. Now DRG (G) and DRG (X) are cascaded by

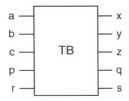

Figure 22.5 Block diagram of a testable block (TB).

connecting the first n outputs of DRG (G) to the first n inputs of DRG (X) in order to form an $(n+1) \times (n+1)$ TRC (G) with two parity outputs P_{oG} and P_{oX}. For any reversible circuit, each gate is to be replaced by TRC (G) to construct a testable reversible circuit. TRC (G) needs a testable cell (TC) to form a testable circuit. Each TRC generates two parity outputs to check whether the circuit is fault free. If the number of TRC (G) is m, $2m+1$ number of TC are required.

22.1.3 Online Testable Circuit Using Online Testable Gate

In this section, a novel 4×4 reversible gate, namely OTG (online testable gate) is used for online testable reversible logic circuits. This procedure is similar to the approach described in subsection 22.1.1. A parity output q is in OTG. OTG is cascaded with a R2 gate to construct an online testable block, where the testable block consists of two parity outputs (s and q) that are used to check whether the block is faulty. It can also be used to design the checker circuit to check parity outputs generated from the online testable block using Toffoli gates and Fredkin gates.

22.1.4 Online Testing of ESOP-Based Circuits

In this section, an approach for online testing of ESOP-based circuit is discussed. ESOP, or exclusive-OR-sum of products is the form of Boolean expressions by which a Boolean function is expressed. There exist some basic ESOP-based reversible logic circuits. The common structure of an ESOP-based reversible circuit is that the control of the Toffoli gates is connected only to input lines and the target is connected only to output lines. To generate a reversible circuit with the required structure for a given function one approach is to create an empty circuit with p input lines and q output lines, and initialize each output line with zero. For each ESOP term, a Toffoli gate is added at the end of the circuit. To convert an ESOP circuit into an online testable circuit, this approach has to add some NOT and CNOT quantum gates. A special representation of the Toffoli gate is extended Toffoli gate (ETG), which can be used to replace the Toffoli gate. A parity line L is required for testing, and it is initialized with a $0's$ (zero). In this approach, an n-bit Toffoli gate of the given circuit is replaced by an $(n+1)$-bit ETG and the last bit of the ETG gate is connected to L. The connections of the first n bits of the ETG are kept the same as that of n-bit Toffoli gate for the required function or expression. Before the required function or expression implemented using Toffoli circuit, each input line is EXORed with parity line L and then both the input and output line also EXORed with parity line L. One extra NOT gate is added on the line parity line L. If any single bit fault occurs in the circuit, the output value of L will be 1; otherwise it will be 0.

22.1.5 Online Testing of General Toffoli Circuit

This approach is very much similar to the approach of online testing using ESOP-based circuits because the Toffoli circuit also consists of only Toffoli gates including NOTs, CNOTs, and negative-control Toffoli gates. Dissimilarity between ESOP and Toffoli circuit is that in general, the input line and output line of a Toffoli circuit may be the same. Therefore, an input line of the Toffoli circuit can be used as an output line. Parity line L is also required to determine whether any single-bit fault occurred. Each n-bit Toffoli gate is replaced by an $(n+1)$-bit ETG and the last bit of the ETG is connected to L. The connections of the first n bits of the ETG are kept the same as that of n-bit Toffoli gate for the required function or expression. Before and after the required function or expression implemented using Toffoli circuit, each input and output line EXORed with parity line L. If any single bit fault occurs in the circuit, the output value of L will be opposite to its input value; otherwise, it will remain the same.

22.2 The Design Approach

It is well known that a parity-preserving gate can detect fault by checking its unique input–output parity. There are some special types of parity-preserving gates whose one or more input lines have no control on other inputs. The double Feynman gate is such a gate for which a universal fault-tolerant gate, namely, UFT is introduced. In the following subsections, this type of input line is renamed as parity line P.

22.2.1 The UFT Gate

In this subsection, a 4×4 fault-tolerant reversible gate namely UFT is introduced and shown in Figure 22.6. The representation of UFT gate and its quantum realization is shown in Figure 22.7 and Figure 22.8, respectively. In Figure 22.8, each dotted rectangle is equivalent to a 2×2 CNOT gate. Thus, the quantum cost of the UFT gate is 6. Truth table of the UFT gate is shown in Table 22.1, and it can be notified from the truth table that the input pattern corresponding to a particular output pattern can be uniquely determined and the parity of the input–output bit is also preserved. In Figure 22.10 it is shown that the last input line has no control on the other inputs. So, any change in the last input cannot affect the others and it can be used as parity line P. So, the UFT can be used to generate an online testable circuit.

Property 22.2.1.1 The UFT gate is a fault-tolerant gate.

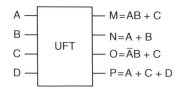

Figure 22.6 Block diagram of UFT gate.

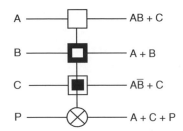

Figure 22.7 Compact representation of a UFT gate.

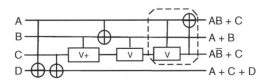

Figure 22.8 Quantum realization of a UFT circuit.

Table 22.1 Truth Table of UFT Gate.

Input				Output			
A	B	C	D	P	Q	R	S
0	0	0	0	0	0	0	0
0	0	0	1	0	0	0	1
0	0	1	0	1	0	1	1
0	0	1	1	1	0	1	0
0	1	0	0	0	1	0	0
0	1	0	1	0	1	0	1
0	1	1	0	1	1	1	1
0	1	1	1	1	1	1	0
1	0	0	0	0	1	1	1
1	0	0	1	0	1	1	0
1	0	1	0	1	1	0	0
1	0	1	1	1	1	0	1
1	1	0	0	1	0	0	1
1	1	0	1	1	0	0	0
1	1	1	0	0	0	1	0
1	1	1	1	0	0	1	1

Proof: The input vector and output vector of UFT gate be $I_v = \{A, B, C, D\}$ and $O_v = \{AB \oplus C, A \oplus B, AB' \oplus C, A \oplus C \oplus D\}$.

$A \oplus B \oplus C \oplus D$ is an input parity by EX-ORing all inputs of the circuit which is as follows:

$$AB \oplus C \oplus A \oplus B \oplus AB' \oplus C \oplus A \oplus C \oplus D$$

$$= B \oplus AB \oplus C \oplus D \oplus AB'$$

$$= A(B \oplus B') \oplus B \oplus C \oplus D$$

$$= A \oplus B \oplus C \oplus D$$

The above parity of the outputs is equal to parity of inputs, which satisfies the property of a fault-tolerant gate. So, from the above explanation, it is shown that the UFT is a fault-tolerant gate.

Property 22.2.2 If no single bit error is occurred in online testable internal circuitry constructed by UFT gate, the parity of the outputs will remain the same as the given value to its input.

Proof: In online testable circuits, all three input and three output lines of a 4×4 UFT gate is EXORed to parity line P. For any online testable circuit using UFT, the input vector is $I_v = \{A, B, C, D\}$ and Output vector is $O_v = \{AB \oplus C, A \oplus B, AB' \oplus C, A \oplus C \oplus P\}$. Therefore, the parity line input line is

$$P_i = A \oplus B \oplus C \oplus P$$

A parity of the output,

$$P_o = P_i \oplus AB \oplus C \oplus A \oplus B \oplus AB' \oplus C \oplus A \oplus C$$

$$= A \oplus B \oplus C \oplus P \oplus AB \oplus C \oplus A \oplus B \oplus AB' \oplus C \oplus A \oplus C$$

$$= P$$

From the above property it is shown that if no single-bit error occurs in online testable internal circuitry, the parity of outputs will remain the same as the given value to its input.

Property 22.2.3 The UFT gate can detect any bit fault concurrently while the gate is performing its operation.

Proof: Here, it is only considered the last input line or the parity line of the UFT gate because this line will detect whether any error occurred. Suppose input vector $\{I_v = I_1, I_2, I_3, I_4\}$ and output vector $O_v = \{O_1, O_2, O_3, O_4\}$.

Property 22.2.2 already proved that if no single-bit error is occurred in an internal circuitry, the parity of the outputs remains the same as the given value to its inputs. Therefore,

$$\text{Parity of inputs } P_i = I_1 \oplus I_2 \oplus I_3 \oplus P$$

$$\text{Parity of outputs } P_0 = O_1 \oplus O_2 \oplus O_3 \oplus O_4$$

If any error occurred such as in O_2, let P_x will be the output of parity line. Then,

$$O_2 \to O_2' = O_2 \oplus 1$$

Therefore,

$$P_x = O_1 \oplus O_2 \oplus O_3 \oplus O_4 \oplus 1$$
$$= I_1 \oplus I_2 \oplus I_3 \oplus P \oplus O_1 \oplus O_2 \oplus O_3 \oplus O_4 \oplus 1$$
$$= P \oplus 1$$
$$= P' \quad [P_i = I_1 \oplus I_2 \oplus I_3 \oplus P_i \oplus O_1 \oplus O_2 \oplus O_3 \oplus O_4 \oplus 1]$$

[From Property 22.2.2]

If any single-bit error is occurred, the value of P must be opposite and it is P'.

From above Property 22.2.3, it is proved that if single-bit error occurred in the circuit, the value of the parity line is opposite to its input value. Any single-bit fault can be detected by checking the parity line.

The UFT gate has one great feature that it can be used to generate any Boolean function such as AND, OR, EX-OR, EX-NOR, NOT, NAND, and NOR. In this subsection, these properties of the UFT gate are shown. Figure 22.9 shows the AND and EX-OR realizations using only UFT gate.

Figure 22.10 shows the OR realization using only UFT gate.

Figure 22.11 shows the NAND and NOT realization using only UFT gate.

Figure 22.12 shows the EX-OR and EX-NOR realizations using only UFT gate.

Figure 22.13 shows the NOR realization using only the UFT gate.

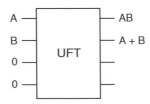

Figure 22.9 AND and EX-OR operations of UFT gate.

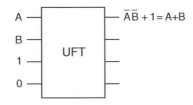

Figure 22.10 OR operation of UFT gate.

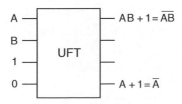

Figure 22.11 NAND and NOT operations of UFT gate.

Algorithm 22.2.1.1 An Approach for Designing Online Testable Reversible Circuits Using UFT Gate

---

1: Begin
2: A Boolean expression uses some basic Boolean operations
3: **if** AND and EX-OR operations have to be done **then**
4:    Give input values to UFT gate such as $I_1 \leftarrow A, I_2 \leftarrow B, I_3 \leftarrow 0, I_4 \leftarrow 0$
5:    Now the output is obtained from the circuit, i.e., $O_1 \rightarrow AB, O_2 \rightarrow A \oplus B$
6: **end if**
7: **if** NAND and NOT operations have to be done **then**
8:    Give input values to UFT gate such as $I_1 \leftarrow A, I_2 \leftarrow B, I_3 \leftarrow 1, I_4 \leftarrow 0$
9:    Now the output is obtained from the circuit, i.e., $O_1 \rightarrow (AB)', O_4 \rightarrow A'$
10: **end if**
11: **if** OR operation has to be done **then**
12:    Give input values to UFT gate such as $I_1 \leftarrow A', I_2 \leftarrow B', I_3 \leftarrow 1, I_4 \leftarrow 0$
13:    Now the output is obtained from the circuit, i.e., $O_1 \rightarrow A + B$
14: **end if**
15: **if** EX-OR and EX-NOR operations have to be done **then**
16:    Give input values to UFT gate such as $I_1 \leftarrow A, I_2 \leftarrow 01, I_3 \leftarrow B, I_4 \leftarrow 1$
17:    Now the output is obtained from the circuit, i.e., $O_3 \rightarrow A \oplus B, O_4 \rightarrow (A \oplus B)'$
18: **end if**
19: **if** NOR operation has to be done **then**
20:    Give input values to UFT gate such as $I_1 \leftarrow A', I_2 \leftarrow B', I_3 \leftarrow 0, I_4 \leftarrow 0$
21:    Now the output is obtained from the circuit, i.e., $O_1 \rightarrow (A + B)'$
22: **end if**
23: End

22.2.2 Analysis of the Online Testable Approach

In this subsection two examples are used to analyze the online testable circuits. The analysis has proved that the online testable approach is better than the previous approaches. Here the same example is discussed using a nontestable approach for better understanding.

Example 22.2.2.1   Consider a Boolean function $f = ab + c'$ can be rewritten as $f = abc \oplus c'$. A nontestable circuit for this function using the basic ESOP-based method is implemented in Figure 22.14.

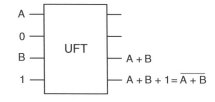

Figure 22.12   EX-OR and EX-NOR operations of UFT gate.

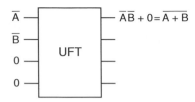

Figure 22.13   NOR operation of the UFT gate.

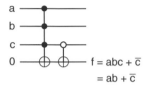

Figure 22.14   Nontestable circuit for $f = abc \oplus c' = ab + c'$.

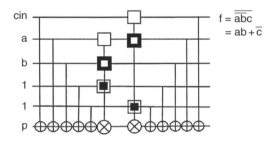

Figure 22.15   Online testable circuit for Example 22.2.2.1.

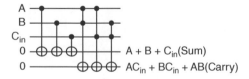

Figure 22.16   Nontestable full adder using ESOP technique.

The design approach of the online testable circuit for Example 22.2.2.1 is shown in Figure 22.15.

Example 22.2.2.2   Figure 22.16 shows a nontestable full adder circuit using the ESOP technique. Let two binary numbers be A and B, and C_{in} be the carry return.

The design approach of the online testable circuit for Example 22.2.2.2 is shown in Figure 22.17.

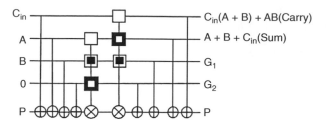

Figure 22.17   Online testable full adder circuit.

22.3   Summary

This chapter presented a technique for online testing of a reversible circuit. In this regard, several approaches with their design issues for online testing of reversible circuits were discussed. Here, a special fault tolerant gate is used to detect whether any single-bit fault occurs in the testable circuit. In addition, a fault-tolerant reversible gate, namely UFT, is also introduced. The UFT gate has two important features that can be used as a universal fault-tolerant gate i.e., any Boolean expression can be generated by this gate and it can also be used in testing a circuit. In this chapter, two examples are used to compare the online testing approach using the UFT gate and other approaches. In both cases, the approaches convert a circuit into an online testable circuit that is synthesized from an ESOP. Though the synthesized circuit can detect any single-bit fault, it can be used in any structure of the reversible circuit for online testability.

23

Applications of Reversible Computing

In the world of computing, heat is the byproduct of advancement and its foe. As per Moore's law, the quantity of parts on a chip of a given size doubles at regular intervals, conceivably multiplying its heat dissipation at a similar rate. In this manner PC engineers have always looked for approaches to lessen the energy utilization of each new age of chips. Workstations utilize circling air and fans to cool their chips, while supercomputers utilize significantly more comprehensive cooling frameworks. These endeavors have satisfied abundantly: Computations per unit of energy have multiplied each 1.6 years since the inception of the primary electronic PCs in 1945. However, even with every one of their developments, chip creators have observed heat to be a noteworthy restriction as they continued looking for quicker PCs. Around 2005, chip producers began restricting clock speed (which controls the rate that calculations are executed) to around 3 gigahertz for each chip on the ground that quicker tickers created heat too quickly and made chips consume more.

To keep pace with Moore's law, the PC equipment industry has expected to look for inventing better approaches to make processors run cooler even as they run quicker. Some anticipate quantum PCs, a best in class innovation that will be normal within 10 years, if innovation proceeds at its present pace. These PCs will take negligible seconds to take care of particular sorts of issues. Quantum PCs store data in the conditions of individual particles, called quantum bits or qubits. Rather than wires, they use quantum-mechanical collisions, for example, photons, superconductivity and superposition to store and impart data between qubits. In any case, quantum circuits are wonderfully sensitive to heat: A little amount of heat can make the ions vibrate excessively and lose stability. The underlying adaptations of quantum PCs depend on circuits. We need quantum circuits that in principle, emit no heat by any means. Figure 23.1 shows a generalized structure of reversible computer.

Standard PC circuits have a shrouded source of heat: the loss of data when bits are deleted. In the middle of a calculation, parts of circuits called logic gates regularly accept two bits as information and produce one piece as output. Each piece is regarded as a burst of energy that initiate movements along wires between gates. A gate that gets two units of input at its sources and conveys one unit of output at its output path must lose a bit of energy. Since energy cannot be made or crushed, that energy must go someplace, so it shows up as heat emitted by the gate.

Reversible and DNA Computing, First Edition. Hafiz Md. Hasan Babu.
© 2021 John Wiley & Sons Ltd. Published 2021 by John Wiley & Sons Ltd.

Figure 23.1 Reversible computer dissipates less heat than a conventional computer.

Why We Need to Use Reversible Circuits

One of the most important recommendations to lessen heat from lost bits was to make PC circuits reversible. A reversible circuit has precisely the same number of outputs as data sources, allocating one output to each input and the other way around. That implies that information can be recreated from the output; no bits are lost, so reversible circuits won't emit heat from bit loss. Besides, reversible circuits can recreate standard computational works and hence it can be utilized in any PC. An inquisitive component of reversible circuits is that, as the name infers, they can really be kept running backward! The circuits keep up a similar capacity regardless of whether we turn around the jobs of the input and output lines. Due to this capacity, reversible PCs are here and there depicted as "PCs that can run in reverse."

Applications of Reversible Computing

There are many applications of reversible circuits. The topic of whether PCs could be worked without spreading heat was the first addressed in the 1950s. Early scientists noticed an intriguing association between data hypothesis and thermodynamics. The two speculations state that entropy increments when data about a framework's state is lost. Thermodynamics says that an expansion of entropy makes heat emit from the framework. Along these lines, if we can abstain from losing data from a framework, we can abstain from emitting that type of heat.

Figure 23.2 Reversible computer has the same number of outputs and inputs.

Reversible circuits take care of that issue, and if different origins of heat can be expelled also – for instance, electrical resistance and the active energy loose when electrons alter course - at that point reversible circuits will work with no heat loss. In 1973, Charles Bennett of IBM acquired the term adiabatic from thermodynamics, where it implies there is no heat trade between a framework and its condition. Adiabatic PCs would be progressive since they would keep on working inconclusively after they were instated and begun, without including or subtracting heat energy. Adiabatic computing can't occur without reversible circuits.

A reversible circuit has precisely the same number of outputs as sources of input. Figure 23.2 expresses the phenomena of a reversible circuit. Information can be reproduced from the output; no bits are lost, so reversible circuits won't radiate heat from bit loss.

23.1 Adiabatic Systems

An all-around protected frictionless vehicle is a decent hypothetical example for an adiabatic PC. When it quickens to a cruising speed, an adiabatic vehicle would lose no energy to friction and radiate no energy into nature, enabling it to travel inconclusively on its energy. It could hinder utilizing 100 percent proficient regenerative braking and afterward accelerate once more, as long as force was preserved and no energy was disseminated by erosion.

Physicist Richard Feynman demonstrated that it is hypothetically conceivable to make an adiabatic reversible PC as shown in Figure 23.3. Feynman checked out reversible computing during the 1970s since he needed to realize whether there is a central lower extreme point how much energy is expected to do a calculation. What sorts of PCs accomplish that limit? He knew about as far as possible, which puts a lower bound on the measure of energy lost to heat when a solitary piece of data is lost. He asked: If we utilize reversible circuits, which

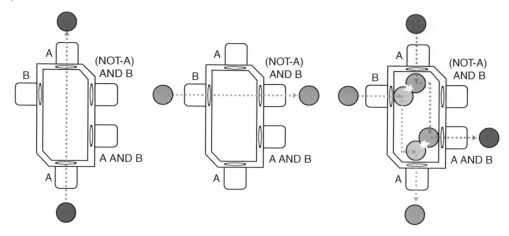

Figure 23.3 Working mechanism of a reversible computer.

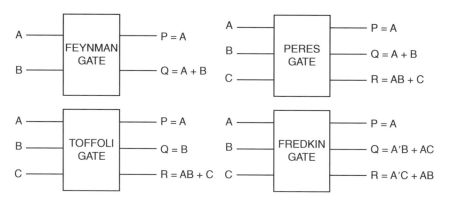

Figure 23.4 Reversible logic gates.

lose no bits, what is the base energy required to complete the calculation? He, in the long run, inferred that there is no hypothetical smallest amount. In 1982 Edward Fredkin and Tommaso Toffoli of the Massachusetts Institute of Technology made reversible computing a stride further by proposing the primary reversible gates as presented in Figure 23.4. A circuit produced using these gates could be unambiguously feedback to its past input. Fredkin and Toffoli likewise demonstrated that their gates are extensive: A computer that worked from them would almost certainly run any program that keeps running on a regular PC.

23.2 Quantum Computing

Reversible computing has accumulated another following lately, in the light of the fact that it underpins energy decrease, but since it is essential for quantum PCs. In 2016 specialists at Griffith University and the University of Queensland in Australia declared that they had manufactured a quantum Fredkin-Toffoli gate utilizing photons of light. What's more, this year an organization called D-wave offered a 2,000-qubit quantum PC that works at close supreme zero temperature as an adiabatic machine.

Quantum PCs use photons of light to convey signals. Since photons do not have mass, they do not produce active heat dissipation in exchanging them. Feynman's alert does not have any significant bearing: Quantum PCs can work a lot quicker than circuits that switch electrons.

23.3 Energy-Efficient Computing

Individuals from an exploration bunch at MIT, driven by Nirvan Tyagi, trust they can overhaul normal irreversible calculations into new, reversible form, which can be communicated in another programming language by utilizing reversible structures. The MIT scientists have structured another dialect, called EEL (energy-efficient language), that limits circles, conditionals, bounces, and capacity calls to reversible structures. They adopt an even-minded strategy and don't demand that each calculation be made reversible; with their language they can gauge how much irreversible code is created when an EEL program is aggregated into language the machine can execute. EEL can enable them to compose programs with generally minor measures of irreversible code. At the point when kept running on regular PCs, EEL projects will spare energy since they don't require the same number of bit-changes as usual calculations.

23.4 Switchable Program and Feedback Circuits

Building programs that can be switched are not new. During the 1970s Brian Randell of Newcastle University called a meeting that considered how to make programming progressively solid. The center of their thought was to construct a few autonomous calculations for a similar capacity into their projects. If one calculation is neglected to give the right result, their framework would back up to the state it was before. Then it attempts to calculate the alternate one. (A comparable adaptation to non-critical failure system utilized in equipment is called N-form figuring.) Interestingly, this strategy increases the likelihood of reversible processing.

In 1976, Tom Anderson and his colleague fabricated a language and working framework around units called recovery squares. These modules of code take inputs, play out a calculation, and test the output for agreeableness. All contributions to all modules could be recovered by moving recovery squares in reverse.

Recovery squares are intended to act naturally contained-or, as the analysts calls it, nuclear. While they execute, they trade no data with their condition. If they weren't nuclear, the failure of an acknowledgment test would have a domino effect, making the framework completely restart and backing up to the start of the program. Just before recovery stops, the working framework saves a duplicate of the framework's state in a recovery store. This makes rollback quick and simple. The framework basically flies in the past state, from the reserve and reestablishes memory to that state, as shown in Figure 23.5.

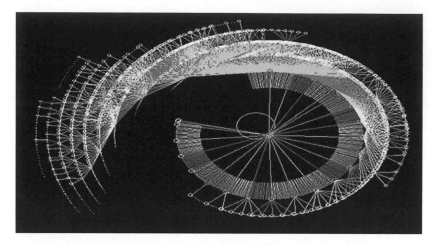

Figure 23.5   Back-up states of a reversible computing system.

23.5   Low-Power CMOS

CMOS transistors, particularly the smallest, cutting-edge ones, spill a lot of current to make exceptionally effective adiabatic circuits. Bigger transistors depend on more established assembling; however they'd be worked gradually to accelerate calculation through parallel tasks. Stacking them in layers could yield conservative and energy-proficient adiabatic circuits, yet right now such 3D creation is still very expensive. Also, CMOS might be a bottleneck regardless.

Luckily, there are some encouraging options. One is to utilize quick superconducting hardware to fabricate reversible circuits, which have just appeared to disseminate less energy per gadget than as far as possible when worked reversibly. Advances in this domain have been made by specialists at Yokohama National University, Stony Brook University, and Northrop Grumman. In the meantime, a group driven by Ralph Merkle at the Institute for Molecular Manufacturing in Palo Alto, California, has planned reversible nanometer-scale subatomic machines, which in principle could consume one-hundred-billionth the energy of the present processing innovation while still exchanging on nanosecond timescales. The message is that the innovation to fabricate such molecularly exact gadgets still ought to be designed.

23.6   Digital Signal Processing (DSP) and Nano-Computing

High-performance chips that endure big amounts of heat impose restrictions on how far we can increase the performance of the system. Reversible circuits will soon offer the only physically possible way to keep improving performance. Reversible computing will also lead to improvement in energy efficiency. Energy efficiency will fundamentally affect the speed of circuits such as digital signal processing (DSP), nano computing, and the speed of most computing applications. To increase the portability of devices, again reversible computing is required. It will let circuit element sizes be reduced to atomic size limits and hence devices will become more portable. Although the hardware design costs incurred in the near future might be high, and the power cost and performance will be more dominant than logic hardware cost in today's computing era, the need for reversible computing cannot be ignored, and thus in the future of nano-computing and DSP, reversible technology will have a great impact.

Part III

DNA Computing

An Overview About DNA Computing

DNA (deoxyribonucleic acid) computing, also known as molecular computing, is a new approach to massively parallel computation. DNA computing was introduced as a means of solving a class of intractable computational problems in which the computing time can grow exponentially with problem size (the NP-complete or nondeterministic polynomial time complete problems). A DNA computer is basically a collection of specially selected DNA strands whose combinations will result in the solution of some problems, depending on the problem at hand. Technology is currently available both to select the initial strands and to filter the final solution. DNA computing is a new computational paradigm that employs molecular manipulation to solve computational problems while at the same time exploring natural processes as computational models.

The DNA found in living cells is composed of four bases, viz. adenine (A), guanine (G), thiamine (T), and cytosine (C). The order of these bases is unique in each individual and determines the unique characteristics of that particular individual. Each base is attached to its neighboring base in the sequence via phosphate bonding. The base, sugar, and phosphate are together called a nucleotide. Two DNA sequences bond with each other via hydrogen bonding between each Watson-Crick complementary base pairs (A with T and C with G) to form a DNA double helix. Each DNA strand has two ends: 5'-end and 3'-end that determine the polarity of the DNA strand. During the formation of a DNA double strand, two complementary single strands bond with each other in anti-parallel fashion.

DNA computing accomplishes the computations using genetic molecules rather than traditional silicon chips. A computation may be assumed as the execution of an algorithm, which itself may be defined as a step-by-step list of well-defined instructions that take some inputs, process it, and yield a result. In DNA computing, information is represented using the four-character genetic alphabet (A, G, T and C), rather than the binary alphabet (1 and 0)

Reversible and DNA Computing, First Edition. Hafiz Md. Hasan Babu.
© 2021 John Wiley & Sons Ltd. Published 2021 by John Wiley & Sons Ltd.

used by traditional computers. This is feasible because short DNA molecules of any arbitrary sequence may be produced to order.

DNA computers use strands of DNA to perform computing operations. The computer consists of two types of strands – the instruction strands and the input data strands. The instruction strands splice together the input data strands to generate the desired output data strand. DNA computing holds out the promise of important and significant connections between computers and living systems, as well as promising parallel computations massively.

The DNA is the major information storage molecule in living cells, and billions of years of evolution have tested and refined both this wonderful informational molecule and highly specific enzymes that can either duplicate the information in DNA molecules or transmit this information to other DNA molecules. Instead of using electrical impulses to represent bits of information, the DNA computer uses the chemical properties of these molecules by examining the patterns of combination or growth of the molecules or strings. DNA can do this through the manufacture of enzymes, which are biological catalysts that could be called the software, used to execute the desired calculation.

Problems are solved in a DNA computer by encoding a problem using *A, G, T,* and *C*. However, a compact DNA computer is constructed by synthesizing corresponding DNA strands that perform several operations on those strands. The advantages of DNA computing over conventional ones are as follows:

1. Massively parallel operation: A single test tube of DNA can contain trillions of DNA strands and all strands respond to the biological operations in parallel.
2. Huge information density: Information density of DNA is huge over silicon. Estimated storage capacity of 490 Exa-bytes per gram of DNA.

DNA computing has drawn the attention of many researchers in recent years for its applicability to solve computationally hard problems. It can perform faster than conventional computers with its inherent massively parallelism nature. Proper synthesis of reversible circuit has been well researched in recent days for its particularly low power consumption (ideally zero) and inherent reversible nature of reversible logic. The optimal synthesis of a reversible truth table means finding the reversible circuit made up of reversible gates satisfying, given truth table with the optimum cost. Nowadays, reversible logic has emerged as a promising computing paradigm having its solicitations in low-power computing, quantum computing, nanotechnology, optical computing, and DNA computing. The classical set of gates such as AND, OR, and EX-OR are not reversible. Recently, it was shown that information can be encoded in DNA and use DNA amplification to implement some reversible gates. Furthermore, in the past reversible gates such as Fredkin gates were constructed using DNA, whose outputs are used as inputs for other reversible gates. Thus, it can be concluded that arbitrary circuits of reversible gates can be constructed using DNA. This has been the driving force leading to the design of reversible adder and multipliers using Fredkin gate.

This part is divided into two types of contents, namely general or irreversible DNA and reversible DNA. In irreversible DNA, some backgrounds and preliminary studies about DNA are given in Chapter 24. In addition, Chapter 25 shows a DNA-based approach of microprocessor design automation. In reversible DNA, Chapter 26 shows some DNA-based reversible gates as well as reversible circuits. Chapter 27 describes the design approaches of a DNA-based reversible addition and subtraction mechanism. Similarly, Chapter 28 shows the design approaches of DNA-based reversible shifting and multiplication techniques. Chapter 29 describes the design approaches of DNA-based reversible multiplexer and arithmetic logic unit (ALU). Chapter 30 shows the design approach of DNA-based flip-flop. Finally, Chapter 31 describes the applications of DNA computing.

24

Background Studies About Deoxyribonucleic Acid

DNA stands for deoxyribonucleic acid, is the hereditary material in humans and almost all other organisms. Nearly, every cell in a person's body has the same DNA. Most DNA is located in the cell nucleus (where it is called nuclear DNA), but a small amount of DNA can also be found in the mitochondria.

24.1 Structure and Function of DNA

The information in DNA is stored as a code made up of four chemical bases: adenine (A), guanine (G), cytosine (C), and thymine (T). Human DNA consists of about 3 billion bases, and more than 99 percent of those bases are the same in all people. The order or sequence of these bases determines the information available for building and maintaining an organism, similar to the way in which letters of the alphabet appear in a certain order to form words and sentences. DNA bases pair up with each other, A with T and C with G, to form units called base pairs. Each base is also attached to a sugar molecule and a phosphate molecule. Together, a base, sugar, and phosphate are called a nucleotide. Nucleotides are arranged in two long strands that form a spiral called a double helix. Figure 24.1 shows the hydrogen bonds of the interior DNA and Figure 24.2 shows the overall DNA structure including (a) a double-helix structure; (b) phosphodiester bonds; and (c) major and minor grooves are binding sites for DNA binding proteins during processes such as transcription (the copying of RNA from DNA) and replication.

The structure of the double helix is somewhat like a ladder, with the base pairs forming the ladders rungs and the sugar and phosphate molecules forming the vertical side pieces of the ladder. An important property of DNA is that it can replicate, or make copies of itself. Each strand of DNA in the double helix can serve as a pattern for duplicating the sequence of bases. This is critical when cells divide because each cell needs to have an exact copy of the DNA present in the old cell. The DNA structure is shown in Figure 24.3.

DNA denaturation is the process of conversion of double stranded DNA into single strands due to heat or alkali induced breakage of hydrogen bonds and Van der Waals force between every two nitrogen bases from two different strands.

Through the process of slow cooling, the denatured single strands can reunite according to the complementarily principle $(A = T; G^G)$, allowing DNA to regain its native double helix; this phenomenon is called denaturation.

Reversible and DNA Computing, First Edition. Hafiz Md. Hasan Babu.
© 2021 John Wiley & Sons Ltd. Published 2021 by John Wiley & Sons Ltd.

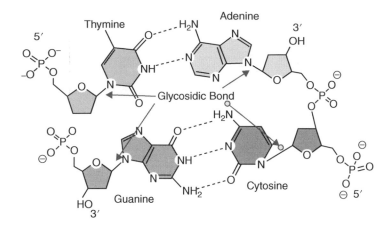

Figure 24.1 Hydrogen bonds of the interior DNA.

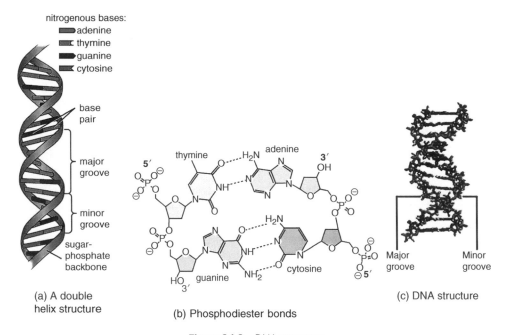

Figure 24.2 DNA structure

Ligation is the chemical process of joining DNA molecules together by forming phosphodiester bonds between two deoxyribose sugars of different terminal nucleotides of those molecules with the activity of ligase enzyme. The ligation process is shown in Figure 24.4.

The polymerase chain reaction (PCR) is a biochemical technology used to amplify a single or a few copies of a piece of DNA across several orders of magnitude, generating thousands to millions of copies of a particular DNA sequence. It relies on thermal cycling, consisting of cycles of repeated heating and cooling of the reaction for DNA melting and

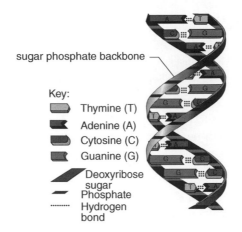

Figure 24.3 Structure of DNA.

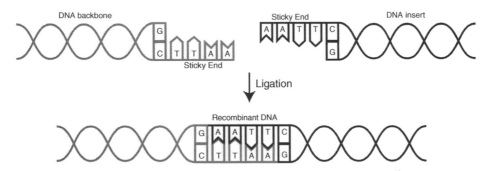

Figure 24.4 Ligation process of DNA.

enzymatic replication of the DNA. Primers containing sequences complementary to the target region along with a DNA polymerase are key components to enable selective and repeated amplification. As PCR progresses, the DNA generated is itself used as a template for replication, setting in motion a chain reaction in which the DNA template is exponentially amplified, and PCR can be extensively modified to perform a wide array of genetic manipulations.

Capillary electrophoresis, also known as capillary zone electrophoresis, can be used to separate ionic species by their charge and frictional forces and hydrodynamic radius. It is the most efficient separation technique available for the analysis of both large and small molecules of DNA.

24.2 DNA Computing

DNA computing is a form of computing that uses DNA, biochemistry, and molecular biology, instead of the traditional silicon-based computer technologies. DNA computing, or, more generally, biomolecular computing, is a fast-developing interdisciplinary area. Research and development in this area concerns theory, experiments, and applications of DNA computing.

24.2.1 Watson-Crick Complementary

By natural of DNA molecule, Watson-Crick complementary plays the most important role in the DNA computing. DNA consists of four bases of nucleic acid: A, G, C, and T. Adenine can only connect with thymine, and cytosine can only connect with guanine (not A always T and not C always equal G).

24.2.2 Adleman's Breakthrough

The first wet experiment that proved DNA and biochemical process could be used as a computing tools to solve of complex computational problem was done by Prof. L.M. Adleman in 1994. In seven days of lab experiments, Adleman successfully solved the Hamiltonian path problem (HPP) of seven cities. HPP is a special case of the traveling salesman problem, obtained by setting the distance between two cities to a finite constants if they are adjacent and to infinity otherwise. In HPP, we can assume G consists of vertices $v_1, v_2, \ldots, v_n$, and v_{start} and v_{end}. One of the main requirements that should be met by HPP is the directed graph if and only if there exists a sequence of compatible one-way edges. $v_1, v_2, \ldots, v_n$, begin with v_{start} and end with v_{end}, and enter other vertex exactly only one time. Figure 24.5 shows the HPP problem that was solved by Adleman in his first wet experiment. To find the unique Hamiltonian path from directed graph, Adleman followed a non deterministic algorithm as presented below:

Algorithm 24.2.2.1 Nondeterministic Algorithm

1: Begin
2: Generate random paths through the graph.Keep only those paths that begin with v_{start} and end with v_{end}.
3: If the graph has n vertices, keep only those paths that enter exactly n vertices.
4: Keep only those paths that enter all of the vertices of the graphs at least once.
5: If any path remains, say Yes, otherwise say No.
6: End

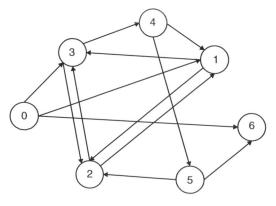

Figure 24.5 HPP on seven vertices.

In Algorithm 24.2.2.1, Adleman randomly generated 20-*mer* of DNA sequences to represent each city denoted as *O*. On the other hand, to represent a vertex that connects between two different cities, Adleman suggested the DNA design a combination of O_i and O_{i-1}. After all DNAs have been synthesized to represent all available vertices, 50 mol of *O* and 50 mol of WC complementary vertices of *O* are added in one test tube for mixing together in single ligation process. This process will randomly produce all possible combinatorial solution for the graph. Only the vertex that passes all cities will be considered as a feasible solution. Therefore, Adleman employed a PCR technique to check whether the strand passes all cities. By using short strands of single-stranded DNA as a primer, the biochemical technique is able to identify related strands and discharge unrelated strands that do not fulfill the requirement. This process is explained in Line 3 in his algorithm. In this way, Prof. Adleman solved the Hamiltonian path problem of seven cities as shown in Figure 24.5.

The output of Line 3 will undergo electrophoresis process. For this purpose, Adleman chose to run electrophoresis process. This process sorts the strands according to their size. Here the number of cities is seven, so only strands with 140-bp (base pair) band (corresponding to double stranded strand) are excited and soaked in double distilled $H_2O(ddH_2O)$. So, only strands that pass seven cities will be extracted in this process to pursue to the next process.

This process is explained in Line 4 in the Adleman algorithm. In order to realize Line 4 in Adleman's algorithm, he employed a magnetic beads separation process. It is most labor-intensive work. In this process, each complementary of city was used to check whether all the cities exist in the strands. The procedure was iterated until the result was obtained. Even though the experiment took almost seven days' of labor-intensive work, the result was a very acceptable arid, and it provides a new approach to solving the most complicated calculations, especially when deal with a huge amount of variables.

24.3 Relationship of Binary Logic with DNA

Several DNA-based designs describe the relationships between 0/1 logic with DNA logic. However, the DNA-based design shows a clever approach to encoding 0/1 logic using two types of single DNA strands, one for input bit and another for operand bit. Also in design, a reversible Toffoli gate is realized using different DNA strands for control inputs and target input of a Toffoli gate where control input signals are constructed using the dinucleotide 5'-*AG* and 5'-*CT* representing bit 1 and 0, respectively. The target input is constructed with 5'-*ψA* and 5'-*UA* as bit 1 and 0, respectively. Similarly, the final output is constructed using a mixer of 3'-*AU*, 3'-*AD*, 3'-*AT*, and 3'-*AX* representing bit 1, 1, 0, and 0, respectively.

24.4 Welfare of DNA Computing

The main advantages of DNA-based circuits over the silicon chips are as follows:

- In double-strand DNA the data density will be one base per square nanometer and the data density will be over one million Gbit per square inch, where in typical high-performance hard drive, the data density is about 7-Gbit per square inch.

- Base pair complement gives a unique error correction mechanism, which works like a RAID 1 array.
- Many copies of the enzyme can work on many DNA molecules simultaneously. This is the power of DNA computing, that it can work in a massively parallel fashion.
- In DNA replication, enzymes start on the second replicated strand of DNA even before they are finished copying the first one. So data rate jumps to two times of initial speed (initially it is 1000 bits/sec). After each replication is finished, the number of DNA strands increases exponentially. For example: After 30 iterations, it increases to *1000-Gbit/sec*. This is beyond the sustained data rates of the fastest hard drives.
- DNA is a stable molecule that never suffers any changes (mutation) unless it faces a harsh (very high temperature, corrosive agents, etc.) environment.
- DNA logic gates can be preserved for a very long time (more than a decade) by maintaining and varying the temperature.
- A tiny energy is required to break the bond when operating DNAs. For example, the energy required to break the bond between *A* and *T* is ≡21 KJ/mol. where *A* denotes adenine and *T* denotes thymine. The same 21 KJ/mol will be gained if a bond between them is formed again. That means, energy is reserved in DNA-based logic circuits.

Figure 24.6 presents the DNA replication process.

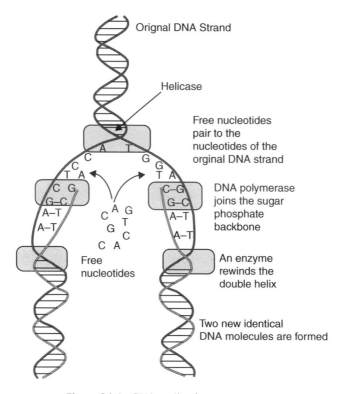

Figure 24.6 DNA replication process.

24.5 Summary

This chapter introduces the concept of DNA, properties of DNA, and how the computing is done using DNA. Some graphical representations corresponding to the structures of DNA are also illustrated. In addition, the relationship of binary logic with DNA is also described. Moreover, the DNA replication process is also presented in this chapter.

25

A DNA-Based Approach to Microprocessor Design

Designing a microprocessor is a costly and time-consuming process, as the huge number of parameters makes it very complicated task. Though many computer aided design (CAD) tools are available to speed up this process, the fundamental design work still lies in the human domain. The terms microprocessor and processor are used interchangeably in this chapter. Several steps are convoluted in designing of a microprocessor. A change at any of the steps will reflect on others. This makes the task of automating design at any one step very difficult and hence any tool established to program the process must take into account all the steps of design. An evolutionary method to automate this process is highly suitable as processor designing is, by itself, evolutionary in nature. The designers tend to build on a microprocessor by adding some features to it and optimizing its features.

25.1   Basics of Microprocessor Design

The different stages involved in designing a microprocessor are as follows:

1. Algorithm stage
2. Architecture stage
3. Logic stage
4. Circuit stage
5. Layout stage

In the algorithm stage, the various functionalities such as addition, multiplication, square rooting etc. estimated from the processor, are fixed. This is done bearing in mind the targeted use of the processor. In the architecture stage, the components obligatory for a specific functionality are found out. If the functionality can be reached by more than a single way, the merits and demerits of each are observed and one of the methods is chosen. In the logic stage, the Boolean equation leading each of the components is evaluated and the corresponding logic circuit (consisting of logical gates) is derived. These circuits are then connected together to form the overall logical circuit of the microprocessor. In the circuit stage, each gate is broken down into transistors using a specific technology. Other electrical components are also fixed up in this stage. By the end of this phase, we can arrive at the circuit diagram for the entire microprocessor. In the layout stage, the actual position of each

Reversible and DNA Computing, First Edition. Hafiz Md. Hasan Babu.
© 2021 John Wiley & Sons Ltd. Published 2021 by John Wiley & Sons Ltd.

and every device and interconnect on the chip is fixed. This is done considering various issues like clock, signal integrity, power distribution, and power dissipation.

The interdependence among various stages of design is a key factor contributing to the complexity of the problem. Modern microprocessors have a very high transistor count. Placement and routing are NP complete problems. Due to the extreme magnitude of the problem, heuristic-based CAD placement and routing tools are used. But even these have severe weaknesses. Human experience plays a very major role in the early placement of modules on the die. In the Deep Sub Micron (DSM) era, new and harder problems related to microprocessor design have crept up. The design paradigm has shifted from the device to the interconnect. With a billion device processors being targeted in the next few years, it is worthwhile to seek out an automation procedure for processor design.

25.2 Characteristics and History of Microprocessors

The Intel family of processors is considered here to show the evolution of processors. Intel 4004 was the first archetype that set the path for the rapid evolution that took place over the next three decades. Intel 4004 was a 4-bit processor supporting clock speeds (the rate, at which the basic gate, inverter, switches) up to 740 kHz. It had a core voltage of 5V, which is high compared to modern processors. Following this came the first 8-bit microprocessor, Intel 8008. Though a little slower in terms of instructions per second than its predecessor, 8008 processed eight data bits at a time and its ability to access more RAM gave it three to four times the processing power of 4-bit chips. Intel's 8085, the next Intel release, boasted an address space of 64 kbytes. It was the first microprocessor to have a register array and special purpose registers.

Intel's next proclamation was the 16-bit microprocessor Intel 8086. This proclamation of 8086 marked the arrival of the innovative concept: pipelining. Pipelining is essentially the process of breaking down a serial process into various stages. Different hardware units can carry out these stages individually and in parallel, thus increasing instruction throughput. A 6-byte prefetching queue was integrated into the 8086 chip for pipelining. The 8086 made the use of four segment registers that helped relative addressing and program relocation. Subsequently, Intel introduced the 80186. One major feature of the 80186 series was the decrease of the number of separate chips required, by including features such as a DMA controller, interrupt controller, timers, and chip select logic. In developing next-generation processors, Intel concentrated on increasing the clock frequency to enhance the overall performance. The Intel 80286 was developed having a clock frequency of 12.5 MHz.

With the initiation of semiconductor technology, more than million transistors got packed into the next generation Intel processor: 80386 and 80486. They supported many landscapes like paging, virtual addressing, multilevel protection, multitasking, and debugging capabilities. To further push forward performance, superscalar processing was familiarized by Intel in the Pentium series of processors. In this series, the clock frequency and device density got significantly increased (60 MHz and 3 million devices). Superscalar processors have the ability to do parallel instruction execution by employing different data paths.

Power issues were conquered in the next release of Intel, the Intel Pentium 2. It functioned at a core voltage of 2.8 V (which is almost half of Pentium) and clock frequency of

233 MHz (which is almost four times that of Pentium). The radical core voltage reduction was possible because of progression in device technologies, like threshold voltage reduction. In Intel's Pentium 3, the number of transistors quadrupled with growth in the device density. The Pentium 3 incorporated dynamic scheduling. The release of Pentium 4 processor, with about 42 million transistors, marked the arrival of the new age processors. It operated at a clock frequency of 3.2 GHz with a core voltage of 1.72 V. Pentium 4 supported hyperthreading, which is the implementation of the simultaneous multithreading technology on the Pentium 4 microarchitecture. Hyper threading provides improved support for multithreaded code, allowing multiple threads to run simultaneously.

This evolution process starting from Intel 4004 to the presentday dual-core Pentium processors, has shown an increase in performance along with increase in power. However, to overcome increased power consumption, yet maintaining performance (having more functionalities), the clock frequency was reduced as in the latest Intel Itanium processor. This is also the first 64-bit processor from Intel. To realize more functionality and computing power, the device count increased. Technology has a significant impact on power, performance, reliability, and cost. Since the devices are packed very close to each other, issues such as hotspot and electromigration have to be considered. Thus, designing a microprocessor is a difficult and laborious task.

25.3 Methodology of Microprocessor Design

There is strong case for the need of an evolutionary methodology in microprocessor design. Various organisms were developed over millions of years. Natural selection enabled inefficient characteristics to be phased out and the efficient characteristics to be enriched and refined. The development of a microprocessor is very similar to that of a species. The characteristics discussed in the previous section were introduced in the microprocessor over a period of time. Some other characteristics were also presented but then rolled back later when they were found to be rather inefficient. Thus, microprocessors too, in this sense, evolve.

As processors become more and more advanced, the complexity in their design growths. This increases the cost of design and the turnaround time of processors. Turnaround time of a processor is well-defined as the time taken during the design phase, design validation phase, and the fabrication and testing phase. This spans from the original commencement of the idea behind the processor, to the actual time when the first processor is produced. Turnaround times have kept increasing. If they become too large, then the processor and the technology it uses might become obsolete by the time its design is complete. Moreover, reduction in turnaround time reduces the cost of production as less man-hours are consumed on the design of a processor. Thus, any methodology that diminishes the turnaround time without reducing the processor's capabilities is widely valued.

Human experience plays a major role in the design of a processor. Experts have unique characters while deciding on critical parameters. This ability, though difficulty to quantify, would be critical for an automation tool developed for designing processors. Every organism is uniquely defined by its genetic code. This code consists of DNA sequences defining the individual traits of the organism. These characteristics are confined in the DNA strands as a sequence of nucleotides. Nucleotides, the basic building blocks of DNA, are chemical mixtures that consist of a heterocyclic base, a sugar and one or more phosphate groups. There are four bases:

1. Adenine: A
2. Thiamine: T
3. Cytosine: C
4. Guanine: G

A string of this A T C G sequence can give a complete genetic description.

An automation tool that would generate a design, given some specifications, is of much interest. A method of finding the genetic sequence for a microprocessor has been shown. This sequence, which completely defines the microprocessor, is then used to progress new microprocessor sequences through combination with other processor sequences. These newly generated sequences can then be converted into the microprocessors they denote. Postprocessing of the new plan is done to rectify any errors that might have occurred. Finally, the new design is tested and evaluated. If the design meets the customized standards that are set by the user, then the design is accepted. In this way, a population of microprocessors is built up. Thus, the algorithm is the first step in the true automation of the microprocessor manipulative process.

25.4   Construction of Characteristic Tree

Figure 25.1 presents an evolutionary method employed. A set of factors that define a processor uniquely (at various design levels) are acknowledged. At the architectural level, around 50 parameters of contemporary processors are used for simulation purposes. Each characteristic is denoted as a node in a directed acyclic graph (DAG). The dependencies that may exist between the parameters (for instance: the critical path of interconnects depends on the technology such as DSM used) are modeled as edge weights between the corresponding parameter nodes. A sample tree is shown in Figure 25.2. The term *tree* is used to refer to a DAG, as it is more applicable.

For programming resolutions an adjacency matrix is used to represent the tree and each node in the tree is a structure containing the parameters, which is shown in Figure 25.3.

The tree is both flexible and reconfigurable to incorporate unforeseen parameters that might ascend as a result of either machinery or architectural advancements.

25.5   Traversal of the Tree

In a DNA sequence more dependent genes are placed closer. Also, highly dependent genes may be associated with one another (a phenomenon called linkage). If the features of the processor need to be encoded as a realistic DNA sequence, then their location becomes very critical. When the adjacency matrix of the tree was symmetrized and a depth-limited depth-first search (DFS) was accomplished on the tree, the placement of the nodes was found to be more realistic. This can be attributed to the fact that DFS, being a recursive

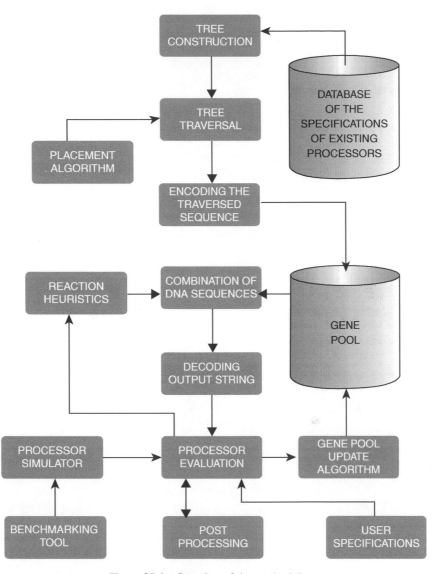

Figure 25.1 Overview of the methodology.

process, first traverses the subtree, with the current node as its root. This ensures that the nodes that are dependent (for example, cache and its dependent characteristics) are placed next to each other as they form a subtree.

This approach, along with the introduction of a fitness factor is used to navigate the tree. The symetrization of the adjacency matrix of the tree converts the DAG into a bidirectional graph. The fitness parameter is a function of the edge weight and specifies the usefulness of placement of a node in the linear DNA sequence. A depth-limited DFS is used to ensure completeness. A further rearrangement is accomplished in the DNA domain to ensure optimal placement. A traversal procedure of the tree is presented in Algorithm 25.5.1.

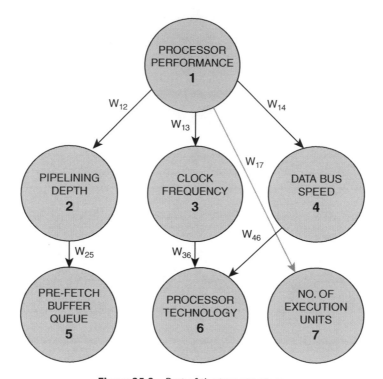

Figure 25.2 Part of the tree structure.

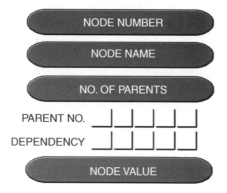

Figure 25.3 Structure of a node.

After performing a DFS on the example graph as shown in Figure 25.2, the following order of traversal is obtained:

Processor Performance → Pipeline Depth → Prefetch Buffer Queue → Clock Speed → Processor Technology → Data Bus Speed → No. of Execution Units

The average effectiveness of placement is: 3.09.

25.6   Encoding of the Traversed Path to the DNA Sequence

A DNA sequence consists of the four prerevealed bases. A group of three bases forms a codon, and a collection of codons makes up a gene that requires a unique characteristic.

Algorithm 25.5.1 Traversal Procedure

1: Begin
2: Symmetrize the adjacency matrix.
3: Compute the maximum depth of traversal to ensure completeness.
4: Traverse the tree using depth-limited DFS. Store the traversed path in an array.
5: Compute the fitness factor for all edges in the tree. Fitness factor $(Node j) = \sum (W_{ij} \times D_{ij}) / \sum W_{ij}$ where W_{ij}: weight between node i and node $j D_{ij}$: distance between node i and node j
6: Rearrange the string in DNA domain using fitness factor.
7: End

This chapter shows a fixed-length encoding scheme (the string length of each character-istic node is constant) to encode the different characteristics of the processor as a DNA sequence. Each processor characteristic has the following fields:

Each field in the above string can be measured as a codon. The value of each field is con-verted to Base 4 system and the corresponding DNA digit (dit) sequence is determined. A single dit can be one of the bases (A or T or C or G). The simplicity of the fixed-length encoding approach confirms less complex decoding policies but has inefficient memory usage.

25.6.1 Gene Pool

The DNA sequences of different processors are stored in a gene pool. Each DNA sequence in the pool has a potency factor linked with it. The eradication of the last 10% (rate of evolution) of the DNA sequence (based on their potency factor) mimics the natural selection process. In the DNA combination, stage two DNA sequences are randomly selected and allowed to combine. The enclosure of the DNA sequence of a new processor (obtained after the evaluation stage) is based on its potency factor.

25.6.2 Potency Factor

The potency factor measures the validity of the DNA sequence and its corresponding pos-sibility and presentation. It is determined during the processor evaluation phase.

25.7 Combination of DNA Sequences

A random combination of DNA sequences produces a large number of invalid resultant DNAs for every valid one. A set of heuristics has been invented for modeling the combina-tion process. These heuristics, apart from collective the yield (percentage of valid processor sequences), also maintain the realism of the combination process. The model for combina-tion is discussed below.

Each node in the tree when mapped to the DNA domain consists of five fields, shown in Figure 25.4. The processor's DNA is a linear combination of these DNA sequences.

NODE NUMBER	NODE NAME	NO. OF PARENTS	PARENT NO. ARRAY	DEPENDENCY ARRAY	NODE VALUE
8	120	8	160	160	120

DIT SIZE

Figure 25.4 DNA sequence of a node for 20 nodes tree.

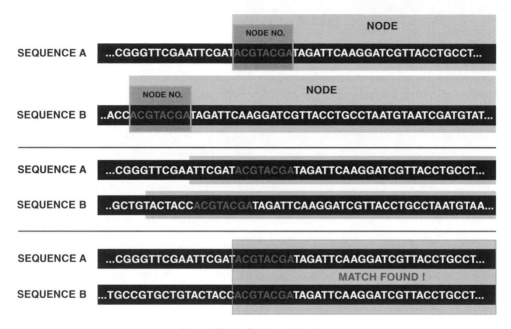

Figure 25.5 DNA combination.

Response between two DNAs comprises a selection of characteristics from both the processors. An example of DNA combination method is shown in Figure 25.5. Two probabilities have been defined as follows:

- P_1: Probability that the current node reacts with a nonhomologous node of another processor's DNA.
- P_2: Probability that the surrounding nodes react.

Another parameter SPREAD is defined as the number of nodes around the current node that might react. Spread varies from 0 to LENGTH/(20 × (NODE SIZE))

LENGTH = size of the DNA sequence in dits.
NODE SIZE = size of one node in dits

Algorithm 25.7.1

25.8 Decoding the Output String

A simple decoding strategy is used to transform the DNA sequence to its corresponding tree structure. A variable length encoding scheme, though, adheres to the natural encoding

---

Algorithm 25.7.1 DNA Combination Method

1: Begin
2: Fix P_1, P_2 and *SPREAD*.
3: $SLIDEDISTANCE = LENGTH \times (1 - P_1)$
4: Slide the 2nd sequence over the first until either a match of the node name is encountered or *SLIDEDISTANCE* becomes 0.
5: Decrement *SLIDEDISTANCE* by *NODESIZE*
6: Allow the current nodes to React*.
7: Let $SPREAD/2$ nodes on both sides of the current node of the first DNA strand to React* with a probability P_2.
8: Continue the process from line 1 until the entire 2nd DNA is exhausted.
9: *React:
10: **if** random $(0,1)<0.9$ **then**
11:    Select values for each field from either of the DNA sequence with equal probability.
12: **else**
13:    Select dit by dit from either node with equal probability.
14: **end if**
15: End

---

process, here found to produce 36% of metadata (field delimiters). This complicates the decoding process. Hence, fixed length encoding is used.

25.9 Processor Evaluation

The resultant processor needs to be estimated and its performance tested before it can be measured for production. This chapter shows how to evaluate the processor by means of an automated simulator. This simulator would reproduce the delays and the power consumption of different functional units and the net list. The interconnects (net list) are modeled using *DIMCIA* (developed at*WARFT*). It is anticipated by using BENSIM, a synthetic bench-marking tool developed at *WARFT*, to benchmark the processor and quantify its performance. Based on these results, post processing is done if necessary.

25.10 Post-Processing

On evaluating the DNA aspect obtained after the DNA combination process, it was perceived that the resultant DNA, when decoded, formed a valid processor with a probability less than 0.2. That is a valid processor that acquired only 20% of the times. It would be a waste of time to remove an offspring sequence (processor) having small discrepancies. These discrepancies can be removed by post-processing (simulated annealing), if the number of discrepancies are found to be below a certain threshold (50% of data dependencies). If not, that offspring is rejected.

Algorithm 25.10.1 Simulated Annealing Technique

1: Begin
2: *S*: user specification (initial solution)
   Class *T*: set of all combinations of deviating characteristics.
3: **repeat**
4:   **while** not yet in equilibrium **do**
5:     *S′* = some random neighboring solution of *S*
       = $C(S') - C(S)[C(x) = \text{cost function}]$
6:     $prob = min(1, e^{-/k}b^T)$
7:     **if** $random(0, 1) <= prob$ **then**
8:       $S = S'$
9:     **end if**
10:   **end while**
11:   Update *T*
12: **until** correlation > 1
13: **return** the best solution
14: End

Two examples of discrepancies follow:

1. Data discrepancy: Getting the value 0.5 for cache levels (cache level must always be integral)
2. Logical discrepancy: Getting the value of 2 for the number of cache levels and 0 for the size of level 1 cache.

Simulated annealing is employed to change the conflicting characteristics, and hence other dependent characteristics, till a preferred correlation between the resultant DNA and the user stipulations is obtained. Simulated annealing, though it takes longer to congregate, is preferable to other probabilistic search algorithms such as game theory as the accuracy of the solution is higher. If more than one parameter in the resultant DNA diverges from user specifications, these parameters are programmed in simulated annealing.

This post-processing advances the overall quality (both validity and potency) of all the processors in the gene pool. This, in return, confirms lesser number of discrepancies in the subsequent offsprings. With time, the discrepancies will be so rare that the DNA combinations most often produce valid offsprings. This would eradicate the need for further post-processing. After this point, the post-processing stage can be removed permanently for future generation of processors. Hence, post-processing is done primarily in the beginning to improve the knowledge represented by the gene pool. It could be able to evolve microprocessors without the post-processing phase (20% of the time).

The general procedure for scheduled simulated annealing technique is given in Algorithm 25.10.1.

25.11 Gene Pool Update

As stated earlier, once the processor is calculated, its performance is quantified as its "potency factor" (ability to produce new processors). Processors with a potency factor greater than "threshold potency" are added to the gene pool. The threshold potency is a extent of both the performance and the validity of the processor. The last 10% (evolution rate) of the gene pool is deleted (based on the potency factor). This models Darwin's natural selection and implements the phenomenon of evolution. The evolution rate is arbitrarily set at 10% and might be changed based on the examination.

25.12 Summary

In this chapter, the characteristics of the processor have been encoded at a higher level of abstraction. A complete depiction of the processor involves encoding not just the functionality of its various functional units but also their connectivity and engagement details. Encoding each and every interconnect is both unrealistic and redundant. Some guidelines about the net list can be encoded. For instance, the number of 10 terminal nets, the number of 5 terminal nets in the cache region can be used as guidelines. A usual method of evolving the exact connectivity matrix during the DNA (deoxyribonucleic acid) combination phase, given these guidelines, is possible. The realization of the connectivity matrix (logical) can be compared (functionally) to the physical realization of interconnections between neurons in the brain. The exact methodology for evolving this connectivity with the use of enzymes is under examination. A straightforward strategy for encoding the processor characteristics onto a DNA sequence is to map each of these characteristics to specific human characters. The development of a new processor would then involve combining two human DNAs in laboratories. The resultant DNA, when mapped back to the processor domain, would give the individualities of the new processor. This can be studied and post-processing can be done when necessary.

26

DNA-Based Reversible Circuits

This chapter explains the reversible logic gates and the composite circuits by using logic gates based on DNA. To make this understand to the readers, necessary figures and expressions are discussed in details.

26.1 DNA-Based Reversible Gates

Let $S = \begin{pmatrix} B \\ \alpha \end{pmatrix}$ and $D = \begin{pmatrix} B \\ B' \end{pmatrix}$, where S represents a single strand, D represents a double strand, B represents base strand and α is an empty base. B and B' are complement to each other.

Let $x = x_1, x_2,...,x_n$, $y = y_1, y_2,...,y_n$ and $u = u_1, u_2,...,u_n$ be n-bit inputs.
Let $z = z_1, z_2, ... , z_m$, $v = v_1, v_2,...,v_m$ and $f = f_1, f_2, ... , f_m$ be m-bit outputs,

where $x_i = s^*+d^*$, $y_i = s^*+D^*$, $f_i = S^*+D^*$, $z = S^*+D^*$, $f = S^*+D^*$, $1< = i< = n$, $n = m$ and

+ represents operation indicating either single strands or double strands; S^* means a set of all single strands including the empty strand, and D^* denotes a set of all double strands including the empty strand. From now on, the above notations are used for defining the DNA-based logic gates.

26.2 DNA-Based Reversible NOT Gate

Let $x = x_1, x_2,...,x_n$ be n-bit input and $z = z_1, z_2,...,z_m$ be m-bit output. Algorithm 26.2.1 presents the construction procedure of a DNA-based reversible NOT gate (DRNG).

Example 26.2.1 Figure 26.1 shows how bits are changed from set to clear and clear to set.

26.3 DNA-Based Reversible Ex-OR Gate

Let $x = x_1, x_2,..., x_n$, and $y = y_1, y_2,...,y_n$ be n-bit inputs. Let $z = z_1,z_2,...,z_m$ and $f = f_1, f_2,...,f_d$ be m-bit outputs. Algorithm 26.3.1 presents the construction procedure of a DNA-based reversible Ex-OR gate (DXRG).

Reversible and DNA Computing, First Edition. Hafiz Md. Hasan Babu.
© 2021 John Wiley & Sons Ltd. Published 2021 by John Wiley & Sons Ltd.

Algorithm 26.2.1 DNA-based reversible NOT gate (DRNG)

1: Begin
2: **for** each strand of input combinations **do**
3:     **if** $x_i \in x$ and $x_i \subset = S * $ **then**
4:        x_i is transformed into Z_i, where $z_i \subset = D *$;
5:     **else**
6:        x_i is transformed into Z_i, where $z_i \subset = S *$
7:     **end if**
8: **end for**
9: End

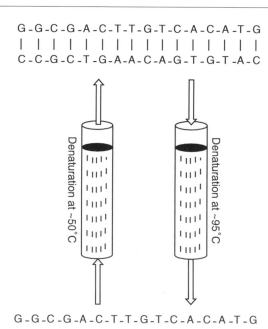

G-G-C-G-A-C-T-T-G-T-C-A-C-A-T-G
| | | | | | | | | | | | | | | |
C-C-G-C-T-G-A-A-C-A-G-T-G-T-A-C

G-G-C-G-A-C-T-T-G-T-C-A-C-A-T-G

Figure 26.1 Operation of DNA-based reversible NOT gate (DRNG).

Example 26.3.1 In this operation, the two DNA sequences representing 1010 and 0110 are used. The output of DRXG is a DNA segment representing the desired signal 1100. The operation is shown in Figure 26.2.

26.4 DNA-Based Reversible AND Gate

Let $x = x_1, x_2, \ldots, x_n$, $y = y_1, y_2, \ldots, y_n$, and $u = u_1, u_2, \ldots, u_n$ be n-bit inputs. Let $z = z_1, z_2, \ldots, z_m$, $v = v_1, v_2, \ldots, v_m$, and $f = f_1, f_2, \ldots, f_m$ be m-bit outputs. Algorithm 26.4.1 presents the construction procedure of a DNA-based reversible AND gate (DRAG).

Example 26.4.1 In this example, the input DNA sequences are 1010 and 0110. The output sequence of DRAG is 0010. The total operation is shown in Figure 26.3.

Algorithm 26.3.1 DNA-Based Reversible Ex-OR Gate (DRXG)

1: Begin
2: **for** every two strands of input combinations **do**
3: **if** $x_i \in x$ and $x_i \mathbb{C}= S *$ **then**
4:   **if** $y_i \in y$ and $y_i \mathbb{C}= S *$ **then**
5:    y_i is transformed into f_i, where $z_i = x_i$ and $f_i \mathbb{C}= S *$;
6:   **else**
7:    y_i is transformed into f_i, where $z_i = x_i$ and $f_i \mathbb{C}= D *$;
8:   **end if**
9: **else**
10:   **if** $y_i \in y$ and $y_i \mathbb{C}= S *$ **then**
11:    y_i is transformed into f_i, where $z_i = x_i$ and $f_i \mathbb{C}= S *$
12:   **end if**
13: **end if**
14: **end for**
15: End

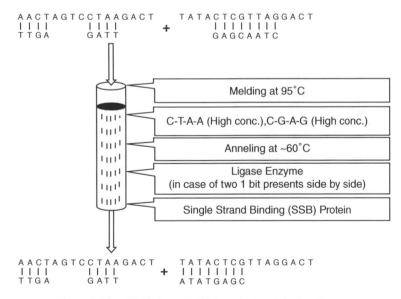

Figure 26.2 Operation of DNA-based reversible Ex-OR gate.

26.5 DNA-Based Reversible OR Gate

Let $x = x_1, x_2,..., x_n$, $y = y_1, y_2,...,y_n$, and $u = u_1, u_2,...,u_n$ be n-bit inputs. Let $z = z_1, z_2,...,z_m$, $v = v_1, v_2,...,v_m$, and $f = f_1, f_2,...f_m$ be m-bit outputs. Algorithm 26.5.1 presents the construction procedure of a DNA-based reversible OR gate (DROG).

Algorithm 26.4.1 DNA-Based Reversible AND Gate (DRAG)

1: Begin

2: **for** every two strands of input combinations **do**

3:    **if** $x_i \in x$ and x_i C= $S * $ **then**

4:       u_i is transformed into f_i, where $z_i = x_i$, $v_i = y_i$ and f_i C= $S *$;

5:    **else if** $y_i \in y$ and y_i C= $S * $ **then**

6:       u_i is transformed into f_i, where $Z_i = x_i$, $v_i = y_i$ and f_i C= $D *$;

7:    **else**

8:       u_i is transformed into f_i, where $z_i = x_i$, $v_i = y_i$ and f_i C= $S *$;

9:    **end if**

10: **end for**

11: End

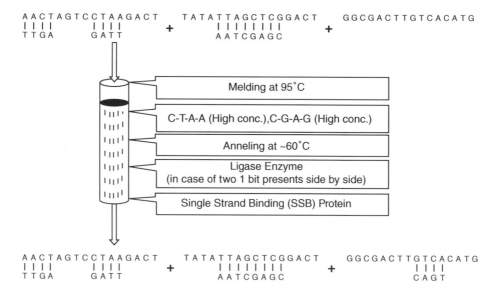

Figure 26.3 Operation of DNA-based reversible AND gate.

Example 26.5.1 Consider 1010 and 0110 are two DNA sequences. These sequences are used to generate the output sequence of DNA-based reversible OR gate (DROG), where the output sequence is 1110. The operation is shown in Figure 26.4.

Property 26.5.1 A DNA logic function is reversible if there is one-to-one mapping between input and output vectors.

Proof: Since the numbers of elements in the input and output vectors of DNA strands are equal and there is a one-to-one mapping between an input DNA strand and an output DNA strand, a DNA logic function is reversible.

Algorithm 26.5.1 DNA-based reversible AND Gate (DRAG)

1: Begin
2: **for** every two strands of input combinations **do**
3:     **if** $x_i \in x$ and $x_i \subset = D *$ **then**
4:         u_i is transformed into f_i, where $z_i = x_i$, $v_i = y_i$ and $f_i \subset = D *$;
5:     **else if** $y_i \in y$ and $y_i \subset = D *$ **then**
6:         u_i is transformed into f_i, where $Z_i = x_i$, $v_i = y_i$ and $f_i \subset = D *$;
7:     **else**
8:         u_i is transformed into f_i, where $z_i = x_i$, $v_i = y_i$ and $f_i \subset = S *$;
9:     **end if**
10: **end for**
11: End

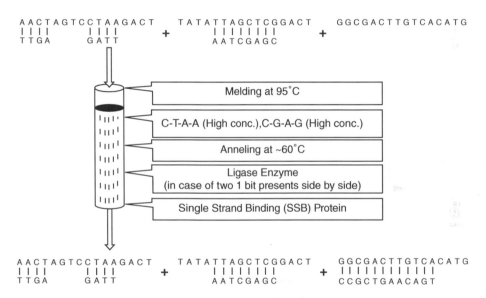

Figure 26.4 Operation of DNA-based reversible OR gate.

Example 26.5.2 In the DNA-based reversible Ex-OR operation in Example 26.3.1, the sizes of the input vectors and the output vectors are equal, i.e., 4. These input and output vectors correspond to the input DNA strands and the output DNA strands, respectively. So, the Ex-OR-based DNA logic function is reversible.

26.6 DNA-Based Reversible Toffoli Gate

A Toffoli gate has three inputs, where the first two are known as control inputs, while the third one is known as a target input. The gate leaves both control inputs unchanged, introducing garbage outputs; flips the target input if both the control inputs are set; and otherwise leaves the target input alone.

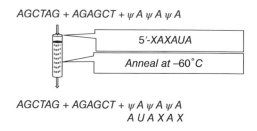

$AGCTAG + AGAGCT + \psi A \psi A \psi A$

| 5'-XAXAUA |
| Anneal at –60°C |

$AGCTAG + AGAGCT + \psi A \psi A \psi A$
$A\,U\,A\,X\,A\,X$

Figure 26.5   Overall procedures of DNA hybridization for a DNA-based Toffoli gate.

To implement a DNA-based reversible Toffoli gate, a sticking system is used. Sticky end of DNA strands are to implement overall computation. Specific temperature helps to anneal DNA bases and additional bases are used to finalize the computation. To construct this gate, two DNA strands are used that act as control inputs and the other one is used as target input. The final output is based on those two control input signals. The final output signal is generated when appropriate DNA segments bind with the target input. Here, the two control inputs and the target input are represent by a string of bits and each bit is represented by a dinucleotide. Both of the control input signals are constructed using the dinucleotide $5'AG$ and $5' - CT$ representing bit 1 and 0, respectively. The target input is constructed with $5' - \psi A$ and $5' - UA$ as bit 1 and 0, respectively. Similarly, the final output is constructed using mixer of $3' - AU$, $3' - AD$, $3' - AT$, and $3' - AX$ representing bit 1, 1, 0, and 0, respectively. Here, the non-natural bases are used to make the gate more efficient in terms of performance and also to modulate the final output signals. For example, whenever there are $3' - AX$ (which is equal to 0) and $3' - AD$ (which is equal to 1), they compel making the output signal in favor of their signal value i.e., 0 and 1, respectively.

Example 26.6.1   In Figure 26.5, there are three inputs 101, 110, and 111 which produce three outputs 101, 110, and 100.

In a sticking system, to make the output signal, it needs to denature and rena-ture the DNA. Here DNA is not necessarily being denatured because it is already in denatured single-stranded form. At first, the predesigned DNA segments represent-ing the control inputs $(5' - AGCTAG = 101$ & $5' - AGAGCT = 110)$ and target input $(5' - \psi A \psi A \psi A - 111)$ are added in test tube system. Then according to desired output signal, a predesigned oligonucleotide $(3' - AUAXAX = 100)$ is added in the test tube. If the temperature is kept low (60°C), then this oligonucleotide forms complementary hydrogen bonds with the target input and produce a double-stranded DNA fragment that ultimate represents the output signal 100. Here, both of the control input signals remain intact making the gate reversible.

26.6.1   Fan-out Technique of a DNA-Based Toffoli Gate

In a DNA-based Toffoli gate, the third output is formed in a double strand. To make any sequential circuit using the gate, it is necessary to form all inputs in a single strand. To fan-out third output of the gate, a unique tag is used in target input. As a result, the third output is formed in a partially double strand where the left part with unique tag is a double strand (representing the output signal) and the right part is a single strand (equivalent to the output signal), which can be used as a fan-out for target input of the gate.

Property 26.6.1.1 The DNA-based Toffoli gate requires $n.2^n$ bases for control inputs, $n.2^n$ bases for target inputs, and 2^n bases for unique tag, which finally produces $2n.2^n + 2^n$ bases for target outputs with other control inputs, where n is the number of bit length.

Proof: The DNA-based Toffoli gate uses two control inputs and one target input as single-stranded DNAs. Depending on the variation of target outputs, different unique tags are used. As each binary value is represented by two bases of DNA, for n different values, it uses 2^n different DNA bases. Thus, it requires $n.2^n$ DNA bases for control inputs. The DNA bases for a unique tag are constant, i.e., 2^n. Similar calculations are applicable for target inputs. Finally, for target outputs, it requires two single-stranded target inputs that are complementary to each other. So, the possible value of DNA bases for target outputs is $2n.2^n + 2^n$.

26.6.2 DNA-Based Reversible NOT Operation

For performing the logical NOT operation, one input variable is fed into the one control input of the Toffoli gate. Another control input and the target input are always constant high. The final output is produced by nipping the target input. Also, two unchanged control inputs are produced, as there are two garbage outputs. Here, the conditional NOT behavior of logical Ex-OR operation is used to perform a logical NOT operation. It means, when one variable is Ex-ORed with logical high, it always produces the logical inversion of that variable.

In Figure 26.6, two control inputs are represented by single-stranded DNAs that are equivalent to binary 1010 and 1111, respectively. The target input is also represented by single-stranded DNA, which is equivalent to binary 1111. The final output is double-stranded DNA, representing binary 0101 with two garbage outputs. Here, there are in total three inputs, where one input is variable and the other two are constant. Also, three outputs are produced in the Toffoli gate, where only one is representing the logical NOT operation with two garbage outputs.

26.6.3 DNA-Based Reversible AND Operation

For performing the logical AND operation using the Toffoli gate, two inputs are fed into two control inputs of the Toffoli gate. The target input contains all zeros. The resulting outputs

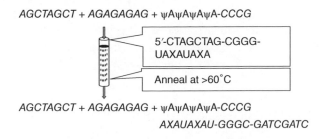

AGCTAGCT + AGAGAGAG + ψAψAψAψA-CCCG

5'-CTAGCTAG-CGGG-UAXAUAXA

Anneal at >60°C

AGCTAGCT + AGAGAGAG + ψAψAψAψA-CCCG

AXAUAXAU-GGGC-GATCGATC

Figure 26.6 DNA-based Toffoli gate as NOT gate.

AGCTAGCT + CTAGAGCT + UAUAUAUA

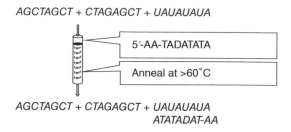

AGCTAGCT + CTAGAGCT + UAUAUAUA
ATATADAT-AA

Figure 26.7   DNA-based Toffoli gate as AND gate.

AGCTAGCT + CTAGAGCT + ψAψAψAψA

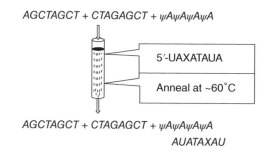

AGCTAGCT + CTAGAGCT + ψAψAψAψA
AUATAXAU

Figure 26.8   DNA-based Toffoli gate as OR gate.

keep the two control inputs unchanged and the changed target output represents the logical AND operation of two given input values.

In Figure 26.7, two control inputs are represented by DNA bases that are equivalent to binary value 1010 and 0110, respectively, and the target input is also represented by DNA bases that are equivalent to binary 0000. The final outputs are one double-stranded DNA representing 0010 with garbage outputs. Here, the DNA-based logical AND operation consists of three inputs in which one is low and, among the three outputs, two outputs are garbage outputs.

26.6.4   DNA-Based Reversible OR Operation

To execute the logical OR using a Toffoli gate, the target input of the Toffoli gate is kept constant high (all ones) and two inverted input variables are fed into two control inputs of the Toffoli gate. The resulting outputs are two control inputs that remain unchanged during operations to keep the logic operation reversible.

In Figure 26.8, two control inputs are represented by single-stranded DNAs, encoded with DNA bases, which are equivalent to binary value 1010 and 0110, respectively; the target input is also represented by single-stranded DNA, which is equivalent to binary 1111, and the final output is one double-stranded DNA signals that is equivalent to 1101. Here, three outputs are produced in which two outputs are garbage outputs to preserve the logical reversibility.

26.6.5   DNA-Based Reversible Ex-OR Operation

Finally, to execute the logical Ex-OR operation, one input is fed into one control input of the Toffoli gate and other input is fed into the target input of the Toffoli gate, and the remaining

AGCTAGCT + AGAGAGAG + UAψAψAUA-GGGC

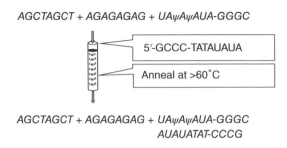

AGCTAGCT + AGAGAGAG + UAψAψAUA-GGGC
AUAUATAT-CCCG

Figure 26.9 DNA-based Toffoli gate as Ex-OR gate.

control input of the Toffoli gate is always kept constant high. The final output is produced by flipping the target input of the Toffoli gate. Also, two unchanged control inputs represent two garbage outputs.

In Figure 26.9, two control inputs are represented by two single-stranded DNAs, which are encoded with different DNA bases representing binary values 1010 and 1111, respectively. Also, the target input is a single-stranded DNA, which is equivalent to binary 0110. One input is fed into the first control input and another is fed into the target input of the Toffoli gate. The final output is a double-stranded DNA that is equivalent to binary 1100 representing the logical Ex-OR operation of two given inputs. Also, two garbage outputs are produced with final output.

26.6.6 Properties of DNA-Based Reversible Toffoli Gate

1. Operational Speed: DNA-based reversible Toffoli gate is faster.
2. Error Correction: It has built-in error correction mechanism.
3. Fan-Out Technique: It is necessary to add extra DNA-bases.
4. Life-Time: It serves longer time by maintaining and varying temperature.
5. Power Consumption: It consumes less power than the Toffoli gate.
6. Parallelism: It provides parallel processing.

26.6.7 DNA-Based Reversible Fredkin Gates

The Fredkin gate is one of the popular reversible gates that is a fault-tolerant and conservation gate. It can be used as a controlled switch based on the logical expression of its outputs. The first input can be used as switch. In this case, if it is logical low, then other two inputs remain unchanged and, on the other hand, if it is high, then other two inputs interchange their positions. Therefore, it can be used to select a single input out of two given inputs. Figure 26.10 shows the realization of a Fredkin gate using DNA. It requires four steps: denaturation, mixing, differential electrophoresis, and annealing.

Initially, one primary input is denatured before being mixed with secondary inputs. Secondary inputs contain input single strand of all primary inputs. So, in differential electrophoresis, if a thicker band is selected, the same input single strand of initially taken primary input will be found because of its high concentration. And, if the thinner band is selected, other input single strand than that of initially taken primary input will be found. Finally, the annealing with suitable operand stand gives the readable double-stranded output.

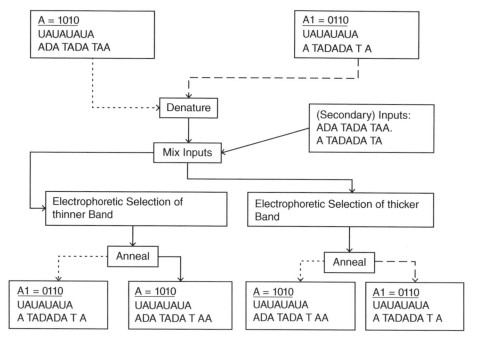

Figure 26.10 Procedures of DNA hybridization for DNA-based Fredkin gate.

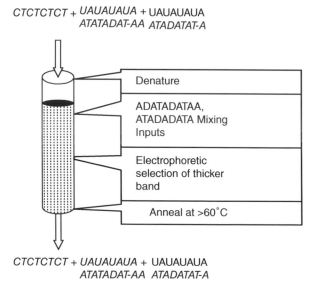

Figure 26.11 DNA hybridization of selection operation between two ANDed products using DNA-based Fredkin gate.

So, in short, after mixing of denatured primary input with the mixture of all input single strands, if a thicker band is selected, the primary input will remain the same, ultimately. And, if the thinner band is selected, the other primary input will be generated through annealing in place of initial one. The hybridization of selection operation using a DNA-based Fredkin gate is shown in Figure 26.11.

26.7 Realization of Reversible DNA-Based Composite Logic

In this section, reversible DNA-based composite logic has been realized by a half-adder circuit using a DNA-based Toffoli gate. There are two Toffoli gates used in this circuit, one for the AND operation and the other for Ex-OR operation. Three inputs have been fed in the AND gate that produces an output of AND operation and two outputs that remain the same as the first two inputs. These two unchanged outputs have been fed to the input of the Ex-OR gate. Finally, the Ex-OR is producing the output, which is similar to the summation output of a half-adder circuit. Figure 26.12 shows the reactions of a DNA-based gate correspond to the logic of the half-adder.

Figure 26.13 shows the realization of reversible DNA-based half-adder circuit. With this realization and reactions, it can be said that other reversible DNA-based composite circuits can also be accomplished.

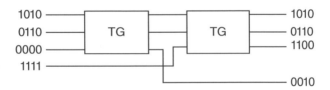

Figure 26.12 Gate-level representation of reversible half-adder using Toffoli gate.

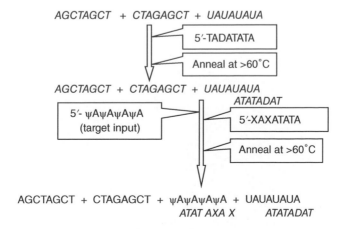

Figure 26.13 Reversible DNA-based half-adder.

26.8 Summary

In this chapter, an approach to design DNA-based Toffoli gate has been realized, where both inputs and outputs are DNA-based and computations are performed using several biochemical operations. The DNA-based Toffoli gate holds the properties of the original one. This study also aims at exploiting some of these bases as well as natural bases to design a DNA-based Toffoli gate. Here, these modified bases are used to augment the complimentary bonding options that can be utilized to represent discrete signals of bits. All of these three bases are complimentary to Adenine and hence, they provide a broad window of complementarities to design the circuit. Also, some of these bases are used to modulate the final output signals. This is one of the key characteristics of the DNA-based Toffoli gate. Finally, a DNA-based reversible composite circuit was shown using the gate, which creates the possibility of stimulating any other composite reversible circuit using this gate. Moreover, DNA-based Toffoli gate can further be used in nano-scale computing, low-power computing modules, and quantum computing.

27

Addition, Subtraction, and Comparator Using DNA

As a powerful material, DNA presents great advantages in the fabrication of molecular devices and higher-order logic circuits. In this chapter, designs of DNA-based adder, subtractor, and comparator are presented. The use of hybridization and displacement of strands DNA-based platform are developed to implement adder, subtractor, and comparator arithmetic processes.

27.1   DNA-Based Adder

In this section design of a DRFA (DNA-based reversible full-adder) has been described. The basic reversible gates have been used to design the DRFA. A DNA-based half-adder circuit is presented in Section 26.7. Likewise the half-adder, the full-adder circuit performs addition operation of A, B, and C, where those three are input variables. So, the equations of summation and carry using full-adder circuit are as follows: Sum $= A \oplus B \oplus C$ and Carry $= AB \oplus AC \oplus BC = C (A \oplus B) \oplus AB$.

The construction procedures DRFA are discussed in this section. At first, two DNA strands are taken as inputs in two test tubes, TA and TB. Then the DNA-based Ex-OR operation is performed using DRXG. Fan-out the Ex-OR output to another test tube and again the DNA-based Ex-OR operation is computed with C_{in}, the output DNA strands represent the sum of DRFA. To get the C_{out}, two inputs A and B are taken and a constant single strand that indicates 0 in a test tube, the output of this operation (AB) is taken to a test tube Tl. Now, again the input A and B are taken in another test tube and with DRXG, it is computed and the output A EB B to a test tube $T2$. Then, repeat the first process for ($AtJJ B$) and C_{in}. Again, the output of this process is taken to a test tube $T3$ and repeat the second process for (A EB B), C_{in} and AB. The final output is the C_{out}. The gate level representation of a reversible full-adder circuit is shown in Figure 27.1.

Property 27.1.1   A composite DNA logic is reversible, if it holds the reversible properties.

Proof: When the individual DNAs of the composition are reversible and their interfaces maintain the reversibility, the composite DNA logic is reversible.

Example 27.1.1   In Figure 27.2, the DNA-based full-adder circuit is reversible, as its individual units consisting of DRXG, DROG, and DRAG and their interfaces are reversible.

Reversible and DNA Computing, First Edition. Hafiz Md. Hasan Babu.

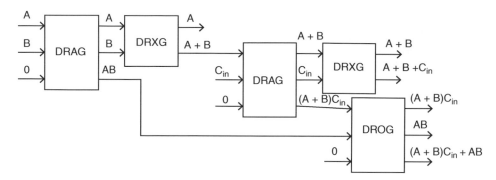

Figure 27.1 Gate-level representation of a reversible full-adder circuit.

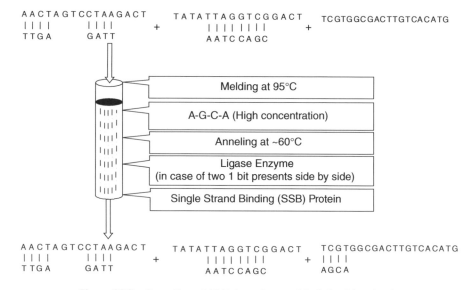

Figure 27.2 Operation of DNA-based reversible full-adder circuit.

The minimized DRFA (DNA-based reversible full-adder) using the DNA-based reversible logic gates has four garbage outputs.

Property 27.1.2 Let R be the highest number of frequency of any element in an output vector. Then, the minimum number of garbage outputs is produced to make DNA logic reversible is $flog3\ R$.

Proof: As the highest number of frequency of any element in an output vector (R) defines the number of garbage outputs at the output level of the DNA function that keep the function reversible, the minimum number of garbage outputs to make DNA logic reversible is $f\ log3\ R$.

Example 27.1.2 In the irreversible two-input AND operation, two garbage outputs are enough to make it reversible as $flog2\ 3 = 2$. It is possible to make DRFA with two garbage outputs that are the minimum garbage outputs for DRFA.

Example 27.1.3 Two DNA strands represent 1010 and 0110 as inputs. Another DNA strand C_{in} represents 00000. The result of the summation is 10000.

27.2 DNA-Based Addition/Subtraction Operations

At first, this section presents how addition and subtraction operations could be combined into one circuit representation. Then, DNA-based representation of addition/subtraction operations is presented with block diagram and biochemical operations. To perform the operations, it uses DNA bases as signal values.

27.2.1 Addition and Subtraction Operations

Unlike adder circuit, a half-subtractor circuit performs the subtraction operation of A and B, where A and B are input variables. So, equations of borrow and difference are as follows: Diff $= A \oplus B$, Borrow $= AB$. Similarly, the full-subtractor circuit performs subtraction operation of A, B, and C; where A, B, and C are input variables. Thus, the equations of borrow and difference are as follows: Diff $= A \oplus B \oplus C$, Borrow $= AB$.

27.2.2 Procedures of DNA-Based Reversible Addition/ Subtraction Operations

From the above discussion on addition and subtraction operations, we can easily construct a circuit that will be capable of performing both addition and subtraction using one circuit-level representation. Figure 27.3 shows the overall gate level diagram of the

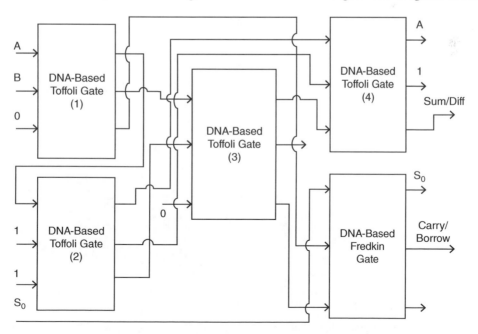

Figure 27.3 DNA-based reversible adder/subtractor circuit.

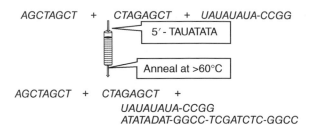

AGCTAGCT + CTAGAGCT + UAUAUAUA-CCGG

5′ - TAUATATA

Anneal at >60°C

AGCTAGCT + CTAGAGCT +
                    UAUAUAUA-CCGG
                    ATATADAT-GGCC-TCGATCTC-GGCC

Figure 27.4 DNA hybridization of logical AND operation using DNA-based Toffoli gate.

combined adder/subtractor circuit. Here, four DNA-based Toffoli gates and one DNA-based Fredkin gate are used to perform addition/subtraction operation. As the logical expressions of summation and difference are the same for both addition and subtraction, one Toffoli gate is enough to do so. Also, one Toffoli gate is used to produce carry operation and another two are used to complement first input and produce borrow operation respectively. Finally, depending on the value of the first control input, the carry or the borrow is selected using the final DNA-based Fredkin gate.

Figure 27.4 shows the DNA hybridization process for logical AND operation to compute carry using DNA-based Toffoli gate. Here, two control inputs of Toffoli gate are represented using DNA bases, which are equivalent to binary 1010 and 0110, respectively. The target input of the Toffoli gate also consists of DNA bases, which is equivalent to binary 1111. The final outputs keep the two control inputs unchanged with flipped double-stranded target output, which represents 0010 in binary. This output is preserved and it will work as a second input of the DNA-based Fredkin gate.

In Figure 27.5, the hybridization process for the DNA-based Toffoli gate is presented where the first input is the output of the previous Toffoli gate. The second and third inputs are single-stranded DNA, which represents all ones in binary. The produced first two outputs are actually unchanged, corresponding to the given two inputs. The other output is double stranded, which represents the complement of the first input and its binary value is 0101.

Another logical AND operation using DNA hybridization process is shown in Figure 27.6 where the first control input of the Toffoli gate is taken from the second output of the first Toffoli gate, the second control input of that gate is also taken from third output of the second Toffoli gate. And the target input is 0000, which is a single-stranded DNA. The produced

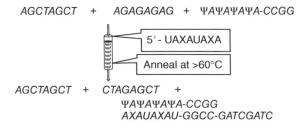

AGCTAGCT + AGAGAGAG + ΨΑΨΑΨΑΨΑ-CCGG

5′ - UAXAUAXA

Anneal at >60°C

AGCTAGCT + CTAGAGCT +
                    ΨΑΨΑΨΑΨΑ-CCGG
                    AXAUAXAU-GGCC-GATCGATC

Figure 27.5 DNA hybridization of logical NOT operation using DNA-based Toffoli gate.

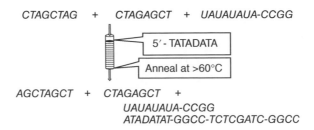

CTAGCTAG + *CTAGAGCT* + *UAUAUAUA-CCGG*

5′ - TATADATA

Anneal at >60°C

AGCTAGCT + *CTAGAGCT* +
                   UAUAUAUA-CCGG
                   ATADATAT-GGCC-TCTCGATC-GGCC

Figure 27.6 DNA hybridization of logical AND operation between complemented and noncomplemented literals using DNA-based Toffoli gate.

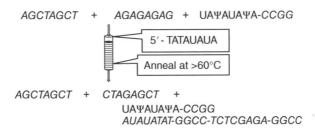

AGCTAGCT + *AGAGAGAG* + *UAΨAUAΨA-CCGG*

5′ - TATAUAUA

Anneal at >60°C

AGCTAGCT + *CTAGAGCT* +
                   UAΨAUAΨA-CCGG
                   AUAUATAT-GGCC-TCTCGAGA-GGCC

Figure 27.7 DNA hybridization of logical Ex-OR operation using DNA-based Toffoli gate.

target output represents AB, where A and B are given inputs. It is 0100 in binary, which is preserved to use as he third input of the DNA-based Fredkin gate. Its first two outputs are also same as given control inputs.

The final Toffoli gate is used to perform the logical Ex-OR operation of two given inputs as shown in Figure 27.7, where two control inputs of Toffoli gate are first two outputs of second Toffoli gate. The target input represents the second given input. The produced outputs are single-stranded two unchanged control inputs with one double-stranded flipped target input, which represents the logical Ex-OR operation two given inputs, where the inputs are 1101 in binary.

Finally, Figure 27.8 shows the selection operation between carry and borrow operations using a DNA-based Fredkin gate, where the first input of the Fredkin gate works as a switch when it is logically low. The second and third inputs remain unchanged when it is logically high. The second and third inputs will interchange their positions. The second input contains carry and the third input contains borrow, which are double-stranded DNAs with extra bases. Here, as the selecting input is asserted low, the second and the third inputs do not change their positions and these inputs thus remain unchanged. So, the carry is propagated for further use.

Algorithm 27.2.2.1 depicts the overall procedures of the DNA-based combined addition and subtraction operations. If n is the total number of bits of given inputs, then the run time complexity of this algorithm is $O(n)$.

Property 27.2.2.1 The design would perform both the DNA-based addition and subtraction operations using single circuit representation.

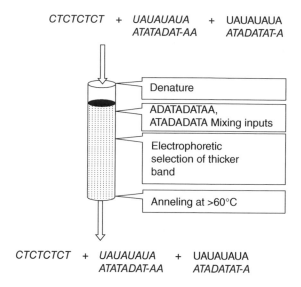

CTCTCTCT + *UAUAUAUA* + *UAUAUAUA*
ATATADAT-AA *ATADATAT-A*

Figure 27.8 DNA hybridization of selection operation using DNA-based Fredkin gate.

Algorithm 27.2.2.1 DNA-Based Reversible Addition/Subtraction Algorithm

---

1: Begin
2: **while** i equals to 1 to n **do**
3:   Carry: DoToffoli(A_i, B_i, 0);
4:   temp«DoToffoli (A_i, 1,1);
5:   Borrow <— DoToffoli (A_i, B_i, 0);
6:   Add/Sub$i$<—DoToffoli (A_i, 1, B_i);
7:   Carry/Borrow i<—DoFredkin (control, $Carry_i$,Borrow)
8: **end while**
9: Procedure DoToffoli (first, second, third)
10: **return** first, second, third;
11: End procedure
12: Procedure DoToffoli (first, second, third)
13: **if** first $= 0$ **then**
14:   **return** second;
15: **else**
16:   **return** third;
17: **end if**
18: End Procedure
19: End

Proof: The design method consists of a DNA-based Fredkin gate and a DNA-based Toffoli gate, which performs binary addition and subtraction operations using one control Signal Add/Sub. However, four DNA-based Toffoli gates perform *Sum(AB)*, *Diff*$(\overline{AB})$ and *Carry/Borrow* ($A \oplus B$), separately. Then, electrophoresis selection is used in the Fredkin gate to select-out one signal between *Sum* and *Diff*. So, if *Add/Sub* is low, the overall circuit uses a Fredkin gate as controlled switch and it feds out only $A \oplus B$ and AB. On the other hand, if Add/Sub is high, it feds out only $A \oplus B$ and $\overline{AB}$. Besides, a DNA-based full-addition/subtraction operation requires only one additional DNA-based Ex-OR gate, (DNA-based Toffoli gate) and two DNA-based half adder/subtractor circuits to perform either an addition operation or a subtraction operation. Add/Sub signal indeed is used as control signal to decide which operations should be performed.

27.3 DNA-Based Comparator

A DNA comparator compares the concentration of the target strand with that of the reference strand. To extend the potentiality of autonomous machines consisting of DNA, the DNA comparator is designed by utilizing a difference of kinetics between hybridizations and branch migrations. A DNA comparator can detect higher concentrations of target strand compared to that of the reference strand.

The DNA comparator is a DNA machine that was developed to compare the DNA concentrations by utilizing hybridization and branch migration reactions such as Seelig's DNA logic gate. The purpose of DNA comparator is to determine whether the concentration of target strand is higher than that of the reference strand. If the concentration of target is higher, the DNA comparator generates an output with a single-strand that may be input strand to the other molecular logic gate.

The DNA comparator consists of three double-strands and two inputs as presented in Figure 27.9. One of two inputs is a target strand whose concentration is the standard. Two of three double-strands have bulge loops in the middle of sequences and toe-hold structures, where two inputs can be hybridized at the 5′-end of the opposite sequences.

A DNA comparator works as follows: First, two inputs such as reference and target hybridize to the toe-holds of blg_comp1 and blg_comp2, respectively. Then, the reference and the blg_comp1 form a double strand, while the target and the blg_comp2 also form a double strand due to the branch migration reaction. As a result, bulge1 and bulge2 are released as single strands and immediately form a double-strand, as the loop domains of bulge1 and bulge2 are complementary to each other. Thus, only the extra amount of bulge2 can hybridize to the toe-hold of BHQ2-out comp due to the branch migration. Now the TRed-output2 is released as the single strand that can be monitored by the fluorescent intensity of Texas Red. The key point is that the hybridization between bulge1 and bulge2 is much faster than the branch migration between TRed-output2 and BHQ2-out comp of

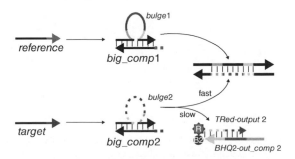

Figure 27.9 The working principle of DNA comparator.

bulge2. Thus, if the concentration of bulge2 is lower than that of bulge1, the almost all bulge2 probably hybridize to bulge1 rather than the toe-hold of BHQ2-out comp. But if the concentration of target is higher than that of reference, the bulge2 can hybridize to the toe-hold of BHQ2-out comp and remove TRed-output2 from BHQ2-out comp due to the branch migration. The full working procedure of a DNA comparator is shown in Figure 27.9.

27.3.1 Sequence Design

DNA sequences are carefully designed as the sequence cannot be hybridized each other except for complementary sequences. The stabilities of every DNA duplexes are calculated by the minimum free energies (ΔG_{min}). The ΔG_{min} is predicted using the program obtained by nucleic acid sequences for DNA computing based on a thermodynamic approach. In addition, the sequences are designed in such a way so that the ΔG_{min} of the toe-hold domains fell into a narrow range in order to uniform the efficiency of branch migration. For the same reason, the sequences are also designed in such a way so that the ΔG_{min} of the duplex domains fell into a narrow range. To prevent the unintended secondary structure, sequences are further designed in such a way so that they do not have continuous repeats of the same base.

The best sequences are searched according to the above criteria by a hill-climbing algorithm. Eventually, the best sequences are chosen among the trials as shown in Table 27.1. The ΔG_{min} of any duplexes except for commentaries are at least -4.65 kcal/mol. The deviation of ΔG_{min} of the toe-hold domains is in the 0.15 kcal/mol range, while that of the duplex

Table 27.1 Sequence list

Sequence name	Sequence[a] (5′ → 3′)
reference	CCAAACTACTTACGTCTTCTAAGCAACTAACTGATG
bulge1	CCAAACTACTTACGTTGAACATACACCGAGGTTTAGTCCAAACTTCT AAGCAACTAA
blg_comp1	CATCAGTTAGTTGCTTAGAAGACGTAAGTAGTTTGG
FAM1-bulge1[b]	F-CCAAACTACTTACGTTGAACATACACCGAGGTTTAGTCCAAACTT CTAAGCAACTAA
BHQ1-blg_comp1[b]	CATCAGTTAGTTGCTTAGAAGACGTAAGTAGTTTGG-B1
target	TATAAGTCAGGTCTCTTTCGTATACCACAATTCCAA
bulge2	TATAAGTCAGGTCTCTTTGGACTAAACCTCGGTGTATGTTCATTTCG TATACCACAA
blg_comp2	TTGGAATTGTGGTATACGAAAGAGACCTGACTTATA
TRed-bulge2[b]	R-TATAAGTCAGGTCTCTTTGGACTAAACCTCGGTGTATGTTCATTTC GTATACCACAA
BHQ2-blg comp2[b]	TTGGAATTGTGGTATACGAAAGAGACCTGACTTATA-B2
TRed-output2	R-TTTGGACTAAACCTCGGTGTA
BHQ2-out_comp2	TGAACATACACCGAGGTTTAGTCCAAA-B2

a) The F, R, B1, and B2 at the end of sequences denote FAM, Texas Red, BHQ1, and BHQ2, respectively.
b) These sequences are used only to estimate the kinetic constants.

domains is in the 0.20 kcal/mol range. Furthermore, the obtained sequences have at most three continuous repeats of the same base.

Thus, a DNA comparator compares the concentration of target with that of reference, which is developed by utilizing the hybridization and branch migration reactions. It can detect at least four times higher concentration of target compared with that of the reference. It is also utilized to release two different outputs according to the difference of concentrations of two targets.

27.3.2 Estimation of Rate Constant

The rate equation can be represented by the following second-order kinetics:

$$I + OC \xrightarrow{k} IC + O \tag{27.3.2.1}$$

where I, O, and C represent the input strand, output strand, and complementary strand to the input, respectively. Here, the output strand represents the single strand produced by the branch migration rather than the output of the DNA comparator. Furthermore, the initial concentration of I is similar to that of OC, which is as follows:

$$I_0 = OC_0 \tag{27.3.2.2}$$

where I_0 and OC_0 are the initial concentrations of I and OC, respectively.

Thus, the rate constant of branch migration is estimated using the following equation:

$$\frac{dO}{dt} = k(\alpha I_0 - O)(\alpha OC_0 - O) \tag{27.3.2.3}$$

$$= k(\alpha I_0 - O)^2 \tag{27.3.2.4}$$

where the parameter α represents the ratio of completion of branch migration. Since all inputs do not complete the branch migration, some double strands with the toe-hold probably remain when a branch migration is completed. The following equation is obtained by solving the above equation:

$$O = \frac{(\alpha I_0)^2}{k} t1 + \alpha I_0 kt \tag{27.3.2.5}$$

To convert the concentration into the fluorescent intensity, the above equation is multiplied by dF/I_0, where the dF is the difference that is calculated by subtracting the fluorescent intensity of quenched state from that of nonquenched state F_0. Therefore, the fluorescent intensities of a function t can be estimated by the following equation:

$$f(t) = F_0 + \frac{dF\alpha^2 I_0 kt}{1 + \alpha I_0 kt} \tag{27.3.2.6}$$

27.4 Summary

This chapter discussed both inputs and outputs of the reversible adder, subtractor, and comparator circuit using DNA bases, and the computations were performed using DNA. This realization aims to exploit some of these bases (X, ψ, and D) as well as natural bases to design a DNA-based composite circuit. These bases are used to augment the complementary bonding options that can be utilized to represent discrete signals of bits, which are complementary to Adenine. Hence, these bases provide a broaden window of the complementary to design the circuit. Also, some of these bases are used to modulate the final output signals, which are mostly prioritized. So, whenever these are represented in the output, the final signal will be the signal that they carry. This is one of the key characteristics of the DNA-based circuit. Unlike reversible logic gate, there is no use of silicon chip, only DNA bases are used in the designs of the circuit.

28

Reversible Shift and Multiplication Using DNA

A multiplication operation works on two numbers: multiplicand and multiplier and it produces final result as product. It usually uses each single bit of multiplier to shift the value of multiplicand and add shifted result with partial product. For an n-bit multiplier, this process iterates n times and finally, the product is produced. The multiplication operation also works using the concept and it uses DNA bases instead of electronic signals as computation signals. It consists of three steps: working on single bit of multiplier, shifting the multiplicand and adding shifted multiplicand with the partial product to produce final product. If n is the number of total bits of multiplier and m is the number of total bits of multiplicand, then the number of total bits of product is $(n+m)$.

The DNA-based multiplication operation will be performed using a DNA-based shifter and a DNA-based adder circuit.

28.1 DNA-Based Reversible Shifter Circuit

A shifter is a circuit that is used to shift a number to left or to right for different kinds of mathematical operations. For example, during multiplication or division of two numbers, the shifter is the most successful approach to execute that operation. The reversible DNA shifter can use any input which is raw DNA strands (represent a bit sequence). Depending on the looping number of shifter, the desired output is generated. An unprecedented mechanism involves intriguing single strands that remain unchanged even after completion of the operation and help fostering the footprint of initial input. From that sense, the circuit can be asserted as reversible inferably.

28.1.1 Procedures of DNA-Based Shifter Circuit

To construct the shifter circuit, the procedure and technical parameters (with slight modifications) of related work have been adopted. As like that study, two single-stranded DNA *(ssDNA)* molecules are used here that act as control inputs and the other one is used as the target input. The final output is based on those two control input signals, and it is generated when the appropriate predesigned *ssDNA* segments bind with the target input and make it double stranded *(dsDNA)*. Here, the slightly modified predesigned DNA segments

Reversible and DNA Computing, First Edition. Hafiz Md. Hasan Babu.

is given with a Poly-A tail that helps this segment to get more molecular weight compared to the target input, and this modification is necessary for the effective execution of shifter functions.

After having the logic gate function performed, there remain two input signals (two *ssDNA* fragments) and an output signal (one *dsDNA* fragment consisting of a smaller DNA control input signal and a larger predesigned segment). This double strand is denatured for separation of the smaller strand by capillary electrophoresis. This smaller strand is ready to be annealed with the modified input strand (it has Poly-A tail and inosines binds with all the bases, which indicates no value after binding with the bases. It is used for blocking one bit). It is important to note that this is the first point of difference between left-shifter and right-shifter function mechanisms. The Poly-A tail for the right shifter is at 5″ end while the tail for the left shifter is at 3′ end. Meanwhile, all other things like the larger strand are transferred to storage for reuse. This double strand is now ligated with another input (two base pairs long double-strand DNA fragment that adds either 1 or 0 bit signal) on the opposite side of flanking Poly-A nucleotide tail. This ligated double strand is the output for the first round of the shifter function.

If another round of shifter function is required, double-stranded output of the first reaction is the input here. It will loop back to the denaturation step for separation of larger strand by capillary electrophoresis again. This larger strand is ready to be annealed with the meaningless shortest input strand (MSIS), when all other things along with smaller strand is transferred to storage. For right shifting, MSIS is $5'IIUUTTT3'$, while the tail for the left shifting, is $5'TTTUUII5'$. The larger strand is now covered by MSIS partially, which is ready to be annealed again with meaningful (one bit) shorter operand strand (MSOS) on the rest larger single-stranded portion. This double strand is now ligated with another input (two base pairs long double strand according to the bases bound to inosines) on the opposite site of the flanking mono nucleotide tail. This ligated double strand is the output of the second round of the consecutive shifter function. For example, suppose an input bit is 1001 $\left(\begin{smallmatrix} U\,A\,U\,A\,U\,A\,U\,A \\ A\,D\,A\,T\,A\,T\,A\,D\,A\,A\,A\,A \end{smallmatrix} \right)$, after the first annealing of operand and input single strand, as shown in Figure 28.1. For the first right shift of the smaller operand strand, *UAUAUAUA* will be separated by capillary electrophoresis and annealed with modified input, where two bases are changed to inosines near the (5′) Poly-A tail, $5'AAAAIITATADA$. After this annealing the input will undergo first right shifting and be converted to $\left(\begin{smallmatrix} U\,A\,U\,A\,U\,A\,U\,A \\ A\,D\,A\,T\,A\,T\,I\,I\,A\,A\,A\,A \end{smallmatrix} \right)$ that means 100.

This is the result for shifter function without rotation. For the rotation, a dinucleotide *dsDNA* fragment. $\left(\begin{smallmatrix} U\,A \\ A\,D \end{smallmatrix} \right)$ (according to the bases that converted to inosines) is needed to be ligated at the opposite side of the flanking Poly-A tail. So, the ultimate output for first right shift with rotation is $\left(\begin{smallmatrix} U\,A\,U\,A\,U\,A\,U\,A \\ A\,D\,A\,T\,A\,T\,I\,I\,A\,A\,A\,A \end{smallmatrix} \right)$ that is 1100. This is the final result of right-shifter function. All sorts of fresh command of right-shifter functions will follow these steps. If the computation requires consecutive right-shifter action, then for second round of shifting, the output of first round is denatured and the larger strand, $5'AAAAIITATADADA$ is selected via capillary electrophoresis. This strand is now annealed with right, shifter-specific MSIS, $5'IIUUTTT$. This partially double-stranded product is then ligated with MSOS, UAUAUA. Ultimately, it is ligated to another dinucleotide, $\left(\begin{smallmatrix} U\,A \\ A\,T \end{smallmatrix} \right)$, which results in the final output of shifter with rotation (0110) of initial signal 1001

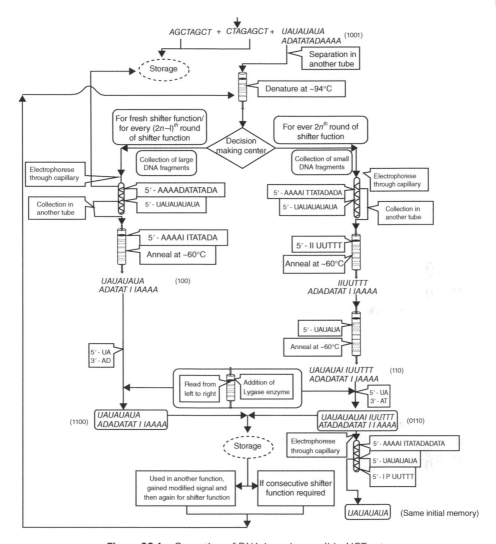

Figure 28.1 Operation of DNA-based reversible NOT gate.

resulting in the formation of $\begin{pmatrix} U\,A\,U\,A\,U\,A\,U\,A\,I\,I\,U\,U\,T\,T \\ A\,T\,A\,D\,A\,D\,A\,T\,A\,T\,I\,I\,A\,A \end{pmatrix}$ without any phosphodiester bond between MSIS and MSOS. This is the output of second round of right shifter. If further successive right-shifter action is required, then it will loop back again to the denaturation step of the first round of shifter mechanism and follow the decision of decision-making center. The fact is, all $(2n-l)^{th}(n=1,2,3,...)$ round of consecutive shifting will follow the first round of reaction and all $(2n)^{th}(n=1,2,3,...)$ round will follow this second round of reaction. It is interesting to note that denaturation and subsequent capillary electrophoresis will give back the same old target input, which is the true application of the reversibility. In a very similar fashion, the left shifter with rotation activity could be carried out in the following reactions, which are depicted in Figure 28.2, but we must keep in mind that the predesigned oligonucleotides need to be adjusted accordingly.

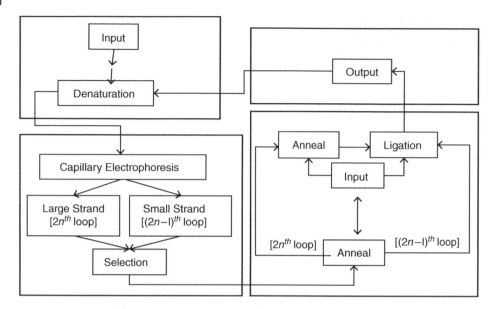

Figure 28.2   The four basic biochemical events of the shifter circuit.

In general, MSIS/MSOS shifter model is composed of four biochemical events such as denaturation, annealing, capillary electrophoresis, and ligation, as shown in Figure 28.2. Denaturation, annealing, and capillary electrophoresis are responsible for shifting, while ligation is introduced to execute rotation. It can shift and rotate any input strand, representing one bit for every two base pairings either from an independent source or from the previously designed added operator.

There is a perfect looping among the steps that can shift with or without rotating. Thus, the desired signal of any length without allowing any error becomes fatal via accumulation.

28.2   DNA-Based Reversible Multiplication Operation

The multiplication operation uses shift operation and adds the partial results. Figure 28.3 shows the operations of multiplication. The DNA strand representing multiplicand (Mt is equal to Ot_1) will pass through the left shift without rotation [L]SnR operation to generate output Ot_2, Ot_3, ..., Ot_p after $p - 1$ looping. Here, p is the number of bits present in the multiplier (Mr). The upper left side of Figure 28.3 shows the operations on multiplicand (Mt). Again, the DNA strand representing multiplier (Mr) will pass through the right shift with rotation R[SR] operation to generate output Or_1, Or_2, ..., Or_p after p looping. The upper right side of Figure 28.3 also shows the operations on multiplier (Mr). In both type of shifter operation, Mt/Mr, the input for one turn will be the output of immediately previous turn after the initial input.

Now after screening, if Or_l is larger than Mr, then DNA strand representing Ot_l will be transferred to temporary storage (TS) and taken as TS_1. And, if Or_l is smaller than Mr, then no DNA strand will be transferred to TS and $TS_1 = 0$, as shown in dashed outline box of

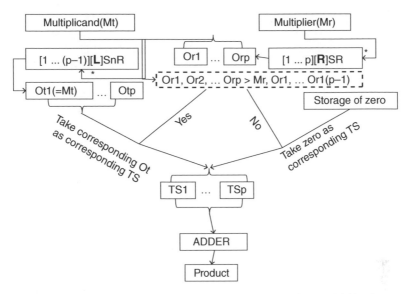

Figure 28.3 The working procedures of DNA-based reversible multiplication operation.

Algorithm 28.2.1 DNA-Based Reversible Multiplication Algorithm

---

1: Begin
2: **while** total p times **do**
3:    $PMr \leftarrow Mr$;
4:    Apply (right) shift with rotation of Mr
5:    Or $\leftarrow$ output (Mr);
6:    Apply (left) shift with not rotation of Mt;
7:    Ot output (Mt)
8:    **if** Or is greater than PMR **then**
9:       $TS \leftarrow Ot$
10:   **else**
11:      $TS' \leftarrow$ Shortage of Zero;
12:   **end if**
13:   Product $\leftarrow$ Addition $(Product, TS)$;
14:   Product is the final result of multiplication;
15: **end while**
16: procedure Addition $(Product, TS)$
17: **if** TS $= 0$ **then**
18:    **return** Product;
19: **else**
20:    **return** DNA-based addition between TS and product;
21: **end if**
22: End

Figure 28.3. Similarly, if Or_2 is larger than Or_1, then $TS_2 = Ot_2$. And, if Or_2 is smaller than Or_1, then $TS_2 = Q$. So, from the above explanation, every output of $[R]SR$ is compared with its input. If the output is larger, then the corresponding output of $[L]SnR$ is taken as temporary storage (TS). On the other hand, if the output is smaller, then it is null. Mathematically if $Orn > Or(n-l)$, then TSn Orn. And if $Orn < Or(n-1)$, then $TSn = Q$. Another case, if the output and input of $[R]SR$ is equal, then the Mr is definitely all 1s or all 0s. After screening, if Mr contains 1, then each Ot is taken for its corresponding TS. And if Mr contains 0, then all TS is 0.

Finally, all contents of temporary storage are passed through the adder circuit and produce the product. So, the addition between the partial result and TS (temporary storage) is done. After n times of addition operation, it produces the overall result of the multiplication operation (product). Here, TS represents as $TS_1, TS_2, ..., TS_p$, which means all TS are passed through the adder to get the product as output of the adder. So, TS is the final product.

Figure 28.3 shows the working procedures of DNA-based multiplier circuit, where $Ot = $ Output (Mt), $Or = $ Output (Mr), $TS = $ Temporary Storage, $p = $ Number of bits in $Mr= $ Number of loops during executing SR/SnR. After initial Mr or Mt, the inputs are $Or_1, Or_2, ..., Or_p$ or $Ot_1, Ot_2, ..., Ot_p$, consecutively, and $Mt = Ot_1$.

The Algorithm 28.2.1 depicts the overall procedures. If Or is equal to previous value of Or, then Mr is screened for 0 or 1. Again, if Mr is equal to 1, then all Ot are taken, otherwise Mr is zero and the product is also zero. The run-time complexity of this algorithm is $O(m)$, where m is total bit of the multiplier.

Example 28.2.1 Figure 28.4 shows an example of the DNA-based multiplication operations with three data paths: multiplicand, multiplier, and temporary storage (TS). When all

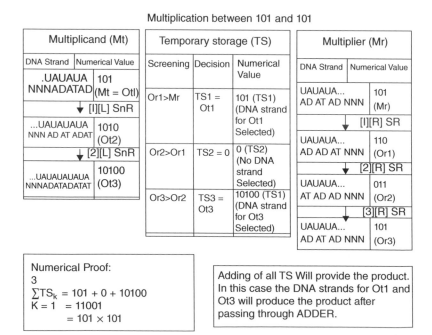

Figure 28.4 An example of DNA-based reversible multiplication operation.

TSs are added, it provides the product. In this case, the DNA strands for Ot_1 and Ot_3 produce the product after passing through adder circuit. Here, $101 \left(\frac{UAUAUA}{NNNADATAD} \right)$ and $\left(\frac{UAUAUA}{ADATADNNN} \right)$ are multiplied together, which produces the final output $11001 \left(\frac{UAUAUAUAUA}{ADADATATAD} \right)$, where $\sum_k TS_k = 101 + 0 + 10100 = 11001 = 101 \times 101$.

28.3 Summary

In this chapter, an approach has been presented to design a DNA (deoxyribonucleic acid) based reversible shifter and multiplier circuit where both inputs and outputs are based on DNA fragments and computations are performed using several analytical biochemical enzymes. Because of unchanging the input sequences, the reusability of DNA strands is achieved. The explanation of this chapter also uses some of the modified bases such as X, ψ, and D, as well as natural bases to design the DNA-based reversible shifter. These modified bases are used to augment the complimentary bonding options that can be utilized to represent discrete signal of bits. All of these modified bases are complimentary to adenine and as a result, they provide a broad window of complementation to design the circuit. Also, some of these bases are used to modulate the final output signals. So, whenever they are represented in output, the final signal will be the signal that they carried. This is one of the key characteristics of the DNA-based reversible shifter circuit. In addition, the multiplier circuits works into three steps: it is working on single bit of multiplier, shifting the multiplicand and adding the partial product to produce the final result. To implement DNA signals, some natural and non-natural (X, ψ, and D) DNA bases have been used for providing a broader window of complementary design. An algorithm is also presented to produce a compact DNA-based multiplier circuit. The run-time complexity of the method is $O(m)$.

29

Reversible Multiplexer and ALU Using DNA

A multiplexer circuit takes multiple inputs and carrys out only one input to the output line. In multiplexer circuits, selector inputs are used to select the particular input placed in the output line. In larger multiplexers, the number of selector inputs is equal to $log_2(n)$ where n is the number of inputs. In addition, the arithmetic unit consists of addition, subtraction and multiplication operations. At first, it performs the addition/subtraction operation according to the select input (S_0) of arithmetic logic unit (ALU). Then, it performs the multiplication operation. The outputs of addition/subtraction (sum/diff) operation and the multiplication operation are fed into multiplexer circuit. The remaining output of addition/subtraction circuit does not feed into the multiplexer and it directly puts into the output signal of ALU. When ALU is used to perform addition/subtraction operation, the above output produces carry/borrow; otherwise it produces garbage output.

29.1 DNA-Based Reversible Multiplexer

The DNA-based multiplexer takes six inputs in total. As a result, it has three selector inputs where $log_2(6)$ is equal to three. The inputs are from outputs of the following operations: basic logical operations, logical Ex-OR operation, addition, subtraction, and multiplication. The multiplexer uses five DNA-based Fredkin gates as building blocks, and it selects one output using three level of selection using three selector inputs. Figure 29.1 shows the overall representation of the multiplexer circuit. At first, it uses two Fredkin gates where four inputs are from the outputs of multiplication operation (MUL) and basic logical operations (AND, OR, NOT) and the select inputs are used as selector input at that level. In next level, two Fredkin gates also are used where four inputs are from operation of addition/subtraction (ADD/SUB). The logical Ex-OR and two selected outputs are taken from previous level. Here, the select input S_1 is used as selector input at this level. Finally, one Fredkin gate is used to select the outputs from the previous level where the select input S_2 is used as the final selector input. The multiplexer has nine inputs, including the three selector inputs and one final output with eight garbage output.

Reversible and DNA Computing, First Edition. Hafiz Md. Hasan Babu.
© 2021 John Wiley & Sons Ltd. Published 2021 by John Wiley & Sons Ltd.

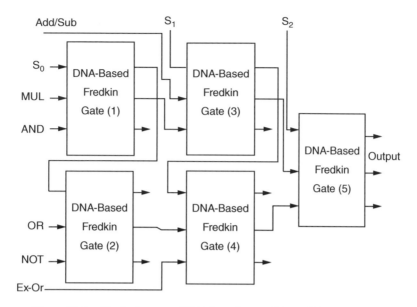

Figure 29.1 Block diagram of DNA-based reversible multiplexer circuit.

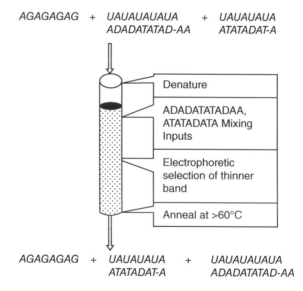

AGAGAGAG + UAUAUAUAUA + UAUAUAUA
ADADATATAD-AA ATATADAT-A

AGAGAGAG + UAUAUAUA + UAUAUAUAUA
ATATADAT-A ADADATATAD-AA

Figure 29.2 DNA hybridization of the first Fredkin gate for selection operation.

29.1.1 The Working Procedures of DNA-Based Multiplexer Circuit

Figure 29.2 to Figure 29.6 show the working procedures for each DNA-based Fredkin gate, respectively. Here, the electrophoretic selection of a thinner band or thicker band is performed based on the selector inputs.

The operations of first DNA-based Fredkin gate are shown in Figure 29.2 where the selector input is presented by the single stranded DNA which is equal to 1111 in binary and two inputs are from the multiplication operation and AND operation which are equal to 11001 and 0010 in binary respectively. At first, the given inputs are renatured and then secondary inputs are mixed with the produced single stranded DNA. Finally, electrophoretic selection of thinner band has taken place to inter change the positions of given two inputs. As a result, the produced outputs are 1111, 0010 and 11001 respectively.

Figure 29.3 shows the procedures of the second DNA-based Fredkin gate of the multiplexer where the selector input is taken from the output of the first Fredkin gate and it reduces the garbage outputs. The remaining two inputs are from the output of the OR and NOT operations, which are equal to 1101 and 0101, respectively. The selection procedures are the same as the previous one. The produced outputs are 1111, 0101, and 1101, respectively.

The previous two Fredkin gates perform first level selection operation. In next level, one Fredkin gate selects one output from the addition/subtraction operation and the first Fredkin gate, whereas the other Fredkin gate selects one output from the second Fredkin gate and EX-OR operation. Figure 29.4 shows the working procedures of the third Fredkin gate where the selector input S_i is represented by single stranded DNA, which is equal to 0000 in binary. The other two inputs are from the output of ADD/SUB operation and NOT operation which are equal to 0010 and 0101 in binary, respectively. Also, the two inputs do not change their values and remain in the same position at the output line due to the value of the selector input.

Figure 29.5 shows the procedures of the fourth Fredkin gate where S_i is used as the selector input that is taken from output line of the third Fredkin gate, but the other two inputs

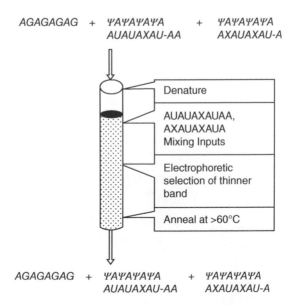

Figure 29.3 DNA hybridization of second Fredkin gate for selection operation.

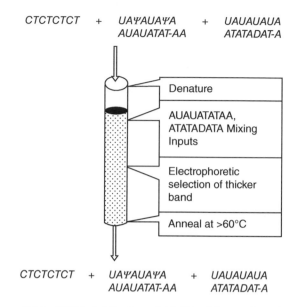

CTCTCTCT   +   *UAΨAUAΨA*   +   *UAUAUAUA*
                   AUAUATAT-AA   *ATATADAT-A*

Figure 29.4   DNA hybridization of third Fredkin gate for selection operation.

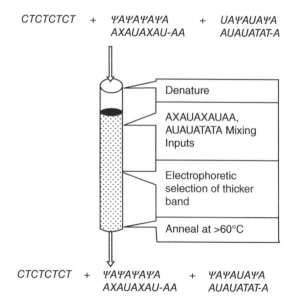

CTCTCTCT   +   *ΨAΨAΨAΨA*   +   *ΨAΨAUAΨA*
                   AXAUAXAU-AA   *AUAUATAT-A*

Figure 29.5   DNA hybridization of the fourth Fredkin gate for selection operation.

are taken from the second Fredkin gate and output of Ex-OR operation (1101). Also, the two do not change their value and remain the same positions at the output line due to the input value of the selector.

Finally, in Figure 29.6, the working procedures of the final Fredkin gate are presented where select input S_2 is used as a selector input of this Fredkin gate. Two inputs are taken

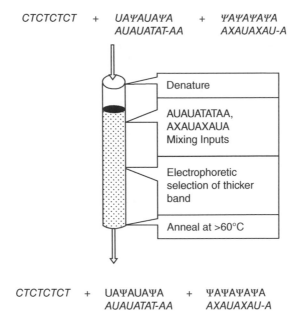

CTCTCTCT + UAΨAUAΨA + ΨAΨAΨAΨA
AUAUATAT-AA   AXAUAXAU-A

CTCTCTCT + UAΨAUAΨA + ΨAΨAΨAΨA
AUAUATAT-AA   AXAUAXAU-A

Figure 29.6   DNA hybridization of the fifth Fredkin gate for selection operation.

from the outputs of the third and the fourth Fredkin gates, which are outputs of ADD/SUB and NOT operations. As the selector input S_2 is equal to 0000, the two given inputs do not change their position and remain at the output line.

29.2 DNA-Based Reversible Arithmetic Logic Unit

In this section, the properties of the ALU are discussed and the analysis of the design methodologies is also described.

29.2.1 Procedures of DNA-Based ALU

The logic unit of the ALU is presented in this subsection. The DNA-based reversible Toffoli gate is used to perform the logical operations. The overall procedures of a DNA-based Toffoli gate is discussed in Chapter 26. As Toffoli gate is universal gate, it is possible to perform any logical operation using this gate. Here, the DNA bases are used as signals of the control input and the target input. The target input is flipped and the final output is preserved based on the value of the control input and again, it is used for further computation. The logic unit of the ALU consists of four logical operations such as AND, OR, NOT, and EX-OR operations.

The logic unit consists of four DNA-based Toffoli gates. Figure 29.7 shows the overall design of the logic unit. At first, the logical AND operation between two inputs A and B are performed. Then the logical EX-OR operation is performed and the two outputs of AND operation are used as two inputs of EX-OR operation to minimize garbage output. Next, the

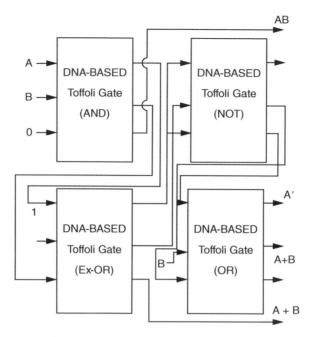

Figure 29.7 Diagram of DNA-based reversible logic unit.

logical NOT of input A is performed and finally, logical OR operation is performed using inverted values of two inputs A and B.

Here, the logic unit is reversible as its building blocks are reversible consisting of four Toffoli gates. The logic unit has six inputs, including two input variable A and B with 0, 1, 1 and inversion of input B, respectively. It produces outputs of basic logical operations such as AND, OR, NOT, and logical EX-OR operations with two garbage outputs A and inversion of B, respectively.

In computing, an ALU is an essential circuit that performs arithmetic and logical operations. The ALU is a fundamental building block of the central processing unit of a computer that also maintains timers. The processors found modern CPUs inside, and CPUs have very powerful and very complex ALUs inside there, where a single component may contain a number of ALUs. Mathematician John von Neumann proposed the ALU concept in 1945, when he wrote a report on the foundations of a computer called the EDVAC. The ALU, which is also capable of performing arithmetic and logical operations, uses DNA signals and biochemical properties. The DNA-based ALU has three modules: The first one is an arithmetic unit that is responsible for performing addition, subtraction, and multiplication operations between two binary numbers; the next one is a logical unit that performs unary and binary logical operation, and finally, the last one is a DNA-based reversible multiplexer that is used to select one output out of many.

Figure 29.8 shows the block diagram of the ALU where input variables A and B are fed into the arithmetic unit first. As every operation of the ALU is reversible, the input variables could be fed into logical unit from the outputs of arithmetic unit and it reduces the garbage

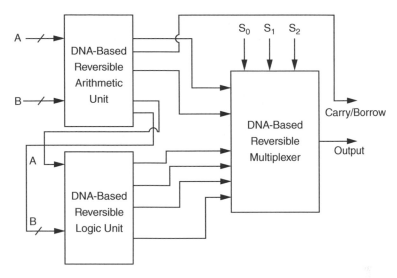

Figure 29.8 Diagram of DNA-based reversible logic unit.

outputs of arithmetic unit. Then, all outputs of the arithmetic unit and the logical unit work as inputs of the reversible multiplexer to select only one output.

29.2.2   Properties of the DNA-Based ALU

The DNA-based arithmetic and logic unit uses a DNA-based Toffoli gate and DNA-based Fredkin gate. In the DNA-based Toffoli gate, there are two types of inputs such as control input and target input. For the produced target output, the DNA bases of the target output is equal to $2n + m + n$ where $2n$ is used for double-stranded DNA, m for unique tag and n for next level input. As a result, it increases the bit density. Since the control inputs remaining unchanged, it is possible to reuse the control inputs again and again.

In a DNA-based Fredkin gate, there are one control input and two given inputs, which are used for the final selection. The control input of Fredkin gate is like the Toffoli gate, which remains unchanged. The given inputs are double-stranded DNA. Here, the secondary inputs are necessary to finish computation. All inputs remained unchanged in the Fredkin gate and the selection is done by electrolysis of thinner or thicker band. The higher degree of parallelism is possible using both gates, where the target input of Toffoli gate is changed; and all other inputs of two gates could be used concurrently with different target and secondary inputs. It preserves sequence by using the unique tag. The copy operation of one signal is actual the replication of one DNA strand.

Some properties DNA-based reversible arithmetic and logic unit are discussed below:

Property 29.2.2.1   The DNA-based ALU circuit is reversible if each individual component is reversible and there is one-to-one mapping between the input and the output vectors.

Proof: Since each component of the DNA-based ALU circuit is reversible and there is one-to-one mapping between the input and the output, it states that the circuit is logically reversible in computation. Also, it maintains the restriction to keep composite circuit reversible such as the fan-out of each signal must be one and there should not be looping or fed backing in the circuit.

Property 29.2.2.2   The arithmetic unit of the ALU performs both addition and subtraction operations using same architecture.

Proof: The adder/subtractor of arithmetic unit consists of several reversible gates, two numbers to perform the desired operation and one control signal with the select input S_i. It uses a Fredkin gate as controlled switch. When the select input S_0 is low, it brings to pass only $A \oplus B$ and AB. Again, when the select input S_0 is high, it performs the inversion of A and it carries out only $A \oplus B$ and AB. The select input S_0 indeed is used as a controller to determine which operations should be performed. This verifies the fulfillment of the described methods.

Property 29.2.2.3   No error at signal level can be retained even over a full loop of MSIS/MSOS shifter, as every round and loop has its auto correct ability, where MSIS is "meaningless shortest input strand" and MSOS is "meaningful (one bit) shorter operand strand."

Proof: In MSIS/MSOS shifter model, signals are readable when the corresponding double strand is present. A single strand is useful for the operation but has no impact on signals. Errors can be generated at signal level only by mismatching the base pairing that will lose its existence when these signals are automatically pass through the denaturation step. As each round and loop starts with a denaturation step, there is no scope for an error to accumulate even after a single loop.

Property 29.2.2.4   It is enough to determine positional value in binary by shifting the positional value with rotation.

Proof: Binary number only contains 0 and 1. It means that right shift any number with rotation produced a number less than the original one if LSB contains zero and vice-versa. It completes the proof.

Property 29.2.2.5   Fredkin gate is sufficient to design any multiplexer circuit.

Proof: A multiplexer circuit takes several inputs and it carries out one input to final output line. One special property of the Fredkin gate is that it can operate as a controlled switch. So, several Fredkin gates are capable of performing multiplexed operation. Therefore, any multiplexer circuit can be realized using only a Fredkin gate.

Property 29.2.2.6   The DNA-based reversible ALU circuit performs operations correctly.

Proof: The ALU circuit consists of arithmetic, logic, and selection modules that are logically reversible and uses DNA bases for representing input and output signals. In arithmetic unit, there are combined adder/subtractor and multiplier circuit where multiplier circuit is a composite of adder circuit and shifter circuit. Also, in logic unit, there are four logical operations. And, in selection module, one multiplexer circuit is used. All individual circuits perform successfully and produce valid results. So, the design methodology is flawless and the ALU circuit works correctly.

29.3 Summary

In this chapter, the logic unit for the ALU is presented where basic logical operations such as AND, OR, NOT, and logical EX-OR are performed. Here, four DNA-based Toffoli gates are used to perform the logical operations. Also, the DNA hybridization procedure individual logical operation is presented. So, the logic unit consists of four Toffoli gates to perform four logical operations. It has also six inputs in which three inputs are constant and six outputs, including two are garbage outputs.

30

Reversible Flip-Flop Using DNA

The reversible logic has been taken as a significant paradigm in low-power computing and it plays an important role in the synthesis of circuits for quantum computing. Numerous applications of synthetic biology require the implementation of scalable and robust biological circuits with information processing capabilities. Basic logic structures, such as logic gates have already been implemented in prokaryotic as well as in eukaryotic cells. Here, in the computational design of edge-triggered D flip-flop in master-slave configuration based on transcriptional logic is described with the counter circuits. The counter can count up to $2n$ cellular events using a sequence of n flip-flops. The state of the counter is changed by edge-triggering either with the synchronization using clock signal or with a pulse, which corresponds to the occurrence of observed event within the cellular environment.

30.1   The Design of a DNA Fredkin Gate

The reversible gates and reversible sequential circuits generate a unique output vector from input to output and output to input, i.e., that is a one-to-one mapping between input and output vectors. The Fredkin gate is a 3×3 reversible gate with three inputs and three outputs. The Fredkin gate is also called a conservative gate, as the Hamming weight (number of logical one) of its inputs are equal to its outputs.

Let $x = (x_1, x_2, x_3)$ be the input vector and $y = (y_1, y_2, y_3)$ be the output vector, where $y_1 = x_1$, $y_2 = (\neg x_1 \wedge x_2) \vee (x_1 \wedge x_3)$ and $y_3 = (x_1 \wedge x_2) \vee (\neg x_1 \wedge x_3)$. The output of a Fredkin gate can be seemed as the input of another Fredkin gate. So, the Fredkin gate is designed in Figure 30.1 as follows:

30.2   Simulating the Fredkin Gate by Sticking System

The Fredkin gate can be implemented using DNA molecules and bio-chemistry operations. Without losing generality, the input and the output of the Fredkin gate are both binary vectors, that is $x_i, y_i = 0$ or $1, i = 1, 2, 3$. So, if a one-to-one mapping between binary vectors and DNA molecules can be constructed and the "$\wedge$", "$\vee$", and "$\neg$" computations of DNA molecules are correspond to the bio-chemistry operations, then a DNA Fredkin gate

Reversible and DNA Computing, First Edition. Hafiz Md. Hasan Babu.

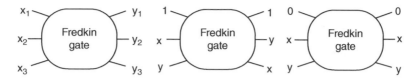

Figure 30.1 Fredkin gate symbol and its working procedure as a conditional switch.

is achieved. In DNA computing, the sticking system was first proposed to encoding the binary information by DNA molecules, so it is considered to simulate the Fredkin gate.

It is known that there are ssDNA (single-strand DNA molecule) and dsDNA (double-strand DNA molecule), both of which consist of bases A, T, C, and G. By applying Watson-Crick's law of complementarity, ddDNA can be formed during a process called hybridization, where two complementary ssDNA stick together by base paring (A to T and C to G). On the other hand, ssDNA can be generated by simply melting the dsDNA. The process is illustrated in Figure 30.2.

In sticking system, a bit of binary information is defined as ssDNA or dsDNA with a fixed length. All but different lengths of DNA molecules, no matter whether they are ssDNA or dsDNA, represent the entire binary information. In computation, if a bit of information is changed from 1 to 0, it corresponds to a process where dsDNA melts to form two ssDNA molecules and vice versa. And the fixed-length DNA strand of each bit is different from each other, which makes the biochemistry operations much easier when the bit changes. Figure 30.3 shows an example for sticking system.

ssDNA A-T-C-G-T-T-A-G-C-C-G-A-T-T-C   hybridizing   A-T-C-G-T-T-A-G-C-C-G-A-T-T-C
ssDNA T-A-G-C-A-A-T-C-G-G-C-T-A-A-G   ⇄ melting | | | | | | | | | | | | | | | dsDNA
                                                      T-A-G-C-A-A-T-C-G-G-C-T-A-A-G

Figure 30.2 Forming dsDNA by hybridizing the two complementary ssDNA and forming two complementary ssDNA by melting the dsDNA.

a. 1st bit is 0 → A-T-C-G   2nd bit is 0 → A-G-C-T   b. x^-(0,1,0)   A-T-C-G--A-G-C-T--T-C-G-T
                                                                                       T-C-G-A
    1st bit is 1 → A-T-C-G
                   T-A-G-C   2nd bit is 1 → A-G-C-T y^-(1,0,1)   A-T-C-G--A-G-C-T--T-C-G-T
    3rd bit is 0 → T-C-G-T   T-C-G-A   T-A-G-C   A-G-G-A

    3rd bit is 1 → T-C-G-T
                   A-G-C-A

c. For example, when inputs are x^-(0,1,0) and y^-(0,1,0), then the output is z^-(,1,0)

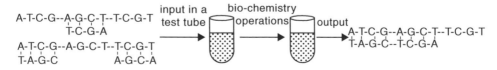

A-T-C-G--A-G-C-T--T-C-G-T   input in a   bio-chemistry
          T-C-G-A   test tube   operations   output
                                                            → A-T-C-G--A-G-C-T--T-C-G-T
A-T-C-G--A-G-C-T--T-C-G-T   T-A-G-C--T-C-G-A
T-A-G-C   A-G-C-A

Figure 30.3 a. A one-to-one mapping between binary bit and DNA strands with fixed length 4. b. A one-to-one mapping between binary information and DNA string by concentrating the DNA strands. c. The simulating of binary information computing by DNA biochemistry operations.

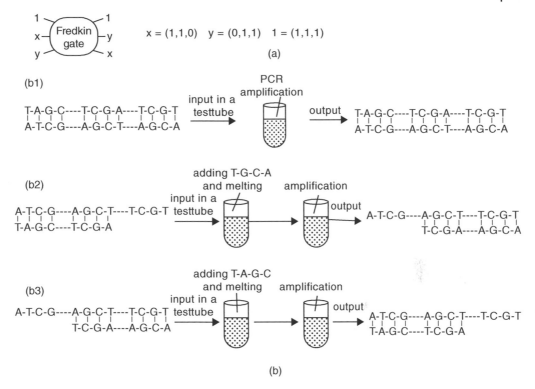

Figure 30.4 a. Fredkin gate with three inputs and three outputs. b. Simulation of the Fredkin gate in detail; b1. When the input is 1, the output is also 1. b2. When the input x and the corresponding biochemistry operations. b3. When the input is y and the corresponding biochemistry operations.

By sticking system, the simulation of the Fredkin gate can easily be achieved by using DNA molecules and biochemistry operations to simulate its inputs, outputs and computation. One of the Fredkin gates in Figure 30.1 is simulated as follows in Figure 30.4. By uniting the test tubes T_{ij}, where $i = 1, 2, 3$, $j = 1, 2$ and the biochemistry operations used in each test tube, we get a system of test tubes and biochemistry operations, it is called as a DNA Fredkin gate based on the sticking system. That seems to be very complexity, but with the development of biochemistry technology, that can be simplified to be a more simple system.

30.2.1 Simulating the Fredkin Gate by Enzyme System

The enzyme system is regarded as another method of simulating Fredkin gate using DNA molecules as the model. This system utilizes the property of restriction enzymes that can only recognize a certain sequence amongst DNA molecules and cut them, producing either a blunt end or a sticky end. As a result, the long DNA molecules can be chopped up into small pieces, which, in turn, represent the binary information mentioned above. For instance, if the input is x and the output is y in computation, it is analogous to the restriction enzyme recognizes a certain sequence on x, cuts the DNA molecule x and produces y as the final product. In another word, if x changes into y in computation, it is

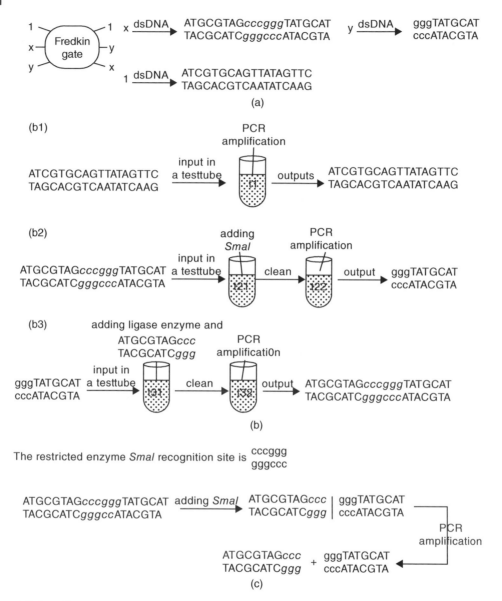

Figure 30.5 a. The Fredkin gate with three inputs and three outputs and three outputs and the one-to-one mapping between the three inputs and the DNA strands. b. Simulation of the Fredkin gate in detail; b1. When the input is 1 using amplification, the output is still 1. b2. When the input is x adding *SmaI*, the output is y. b3. When the input is y is adding ligase enzyme, the output is x. c. The progress of cutting x into two parts using *SmaI*.

very likely that x DNA molecules are partially cut by restriction enzymes; and if y changes into x, piece of x DNA molecules are ligated using ligase. Figure 30.5 illustrates an example. By uniting the test tubes t_{ij} (where $i = 1, 2, 3,$ and $j = 1, 2$) and performing the biochemistry operations in each test tube, the DNA Fredkin gate is achieved using enzyme system.

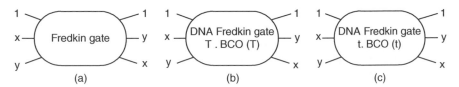

Figure 30.6 a. The Fredkin gate with three inputs and three outputs. b. The DNA Fredkin gate based on the sticking system. c. The DNA Fredkin gate based on the enzyme system.

In this section, two methods are used in simulating the Fredkin gate to do two tasks:

(i) To construct a one-to-one mapping between the binary information and the DNA strands, and

(ii) To compute the binary bits which corresponds to the biochemistry operations. These both can be computed as the Fredkin gate which generates an unique output from each input. So the DNA Fredkin gate based on sticking system and enzyme system, generates an unique DNA strand from each input DNA strand. The test tubes and the biochemistry operations used in the sticking system are denoted as T and BCO (T), respectively. And the test tubes and the biochemistry operations used in the enzyme system are noted as t and BCO (t), respectively. The two DNA Fredkin gates of which each can be designed using sticking system and enzyme system separately are shown in Figure 30.6. These two DNA Fredkin gates are used to simulate the DNA reversible circuits.

30.3 Simulation of the Reversible D Latch Using DNA Fredkin Gate

It can be seen that the reversible D latch is highly optimized in terms of the numbers of reversible gates and garbage outputs. The D latch and the DNA D latch are demonstrated in Figure 30.7. In Figure 30.7 (a), D. E, 0 and 1 are binary information, but in Figure 30.7 (b) and (c), D', E', 0 and 1 are DNA strands.

30.3.1 Simulation of the Reversible Sequential Circuit Using DNA Fredkin Gate

The D latch has already been simulated by the two DNA Fredkin gates as mentioned above. So it is possible to simulate the reversible sequential circuit based on the D latch. That

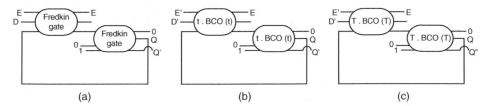

Figure 30.7 a. The D latch based on Fredkin gate. b. The DNA D latch based on sticking system. c. The DNA D latch based on enzyme system.

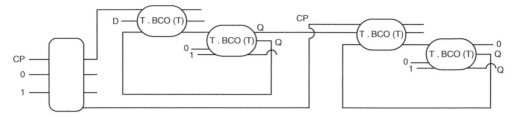

Figure 30.8 DNA reversible master–slave D flip-flop with the CP clock pulse, which can be simulated by DNA strands.

might be a new computing model for both DNA computing and quantum computing. With the development of biochemistry and quantum theory, the DNA-based reversible sequential circuit will play an important role in nano-scale computing and low-power computing in the future. A reversible sequential circuit is shown in Figure 30.8, which is noted as a reversible master–slave D flip-flop, and it is simulated by enzyme system without having a lot of changes. Many other reversible sequential circuits can also be simulated by DNA strands using sticking system and enzyme system easily.

30.4 DNA-Based Biochemistry Technology

BCO (T) and BCO (t) are used in simulating the Fredkin gates based on the sticking system and enzyme system. There are some reminders of the biochemistry operations used in these systems that are very common in biochemistry lab. The ssDNA can be linked together in two different ways: connection and hybridization. Two short ssDNA can be connected by ligase enzyme to form a long ssDNA. And two complementary ssDNA can hybridize each other at an appropriate temperature. Also the dsDNA can melt two complementary ssDNA at an appropriate temperature. In general, the hybridization temperature is 98°, and the melting temperature is 78°. The dsDNA can be cut into two parts by restriction enzyme with the blunt or sticking end, as in Figure 30.5. A kind of restriction enzyme can only cut the DNA strands that have its recognition site, i.e., it does nothing to the stands without the recognition site. By restriction enzymes and by ligasing and hybridizing a DNA strand can be shortened and lengthened easily. There are hundreds of known restriction enzymes, which can be used as a computation in the reversible gate. The technique called PCR (polymerase chain reaction) can amplify the dsDNA strands in huge number within a very short time and, i.e., we can increase the numbers of one dsDNA by PCR. There are three steps for PCR: melting, adding primer, and amplifying. Also, we can separate and measure the DNA strands in length by gel electrophoresis, which is often used in the lab. A DNA strand based on the polymerase action can extend a primed single strand template which is useful in determining an output of a reversible gate or a reversible sequential circuit. With the development of biochemistry technique such as PNA technology, real-time PCR, DNA self-assembling and so on, the DNA computing models are much more simple and efficient than before. These also can be more helpful for DNA reversible gates and DNA reversible sequential circuits.

30.5   Summary

In this chapter, the DNA Fredkin gate is simulated as a reversible logic using sticking system and enzyme system. The binary inputs and outputs of the Fredkin gate are represented by the DNA strands, and the computations of these binary information correspond to the biochemistry operations. So, the sticking system and the enzyme system can be used to compute inputs and outputs of a Fredkin gate. The computing models have the advantages of both DNA computing and quantum computing. This dimension of computing might be a new direction of the nano-scale computing and low-power computing models.

31

Applications of DNA Computing

For the last few decades, engineers have faced the gospel truth of material science in the quest for more powerful computers: transistors, the on-off switches that control the microprocessor, cannot be made any smaller than their present size. There are some alternatives of silicon chip, an instinctive option is right now being created utilizing DNA to play out similar sorts of complex computations that silicon transistors do now. Figure 31.1 represents a DNA double helix. So in this chapter the discussion will be on DNA processing, the applications of DNA computing, and the reasons for which it is a major deal.

How DNA Computing Works

The issue with transistors is that they presently exist at the size of a couple of nanometers in size, which is just a couple of silicon particles thick. They can not be made any smaller than their current size for all intents and purposes.

If they get any smaller, the electrical current-flowing through the transistor effectively leaks out into different segments close by or disfigures the transistor because of heat, rendering it useless. We need a base number of molecules to make the transistor work and we have practically achieved that limit.

Designers have discovered some workarounds for this issue by utilizing multicore and multiprocessing frameworks to increment computational power without shrinking the transistors further, yet this also accompanies programming difficulties and power prerequisites, so another arrangement is required if we want to see more powerful computers later on.

While quantum computing is getting a ton of media, DNA computing can be similarly or significantly ground-breaking than quantum computing. Moreover, it does not have a large number of limitations that quantum computing has. Also, we ourselves are living examples of the information stockpiling and computational intensity of DNA computing.

Applications of DNA Computing

There are many applications of this technology that will enhance our life in future. And moreover, there are many researchers who are giving their utmost to make a better DNA computer.

Reversible and DNA Computing, First Edition. Hafiz Md. Hasan Babu.
© 2021 John Wiley & Sons Ltd. Published 2021 by John Wiley & Sons Ltd.

Figure 31.1   DNA double helix.

31.1   Solving the Optimization and Scheduling Problems Like the Traveling Salesman Problem

As the traveling salesman issue characterizes it, an organization has a salesman who must visit *n* numbers of urban communities making calls and can just visit every city once. What should be the arrangement for visiting the urban communities that will give the most cost-effective and shortest way?

At the point when *n* rises to 5, the issue can be worked out by hand on a paper and a computer can test each conceivable way rapidly. Consider the possibility that *n* rises to 20. Finding the most limited way through 20 urban areas turns out to be substantially more computationally difficult, and it would take a traditional PC exponentially longer to discover the appropriate response. An example of shortest path problem is presented in Figure 31.2.

Endeavor to locate the shortest way between 500 urban areas and it would take a traditional PC longer than the whole lifetime of the universe to locate the shortest way since the best way to confirm that we have discovered the shortest way is to check each and every change of urban areas. A few calculations exist utilizing dynamic algorithm that can hypothetically decrease the quantity of checks required (and the genuine Hamilton path issue does not require checking each hub in a diagram); however, that may save a couple of million years off the top; the issue at present computationally unthinkable on a traditional computer. An example of traveling salesman problem is shown in Figure 31.3.

DNA can be collected so that a test tube loaded with DNA squares could gather them to encode the majority of the potential ways to solve the traveling salesman problem as shown in Figures 31.2 and 31.3.

In DNA, genetic coding is represented by four distinct atoms, called *A, T, C*, and *G*. These four "bits" when fastened together can hold a staggering measure of information. All things considered, the human genome is encoded in something that can be stuffed into a solitary core of a cell.

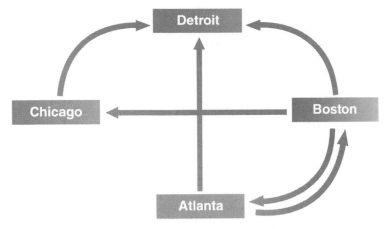

CITY	DNA NAME	COMPLEMENT
ATLANTA	ACTTGCAG	TGAACGTC
BOSTON	TCGGACTG	AGCCTGAC
CHICAGO	GGCTATGT	CCGATACA
DETROIT	CCGAGCAA	GGCTCGTT

FLIGHT	DNA FLIGHT NUMBER
ATLANTA - BOSTON	GCAGTCGG
ATLANTA - DETROIT	GCAGCCGA
BOSTON - CHICAGO	ACTGGGCT
BOSTON - DETROIT	ACTGCCGA
BOSTON - ATLANTA	ACTGACTT
CHICAGO - DETROIT	ATGTCCGA

Figure 31.2 DNA computer solving a shortest path problem.

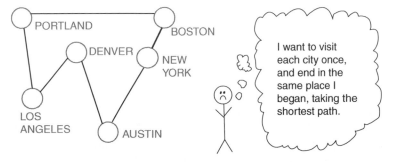

Figure 31.3 Traveling salesman problem.

By blending these four atoms into a test tube, the particles normally amassed themselves into strands of DNA. If a mix of these atoms speaks to a city and a flight way, each strand of DNA could speak to an alternate flight way for the salesman, all being determined immediately in the blend of the DNA strands amassing themselves in parallel.

At that point, it would essentially involve examining through the more drawn out ways until you have just the shortest way left. There is a demonstration on how this should be possible with seven urban communities and the answer for the issue would be encoded when the DNA strands were blended.

This creates energy because that DNA structures are shabby, moderately simple to deliver, and versatile. There is no restriction to the power that DNA processing can hypothetically have since its capacity expands the more atoms you add to the condition and not at all like silicon transistors that can play out a single legitimate activity at any given moment, these DNA structures can hypothetically execute the same number of computation at once as expected to take care of an issue and do it at the same time.

31.2   Parallel Computing

The flexibility of DNA computing over established computing and even quantum computing to a degree is that it can perform uncountable calculation in parallel. This thought of parallel computing is not new and has been imitated in traditional processing for a considerable length of time. In Figure 31.4, a parallel computer is presented.

When two applications are running on a PC in the meantime, they aren't really running simultaneously; at some random time, just a single instance is being completed. If a music program and shopping web-based utilizing a program both are running in the PC, the PC is really utilizing something referred to as context switching to give the presence of simultaneousness.

It runs instance for one program, spares the condition of that program after the processing is done, and expels the program from dynamic memory. At that point it stacks up the recently spared condition of the second program, runs its next instance, spares its new state, and after that empties it from dynamic memory. At that point it reloads the principal program to complete its next iteration.

By making a great many stages of context switching, the presence of simultaneousness is accomplished, yet nothing is ever really being kept running in parallel. DNA computing can really complete these great many tasks in the meantime. More than 10 trillion DNA particles can be crushed into a solitary cubic centimeter. This cubic centimeter of material could hypothetically perform 10 trillion estimations without a moment's delay and hold as much as 10 terabytes of information.

DNA computing is the best idea as a supplement of quantum computing, with the goal that when combined together and driven by an established computer, the sorts of sensational increments in computational power that individuals want to find can really become an important step toward next-generation computing.

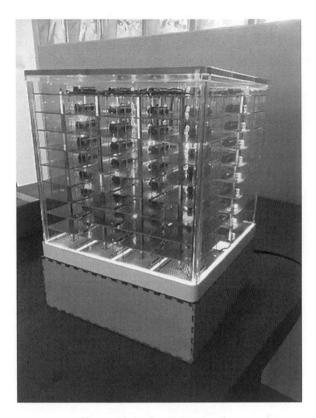

Figure 31.4   Parallel computing.

31.3   Genetic Algorithm

A genetic algorithm (GA) is a sort of delicate computing with a genetic system in life forms, looks ideal qualities when it accepts various control designs in rehashing accepts GA restoration. Genetic calculation is utilized to solve the control of lifts. The setting of parameters in a control framework of lifts is likewise difficult to oversee on a manual premise. DNA computing can solve the genetic algorithms very efficiently than regular computers.

31.4   Neural System

In the field of neural networks, the optimization of an evaluation function is done by back-propagation learning depending on past occurrences, the easiness of updating rules is welcome in comparison with GAs and neural networks. Neural networks can be enhanced by using DNA calculations.

31.5 Fuzzy Logic Computation and Others

It is not only limited in the fields of artificial intelligence, genetic algorithms, neural networks and fuzzy logic computation but also such soft computing methods as enhanced learning evolutionary strategy and genetic network programming are employed in optimizing group control scenarios.

31.6 Lift Management System

Different lifts are usually utilized in tall structures. Powerful control of such different lifts is fundamental. The general point in controlling a gathering of lifts is to fulfill the time imperatives all things considered, DNA computing and its application giving the most proficient framework. The essential issue is to choose which lift should stop at a specific floor where travelers are hanging tight to go up or down. Indeed, even in pinnacle (surge) hours, it is conceivable to discover all lifts moving in a similar course, or then again all lifts arriving all the while at a similar floor. So as to determine such circumstances, all lifts should be ideally allocated to travelers, paying little respect to the last's changing landing times at the different floors in the multistory building.

The gathering control framework chooses lift development designs as indicated by irregular changes in rush hour gridlock volumes as well as driving administration, or for the situation of a mishap. Such gathering control acknowledges agreeable, protected, and efficient the board of lifts by utilizing DNA calculations.

31.7 DNA Chips

DNA chip or DNA microarrays is a collection of DNA that can be used to calculate a large number of calculations simultaneously. This chip can be used to measure how concurrent operations perform. Especially in genome sequencing and DNA mapping, this field has huge consequences as it is very dynamic for genetic applications; for example, medicate plan and difficulty characterization. So by using the power of these microarrays, a huge opportunity arrives. The field of machine learning can be hugely affected by using DNA chips as this would calculate millions of calculation at a time that can lead to an improved outcome than the regular computers. The issue of over fitting is the main problem with AI solutions to deal with DNA chip information. This beneficial information is portrayed by class unevenness, non-straight reaction, high disorder, and huge quantities of traits. We can see the concept of DNA chip in Figure 31.5.

Moreover, DNA chips are used in medical diagnostic tests. Sometimes they are also used to decide which medication is best for a patient, because DNAs determine how humans react to any medicine. With the arrival of new DNA sequencing technologies, it is becoming easier to access this technology.

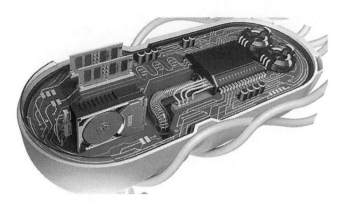

Figure 31.5   Concept of DNA chips.

31.8   Swarm Intelligence

Swarm intelligence is a branch of intelligence that discusses about the group behavior by agents together. These agents are usually decentralized and they are self-organized. In the nature this self-organizing capability can be found in many situations. The ants, bees, flocks of birds, fireflies, etc. depict a collective behavior that is not controlled by a single centralized authority but in a decentralized manner. So this is really a useful intelligent scheme in nature.

Every individual may not be smart, however together they perform complex community practices. This swarm knowledge can help to the investigation of human social conduct by watching other social creatures and to take care of different rearrangement issues. There are three principle kinds of swarm knowledge strategies: models of winged animal rushing, the ant colony optimization (ACO) calculation, and particle swarm optimization (PSO) calculation.

These artificial intelligence algorithms can be used to map many types of group problems that can be faced in regular life. These problems are hard to solve in regular computers, as many times regular computers need to calculate simultaneous operations beyond its limit. So it takes a lot of time even for a powerful computer. But with the use of DNA computer this calculations are easier as DNA computing uses simultaneous approaches that are much more concurrent than the regular systems. That is the reason the researchers are interested in using DNA computers for solving many artificial intelligence problems like swarm intelligence etc. An example of swarm intelligence is shown in Figure 31.6.

31.9   DNA and Cryptography Systems

Nowadays, DNA computing fills in as a standout amongst the most significant cryptosystems in view of its restrictive parallelism and huge information thickness attributes. The greater part of the cryptosystems depends on numerical conditions.

Figure 31.6 Swarm intelligence in nature.

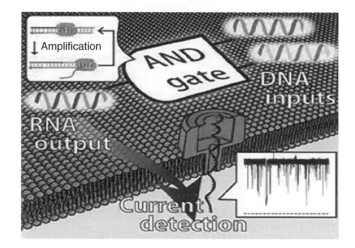

Figure 31.7 Swarm intelligence in nature.

The harder are these scientific conditions, the harder the aggressor can approach the framework. There are numerous cryptographic frameworks utilized these days. A standout amongst the most known cryptographic frameworks is the DES, which can be utilized to deliver 64-bit figure content from 64-bit plain content by utilizing 56-bit mystery key. Utilizing the parallel handling of DNA computing is further developed than what DES is doing. Utilizing a mystery key, effective DNA technique had the option to break the DES key inside a day. The equivalent goes for the RSA cryptographic strategy, which has a dependability issue. All this work can be done using DNA logic gates. We will see a DNA logic gate in Figure 31.7.

31.10 Monstrous Memory Capacity

The other certain impact of DNA computer is its enormous memory limit. Putting away data in atoms of DNA takes into consideration a data thickness of roughly 1 bit for each

Figure 31.8 Huge memory capacity of DNA.

cubic nanometer. The bases (moreover known as nucleotides) of DNA atoms, which speak to the limit unit of data in DNA computers, are separated each 0.34 nanometers along the DNA particle, giving DNA a surprising information thickness of almost 18 megabits for every inch. Figure 31.8 shows the huge memory capacity of a DNA. In two measurements, in the event that we expect one base for every square nanometer, the information thickness is more than one million gigabits for each square inch. Contrast this with the information thickness of an average superior hard driver, which is around 7 gigabits for every square inch a factor of more than 100,000 times smaller.

31.11 Low-Power Dissipation

The capability of DNA-based calculation lies in the way that DNA has a huge memory limit and furthermore in the way that the biochemical activities disperse close to noenergy.

DNA PCs can perform 2×1019 activities for every joule. This is astonishing, taking into account that the second law of thermodynamics manages a hypothetical limit of 34×1019 (irreversible) tasks per joule (at 300K). Existing supercomputers are not very energy proficient, executing a limit of 109 tasks for each joule. Simply consider the energy could be truly important in future. In this way, this characteristic of DNA PCs can be significant. Figure 31.9 shows a low-power DNA computer. According to the researchers, A DNA computer needs less than one percent of the energy needed by a regular transistor, which is a very good prospect. These attributes make the DNA computers more viable than electrical supercomputers, as the regular super computers are power hungry and thus require a lot of power support. In some cases, they need their own power station to run huge operations.

31.12 Summary

The primary advantage of utilizing DNA PCs is to solve complex issues, where various potential arrangements are made all at once and in a parallel design. General electronic

Figure 31.9 Low-power DNA computers.

computers must endeavor to tackle the issue one procedure at a period. DNA itself gives the additional advantages of being energy effective asset. The expanding capacity to plan complex atoms and frameworks makes these models of calculation progressive for nanotechnology and the future of computing.

Conclusion

We are living in a world with an ever-increasing hunger for energy. Human energy consumption continues to grow. Therefore, huge exertions are being made to satisfy our hunger for energy in a manner that brings as little damage to our environment as possible. Besides an ever-increasing thirst for energy, our society also displays an ever-increasing demand for information. People are constantly harvesting, transporting, and processing information. Little effort has been spent to make the computation processes reversible. Recent computer hardware and software are designed for performing computations in a definite direction from input information to the output information and from questions to answers. However, there are systems that can compute either forward direction or backward direction, where the direction of computation can be chosen by the user. Reversible computing models allow exploration of the fundamentals of computation, which is independent of any particular choice of computer hardware and computer language.

Reversible computing is a paradigm in which reversible computers are hierarchically constructed based on reversible physical phenomena and reversible operations. The reversible logic has promising applications in emerging computing paradigms such as quantum computing, quantum-dot cellular automata, optical computing, etc. In the first part of the book, reversible arithmetic units such as adders, subtractors, and multipliers are demonstrated that form the essential component of a computing system. In addition, reversible binary comparator, sequential and encoder circuits, barrel shifter, programmable logic devices, RAM, ROM, ALU, and control unit were presented with the applications.

As reducing the failure probability and increasing reliability is a goal of electronic systems, designers have been developing the first components ever since. No matter how much care is taken designing and building an electronic system, sooner or later an individual component or device will fail. Fault tolerance is the property that permits a system to continue functioning properly in the event of the failure of some of its constituents. If the system itself is made of fault-tolerant constituents, then the detection and correction of faults become easier and simple. In order to achieve reliable circuits in reversible computing, provision of fault tolerance is necessary. To achieve fault tolerance, the first stage is to identify occurrence of fault during the application of circuits. The second part of the book, properties of fault tolerance, as well as fault-tolerant, such as adders, multipliers, divisions, decoders, and the arithmetic logic unit. Reversible hardware computation with fault-tolerant support, that is, performing logic signal transformations in a way that allows the original input signals to be recovered from the produced outputs that are helpful in diverse areas such as

Reversible and DNA Computing, First Edition. Hafiz Md. Hasan Babu.
© 2021 John Wiley & Sons Ltd. Published 2021 by John Wiley & Sons Ltd.

quantum computing, low-power design, nanotechnology, optical information processing, and bioinformatics.

Part One and Part Two of this book has covered by the important topic of modern computing. Reversible and fault tolerant computing are the basic of quantum computing. The reversible logic has extensive applications in emerging technologies such as quantum computing. A quantum computer will be viewed as a quantum network (or a family of quantum networks) composed of quantum logic gates, where each gate is performing an elementary unitary operation on one, two, or more two-state quantum systems called qubits. Quantum networks must be built from reversible logical components. The reversible logic satisfies the quantum computing property by maintaining one-to-one mapping between the input and output vectors.

In the last part or Part Three of this book, the most promising computing technology such as DNA computing is presented with its circuit level applications. With just a little bit of DNA and some basic laboratory techniques, a DNA computer can be created to provide an enormous range of useful applications. This technology works through some interesting computer hardware known as logic gates that are made of DNA. In fact, this development has led us to place trust in the DNA computer technology, which relies on DNA codes to operate instead of the usual electrical signals. Through DNA codes, the computer can successfully complete logic-based operations. DNA computers could help researchers to answer complicated mathematical problems that other types of computers have thus far been unable to solve. It is hoped that DNA computers will be able to mimic the way that our current electronic computers think and perform.

Copyright Permission of Third-Party Materials

- Design of reversible sequential circuits optimizing quantum cost, delay, and garbage outputs, by Himanshu Thapliyal. Copyright Jan. 2010 by IEEE. Reproduced with permission of IEEE via Copyright Clearance Center.
- Design of reversible multiplexer/de-multiplexer, by Lenin Gopal. Copyright Nov. 2014 by IEEE. Reproduced with permission of IEEE via Copyright Clearance Center.
- Minimum Cost Fault Tolerant Adder Circuits in Reversible Logic Synthesis, by Sajib Kumar Mitra. Copyright Jan. 2012 by IEEE. Reproduced with permission of IEEE via Copyright Clearance Center.
- DNA Based Evolutionary Approach for Microprocessor Design Automation, by Nagarajan Venkateswaran. Copyright Jan. 2007 by Springer Nature. Reproduced with permission of Springer Nature via Copyright Clearance Center.
- The design of reversible gate and reversible sequential circuit based on DNA computing, by Tao Song. Copyright Nov. 2008 by IEEE. Reproduced with permission of IEEE via Copyright Clearance Center.

Reversible and DNA Computing, First Edition. Hafiz Md. Hasan Babu.
© 2021 John Wiley & Sons Ltd. Published 2021 by John Wiley & Sons Ltd.

Bibliography

LM Adelman. On constructing a molecular computer. DNA based computers. RL A. E. Baum. *American Mathematical Society*, pages 1–21, 1996.

Leonard M Adleman. Molecular computation of solutions to combinatorial problems. *Science*, pages 1021–1024, 1994.

Ramesh C Agarwal, Fred G Gustavson, and Martin S Schmookler. Series approximation methods for divide and square root in the power3/sup tm/processor. In *Proceedings 14th IEEE Symposium on Computer Arithmetic (Cat. No. 99CB36336)*, pages 116–123. IEEE, 1999.

Abhinav Agrawal and Niraj K Jha. Synthesis of reversible logic. In *Proceedings Design, Automation and Test in Europe Conference and Exhibition*, volume 2, pages 1384–1385. IEEE, 2004.

Tanvir Ahmed, Ankur Sarker, Mohd Istiaq Sharif, SM Mahbubur Rashid, Md Atiqur Rahman, and Hafiz Md Hasan Babu. A novel approach to design a reversible shifter circuit using DNA. In *2013 IEEE International SOC Conference*, pages 256–261. IEEE, 2013.

Ehsan Pour Ali Akbar, Majid Haghparast, and Keivan Navi. Novel design of a fast reversible Wallace sign multiplier circuit in nanotechnology. *Microelectronics Journal*, 42(8):973–981, 2011.

Anas N Al-Rabadi. Closed-system quantum logic network implementation of the viterbi algorithm. *Facta universitatis-series: Electronics and Energetics*, 22(1):1–33, 2009.

Franz L Alt. *Advances in computers*, volume 2. Academic Press, 1961.

HV Ravish Aradhya, BV Praveen Kumar, and KN Muralidhara. Design of control unit for low power AU using reversible logic. *Procedia Engineering*, 30: 631–638, 2012.

Somayeh Babazadeh and Majid Haghparast. Design of a nanometric fault tolerant reversible multiplier circuit. *Journal of Basic and Applied Scientific Research*, 2(2):1355–1361, 2012.

Hafiz Md Hasan Babu and Ahsan Raja Chowdhury. Design of a compact reversible binary coded decimal adder circuit. *Journal of Systems Architecture*, 52(5):272–282, 2006.

Hafiz Md Hasan Babu, Md Rafiqul Islam, Ahsan Raja Chowdhury, and Syed Mostahed Ali Chowdhury. Reversible logic synthesis for minimization of full-adder circuit. In *Euromicro Symposium on Digital System Design, 2003. Proceedings.*, pages 50–54. IEEE, 2003.

Hafiz Md Hasan Babu, Md Rafiqul Islam, Syed Mostahed Ali Chowdhury, and Ahsan Raja Chowdhury. Synthesis of full-adder circuit using reversible logic. In *17th International Conference on VLSI Design. Proceedings*, pages 757–760. IEEE, 2004a.

Reversible and DNA Computing, First Edition. Hafiz Md. Hasan Babu.
© 2021 John Wiley & Sons Ltd. Published 2021 by John Wiley & Sons Ltd.

Hafiz Md Hasan Babu, Moinul Islam Zaber, Md Mazder Rahman, and Md Rafiqul Islam. Implementation of multiple-valued flip-flops using pass transistor logic. In *Euromicro Symposium on Digital System Design, 2004. DSD 2004*, pages 603–606. IEEE, 2004b.

Thomas Back. *Evolutionary algorithms in theory and practice: evolution strategies, evolutionary programming, genetic algorithms*. Oxford: Oxford University Press, 1996.

Jean-Loup Baer. *Computer Systems Architecture*, volume 11. Rockville, MD: Computer Science Press, 1980.

Rohani Binti Abu Bakar, Junzo Watada, and Witold Pedrycz. DNA approach to solve clustering problem based on a mutual order. *Biosystems*, 91(1):1–12, 2008.

Alexander A Balandin and Kang L Wang. Implementation of quantum controlled-not gates using asymmetric semiconductor quantum dots. In *NASA International Conference on Quantum Computing and Quantum Communications*, pages 460–467. Springer, 1998.

Anindita Banerjee and Anirban Pathak. On the synthesis of sequential reversible circuit. *arXiv preprint arXiv:0707.4233*, 2007.

Charles R Baugh and Bruce A Wooley. A two's complement parallel array multiplication algorithm. *IEEE Transactions on Computers*, 100(12): 1045–1047, 1973.

Parhami Behrooz. *Computer arithmetic: Algorithms and hardware designs*. Oxford: *Oxford University Press*, 19:512583–512585, 2000.

Richard Bellman. Dynamic programming treatment of the travelling salesman problem. *J. ACM*, 9(1):61–63, January 1962. ISSN 0004-5411. doi: 10.1145/321105.321111. URL http://doi.acm.org/10.1145/321105.321111.

Charles H Bennett. Logical reversibility of computation. *IBM Journal of Research and Development*, 17(6):525–532, 1973.

Charles H Bennett. Notes on the history of reversible computation. *IBM Journal of Research and Development*, 32(1):16–23, 1988.

Charles H Bennett, Ethan Bernstein, Gilles Brassard, and Umesh Vazirani. Strengths and weaknesses of quantum computing. *SIAM Journal on Computing*, 26(5):1510–1523, 1997.

Sidney W Benson. Iii-bond energies. *Journal of Chemical Education*, 42(9):502, 1965.

Gary H Bernstein, Wenchuang Hu, Qingling Hang, Koshala Sarveswaran, and Marya Lieberman. Electron beam lithography and liftoff of molecules and dna rafts. In *4th IEEE Conference on Nanotechnology, 2004*, pages 201–203. IEEE, 2004.

Gary W Bewick. *Fast multiplication: algorithms and implementation*. PhD thesis, Department of Electrical Engineering. Stanford University, 1994.

Ashis Kumer Biswas, Md Mahmudul Hasan, Ahsan Raja Chowdhury, and Hafiz Md Hasan Babu. Efficient approaches for designing reversible binary coded decimal adders. *Microelectronics Journal*, 39(12):1693–1703, 2008.

Mark T Bohr. Interconnect scaling-the real limiter to high performance ULSI. In *Proceedings of International Electron Devices Meeting*, pages 241–244. IEEE, 1995.

Terence A Brown. *Gene cloning and DNA analysis: an introduction*. Hoboken, NJ: John Wiley & Sons, 2016.

JW Bruce, Mitchell A Thornton, L Shivakumaraiah, PS Kokate, and X Li. Efficient adder circuits based on a conservative reversible logic gate. In *Proceedings IEEE Computer Society Annual Symposium on VLSI. New Paradigms for VLSI Systems Design. ISVLSI 2002*, pages 83–88. IEEE, 2002.

K Buch. Low-power fault tolerant state machine design using reversible logic gates. In *Military and Aerospace Programmable Logic Devices Conference*, 2008.

J. Burr and A. Peterson. Ultra low-power CMOS technology. 1991. URL https://academic .microsoft.com/paper/1680657443.

Mario Cannataro, Rodrigo Weber dos Santos, Joakim Sundnes, and Pierangelo Veltri. Advanced computing solutions for health care and medicine, 2012.

Amir Mokhtar Chabi, Samira Sayedsalehi, Shaahin Angizi, and Keivan Navi. Efficient QCA exclusive-or and multiplexer circuits based on a nanoelectronic-compatible designing approach. *International scholarly research notices*, 2014, 2014.

Leland Chang, David J Frank, Robert K Montoye, Steven J Koester, Brian L Ji, Paul W Coteus, Robert H Dennard, and Wilfried Haensch. Practical strategies for power-efficient computing technologies. *Proceedings of the IEEE*, 98(2): 215–236, 2010.

Mark Chee, Robert Yang, Earl Hubbell, Anthony Berno, Xiaohua C Huang, David Stern, Jim Winkler, David J Lockhart, MacDonald S Morris, and Stephen PA Fodor. Accessing genetic information with high-density DNA arrays. *Science*, 274(5287):610–614, 1996.

Ahsan Raja Chowdhury, Rumana Nazmul, et al. A new approach to synthesize multiple-output functions using reversible programmable logic array. In *19th International Conference on VLSI Design held jointly with 5th International Conference on Embedded Systems Design (VLSID'06)*, pages 6–pp. IEEE, 2006.

Malgorzata Chrzanowska-Jeske. Architecture and synthesis issues in FPGAS. In *Proceedings of NORTHCON'93 Electrical and Electronics Convention*, pages 102–105. IEEE, 1993.

Min-Lun Chuang and Chun-Yao Wang. Synthesis of reversible sequential elements. *ACM Journal on Emerging Technologies in Computing Systems (JETC)*, 3(4):4, 2008.

Hauke Clausen-Schaumann, Matthias Rief, Carolin Tolksdorf, and Hermann E Gaub. Mechanical stability of single DNA molecules. *Biophysical Journal*, 78 (4):1997–2007, 2000.

Luigi Dadda. Multioperand parallel decimal adder: A mixed binary and BCD approach. *IEEE Transactions on Computers*, 56(10):1320–1328, 2007.

Faraz Dastan and Majid Haghparast. A novel nanometric fault tolerant reversible divider. *International Journal of Physical Sciences*, 6(24): 5671–5681, 2011.

Faraz Dastan and Majid Haghparast. A novel nanometric reversible signed divider with overflow checking capability. *Research Journal of Applied Sciences, Engineering and Technology*, 4(6):535–543, 2012.

F De Santis and G Iaccarino. A DNA arithmetic logic unit. *WSEAS Transactions on Biology and Biomedicine*, 1(4):436–440, 2004.

Alexis De Vos and Yvan Van Rentergem. Power consumption in reversible logic addressed by a ramp voltage. In *International Workshop on Power and Timing Modeling, Optimization and Simulation*, pages 207–216. New York: Springer, 2005.

Narsingh Deo. *Graph theory with applications to engineering and computer science*. Courier Dover Publications, 2017.

Bart Desoete and Alexis De Vos. A reversible carry-look-ahead adder using control gates. *Integration, the VLSI Journal*, 33(1-2):89–104, 2002.

David Elieser Deutsch. Quantum computational networks. *Proceedings of the Royal Society of London. A. Mathematical and Physical Sciences*, 425(1868): 73–90, 1989.

Akanksha Dixit and Vinod Kapse. Arithmetic & logic unit (ALU) design using reversible control unit. *Development*, 32:16–23, 1998.

Harrison Echols and Myron F Goodman. Fidelity mechanisms in DNA replication. *Annual Review of Biochemistry*, 60(1):477–511, 1991.

Milos D Ercegovac and Tomas Lang. *Digital arithmetic*. Amsterdam: Elsevier, 2004.

Z Ezziane. DNA computing: applications and challenges. *Nanotechnology*, 17 (2):R27, 2005.

Mike J Farabee. Transport in and out of cells. 2007.

Navid Farazmand, Masoud Zamani, and Mehdi B Tahoori. Online fault testing of reversible logic using dual rail coding. In *2010 IEEE 16th International On-Line Testing Symposium*, pages 204–205. IEEE, 2010.

Kenneth Fazel, Mitchell A Thornton, and JE Rice. ESOP-based Toffoli gate cascade generation. In *2007 IEEE Pacific Rim Conference on Communications, Computers and Signal Processing*, pages 206–209. IEEE, 2007.

Richard P Feynman. Quantum mechanical computers. *Foundations of Physics*, 16(6):507–531, 1986.

Harold Fleisher and Leon I. Maissel. An introduction to array logic. *IBM Journal of Research and Development*, 19(2):98–109, 1975.

Michael P Frank. The physical limits of computing. *Computing in Science & Engineering*, 4(3):16, 2002.

Michael P Frank. Approaching the physical limits of computing. In *35th International Symposium on Multiple-Valued Logic (ISMVL'05)*, pages 168–185. IEEE, 2005a.

Michael P Frank. Introduction to reversible computing: motivation, progress, and challenges. In *Proceedings of the 2nd Conference on Computing Frontiers*, pages 385–390. ACM, 2005b.

Michael P Frank. Foundations of generalized reversible computing. In *International Conference on Reversible Computation*, pages 19–34. New York: Springer, 2017.

Edward Fredkin and Tommaso Toffoli. Conservative logic. *International Journal of Theoretical Physics*, 21(3-4):219–253, 1982.

Stefan Frehse, Robert Wille, and Rolf Drechsler. Efficient simulation-based debugging of reversible logic. In *2010 40th IEEE International Symposium on Multiple-Valued Logic*, pages 156–161. IEEE, 2010.

MH Fulekar. Nanotechnology-in relation to bioinformatics. In *Bioinformatics: Applications in Life and Environmental Sciences*, pages 200–206. New York: Springer, 2009.

Michael R Garey and David S Johnson. *Computers and Intractability*, volume 29. New York: WH Freeman, 2002.

Lenin Gopal, Adib Kabir Chowdhury, Alpha Agape Gopalai, Ashutosh Kumar Singh, and Bakri Madon. Reversible logic gate implementation as switch controlled reversible full adder/subtractor. In *2014 IEEE International Conference on Control System, Computing and Engineering (ICCSCE 2014)*, pages 1–4. IEEE, 2014a.

Lenin Gopal, Nikhil Raj, Alpha Agape Gopalai, and Ashutosh Kumar Singh. Design of reversible multiplexer/de-multiplexer. In *2014 IEEE International Conference on Control System, Computing and Engineering (ICCSCE 2014)*, pages 416–420. IEEE, 2014b.

Saeid Gorgin and Amir Kaivani. Reversible barrel shifters. In *2007 IEEE/ACS International Conference on Computer Systems and Applications*, pages 479–483. IEEE, 2007.

Daniel Große, Robert Wille, Gerhard W Dueck, and Rolf Drechsler. Exact synthesis of elementary quantum gate circuits for reversible functions with don't cares. In *38th International Symposium on Multiple Valued Logic (ismvl 2008)*, pages 214–219. IEEE, 2008.

Daniel Große, Robert Wille, Gerhard W Dueck, and Rolf Drechsler. Exact multiple-control toffoli network synthesis with SAT techniques. *IEEE Transactions on Computer-Aided Design of Integrated Circuits and Systems*, 28(5):703–715, 2009.

Zhijin Guan, Wenjuan Li, Weiping Ding, Yueqin Hang, and Lihui Ni. An arithmetic logic unit design based on reversible logic gates. In *Proceedings of 2011 IEEE Pacific Rim Conference on Communications, Computers and Signal Processing*, pages 925–931. IEEE, 2011.

Pallav Gupta, Abhinav Agrawal, and Niraj K Jha. An algorithm for synthesis of reversible logic circuits. *IEEE Transactions on Computer-Aided Design of Integrated Circuits and Systems*, 25(11):2317–2330, 2006.

Vineet Gupta, Srinivasan Parthasarathy, and Mohammed Zaki. Arithmetic and logic operations with DNA. 1997.

Majid Haghparast and Keivan Navi. A novel reversible BCD adder for nanotechnology based systems. *American Journal of Applied Sciences*, 5(3): 282–288, 2008.

Majid Haghparast, Somayyeh Jafarali Jassbi, Keivan Navi, and Omid Hashemipour. Design of a novel reversible multiplier circuit using HNG gate in nanotechnology. In *World Appl. Sci. J.* Citeseer, 2008.

Majid Haghparast, Majid Mohammadi, Keivan Navi, and Mohammad Eshghi. Optimized reversible multiplier circuit. *Journal of Circuits, Systems, and Computers*, 18(02):311–323, 2009.

Hanadi Ahmed Hakami, Zenon Chaczko, and Anup Kale. Review of big data storage based on DNA computing. In *2015 Asia-Pacific Conference on Computer Aided System Engineering*, pages 113–117. IEEE, 2015.

Siva Kumar Sastry Hari, Shyam Shroff, Sk Noor Mahammad, and V Kamakoti. Efficient building blocks for reversible sequential circuit design. In *2006 49th IEEE International Midwest Symposium on Circuits and Systems*, volume 1, pages 437–441. IEEE, 2006.

Irina Hashmi and Hafiz Md Hasan Babu. An efficient design of a reversible barrel shifter. In *2010 23rd International Conference on VLSI Design*, pages 93–98. IEEE, 2010.

John P Hayes. *Computer architecture and organization*. New York: McGraw-Hill, Inc., 2002.

Hiroshi Hiasa and Kenneth J Marians. Primase couples leading-and lagging-strand DNA synthesis from oric. *Journal of Biological Chemistry*, 269 (8):6058–6063, 1994.

Hiroshi Hiasa and Kenneth J Marians. Initiation of bidirectional replication at the chromosomal origin is directed by the interaction between helicase and primase. *Journal of Biological Chemistry*, 274(38):27244–27248, 1999.

KSV Hornweder. An overview of techniques and applications of DNA nanotechnology. *Tech Reports*, pages 1–20, 2011.

Nayeemul Huda. On the implementation of reversible random access memory.

William NN Hung, Xiaoyu Song, Guowu Yang, Jin Yang, and Marek Perkowski. Quantum logic synthesis by symbolic reachability analysis. In *Proceedings of the 41st annual Design Automation Conference*, pages 838–841. ACM, 2004.

William NN Hung, Xiaoyu Song, Guowu Yang, Jin Yang, and Marek Perkowski. Optimal synthesis of multiple output boolean functions using a set of quantum gates by symbolic reachability analysis. *IEEE Transactions on Computer-Aided Design of Integrated Circuits and Systems*, 25(9):1652–1663, 2006.

James A Ibers. Molecular structure. *Annual Review of Physical Chemistry*, 16 (1):375–396, 1965.

Md Islam, Zerina Begum, et al. Reversible logic synthesis of fault tolerant carry skip BCD adder. *arXiv preprint arXiv:1008.3288*, 2010a.

Md Islam, Muhammad Mahbubur Rahman, Zerina Begum, Mohd Hafiz, et al. Fault tolerant variable block carry skip logic (VBCSL) using parity preserving reversible gates. *CoRR*, abs/1009.3819, 2010b. URL http://arxiv.org/abs/1009.3819.

Md Islam, Muhammad Mahbubur Rahman, Mohd Hafiz, et al. Efficient approaches for designing fault tolerant reversible carry look-ahead and carry-skip adders. *arXiv preprint arXiv:1008.3344*, 2010c.

Md Saiful Islam, Muhammad Mahbubur Rahman, Zerina Begum, and Mohd Zulfiquar Hafiz. Fault tolerant reversible logic synthesis: Carry look-ahead and carry-skip adders. In *2009 International Conference on Advances in Computational Tools for Engineering Applications*, pages 396–401. IEEE, 2009a.

Md Saiful Islam, Muhammad Mahbubur Rahman, Zerina Begum, Mohd Zulfiquar Hafiz, and Abdullah Al Mahmud. Synthesis of fault tolerant reversible logic circuits. In *2009 IEEE Circuits and Systems International Conference on Testing and Diagnosis*, pages 1–4. IEEE, 2009b.

Lafifa Jamal and Hafiz Md Hasan Babu. Efficient approaches to design a reversible floating point divider. In *2013 IEEE International Symposium on Circuits and Systems (ISCAS2013)*, pages 3004–3007. IEEE, 2013.

Lafifa Jamal, Md Shamsujjoha, and HM Hasan Babu. Design of optimal reversible carry look-ahead adder with optimal garbage and quantum cost. *International Journal of Engineering and Technology*, 2(1):44–50, 2012.

Rekha K James, K Poulose Jacob, Sreela Sasi, et al. Fault tolerant error coding and detection using reversible gates. In *TENCON 2007-2007 IEEE Region 10 Conference*, pages 1–4. IEEE, 2007.

Claude-Pierre Jeannerod, Hervé Knochel, Christophe Monat, Guillaume Revy, and Gilles Villard. A new binary floating-point division algorithm and its software implementation on the st231 processor. In *2009 19th IEEE Symposium on Computer Arithmetic*, pages 95–103. IEEE, 2009.

William Kahan. A brief tutorial on gradual underflow. *Available as a PDF file at* www.cs .berkeley.edu/~wkahan/ARITH_17U.pdf, 2005.

Chris A Kaiser, Monty Krieger, Harvey Lodish, and Arnold Berk. *Molecular cell biology*. New York: WH Freeman, 2007.

Lila Kari, Gheorghe Păun, Grzegorz Rozenberg, Arto Salomaa, and Sheng Yu. DNA computing, sticker systems, and universality. *Acta Informatica*, 35 (5):401–420, 1998.

Lila Kari, Mark Daley, Greg Gloor, Rani Siromoney, and Laura F Landweber. How to compute with DNA. In *International Conference on Foundations of Software Technology and Theoretical Computer Science*, pages 269–282. New York: Springer, 1999.

Lila Karl. DNA computing: arrival of biological mathematics. *The Mathematical Intelligencer*, 19(2):9–22, 1997.

Pawel Kerntopf, Marek A Perkowski, and Mozammel HA Khan. On universality of general reversible multiple-valued logic gates. In *Proceedings. 34th International Symposium on Multiple-Valued Logic*, pages 68–73. IEEE, 2004.

Robert W Keyes and Rolf Landauer. Minimal energy dissipation in logic. *IBM Journal of Research and Development*, 14(2):152–157, 1970.

Mozammel HA Khan. Logic synthesis with cascades of new reversible gate families. *Reed Muller, 2003*, 2003.

Mozammel HA Khan and Marek Perkowski. Synthesis of reversible synchronous counters. In *2011 41st IEEE International Symposium on Multiple-Valued Logic*, pages 242–247. IEEE, 2011.

A Khazamipour and Katarzyna Radecka. Adiabatic implementation of reversible logic. In *48th Midwest Symposium on Circuits and Systems, 2005*, pages 291–294. IEEE, 2005.

A Khazamipour and Katarzyna Radecka. A new architecture of adiabatic reversible logic gates. In *2006 IEEE North-East Workshop on Circuits and Systems*, pages 233–236. IEEE, 2006.

Andrei B Khlopotine, Marek A Perkowski, and Pawel Kerntopf. Reversible logic synthesis by iterative compositions. In *Iwls*, pages 261–266, 2002.

Scott Kirkpatrick, C Daniel Gelatt, and Mario P Vecchi. Optimization by simulated annealing. *Science*, 220(4598):671–680, 1983.

Joshua P Klein, Thomas H Leete, and Harvey Rubin. A biomolecular implementation of logically reversible computation with minimal energy dissipation. *Biosystems*, 52(1-3):15–23, 1999.

Emanuel Knill, Raymond Laflamme, and Gerald J Milburn. A scheme for efficient quantum computation with linear optics. *Nature*, 409(6816):46, 2001.

Donald Ervin Knuth. *The art of computer programming*, volume 3. Pearson Education, 1997.

Mel Krajden, James M Minor, Oretta Rifkin, and Lorraine Comanor. Effect of multiple freeze-thaw cycles on hepatitis B virus DNA and hepatitis c virus RNA quantification as measured with branched-DNA technology. *Journal of Clinical Microbiology*, 37(6):1683–1686, 1999.

Taek-Jun Kwon and Jeffrey Draper. Floating-point division and square root using a taylor-series expansion algorithm. *Microelectronics Journal*, 40(11): 1601–1605, 2009.

Sam J La Placa, Walter Clark Hamilton, James A Ibers, and Alan Davison. Nature of the metal-hydrogen bond in transition metal-hydrogen complexes. neutron and x-ray diffraction studies of beta-pentacarbonylmanganese hydride. *Inorganic Chemistry*, 8(9):1928–1935, 1969.

AV Anantha Lakshmi and GF Sudha. Design of a reversible single precision floating point subtractor. *SpringerPlus*, 3(1):11, 2014.

Parag K Lala. *Self-checking and fault-tolerant digital design*. Morgan Kaufmann, 2001.

PK Lala, JP Parkerson, and P Chakraborty. Adder designs using reversible logic gates. *WSEAS Transactions on Circuits and Systems*, 9(6):369–378, 2010.

Rolf Landauer. Irreversibility and heat generation in the computing process. *IBM journal of research and development*, 5(3):183–191, 1961.

Laura F Landweber and Lila Kari. The evolution of cellular computing: nature's solution to a computational problem. *Biosystems*, 52(1-3):3–13, 1999.

R Laundauer. Irreversibility and heat generation in the computational process. *IBM Journal of Research and Development*, 5:183–191, 1961.

Robert Steven Ledley. *Digital computer and control engineering*. New York: McGraw-Hill, 1960.

M Lehman. High-speed digital multiplication. *IRE Transactions on Electronic Computers*, (3):204–205, 1957.

Charles Eric Leiserson, Ronald L Rivest, Thomas H Cormen, and Clifford Stein. *Introduction to algorithms*, volume 6. Cambridge, MA: MIT Press, 2001.

Craig S Lent, P Douglas Tougaw, Wolfgang Porod, and Gary H Bernstein. Quantum cellular automata. *Nanotechnology*, 4(1):49, 1993.

Sam Fong Yau Li. *Capillary electrophoresis: principles, practice and applications*, volume 52. Amsterdam: Elsevier, 1992.

Xiaoqin Li, Duncan Steel, Daniel Gammon, and Lu J Sham. Quantum information processing based on optically driven semiconductor quantum dots. *Optics and Photonics News*, 15(9):38–43, 2004.

Albert A Liddicoat and Michael J Flynn. High-performance floating point divide. In *Proceedings Euromicro Symposium on Digital Systems Design*, pages 354–361. IEEE, 2001.

Joonho Lim, Dong-Gyu Kim, and Soo-Ik Chae. A 16-bit carry-lookahead adder using reversible energy recovery logic for ultra-low-energy systems. *IEEE Journal of Solid-State Circuits*, 34(6):898–903, 1999.

Richard J Lipton. DNA solution of hard computational problems. *Science*, 268 (5210):542–545, 1995.

Bin Liu, Mei Lie Wong, and Bruce Alberts. A transcribing RNA polymerase molecule survives DNA replication without aborting its growing RNA chain. *Proceedings of the National Academy of Sciences*, 91(22):10660–10664, 1994.

Stine Lund and Jørgen Dissing. Surprising stability of DNA in stains at extreme humidity and temperature. In *International Congress Series*, volume 1261, pages 616–618. Elsevier, 2004.

Xiaojun Ma, Jing Huang, Cecilia Metra, and Fabrizio Lombardi. Reversible gates and testability of one dimensional arrays of molecular QCA. *Journal of Electronic Testing*, 24(1-3):297–311, 2008.

Sk Noor Mahammad and Kamakoti Veezhinathan. Constructing online testable circuits using reversible logic. *IEEE Transactions on Instrumentation and Measurement*, 59(1):101–109, 2009.

Madhusmita Mahapatro, Sisira Kanta Panda, Jagannath Satpathy, Meraj Saheel, M Suresh, Ajit Kumar Panda, and MK Sukla. Design of arithmetic circuits using reversible logic gates and power dissipation calculation. In *2010 International Symposium on Electronic System Design*, pages 85–90. IEEE, 2010.

Md Mamun, Selim Al, and Syed Monowar Hossain. Design of reversible random access memory. *arXiv preprint arXiv:1312.7354*, 2013.

Md Mamun, Selim Al, and David Menville. Quantum cost optimization for reversible sequential circuit. *arXiv preprint arXiv:1407.7098*, 2014.

M Morris Mano. *Digital design*. Englewood Cliffs, NJ: Prentice Hall. 7632: 119–125, 1984.

Dmitri Maslov and Gerhard W Dueck. Reversible cascades with minimal garbage. *IEEE Transactions on Computer-Aided Design of Integrated Circuits and Systems*, 23(11):1497–1509, 2004.

Dmitri Maslov and D Michael Miller. Comparison of the cost metrics for reversible and quantum logic synthesis. *arXiv preprint quant-ph/0511008*, 2005.

Dmitri Maslov, Gerhard W Dueck, and D Michael Miller. Synthesis of Fredkin-Toffoli reversible networks. *IEEE Transactions on Very Large Scale Integration (VLSI) Systems*, 13(6):765–769, 2005.

Jimson Mathew, Jawar Singh, Anas Abu Taleb, and Dhiraj K Pradhan. Fault tolerant reversible finite field arithmetic circuits. In *2008 14th IEEE International On-Line Testing Symposium*, pages 188–189. IEEE, 2008.

James Clerk Maxwell and Peter Pesic. *Theory of heat*. Courier Corporation, 2001.

DN Mehta and PN Sanghavi. Intel's 45 nm CMOS technology performance parameters in vlsi design. *Int. J. Electron. Commun. Comput. Eng., REACT-2013*, 4(2), 2013.

Giovana Mendes. Battle against intel CPU monopoly moves from lower costs to faster ia-32 processors.

Ralph C Merkle. Reversible electronic logic using switches. *Nanotechnology*, 4 (1):21, 1993a.

Ralph C Merkle. Two types of mechanical reversible logic. *Nanotechnology*, 4 (2):114, 1993b.

Ralph C Merkle and K Eric Drexler. Helical logic. *Nanotechnology*, 7(4):325, 1996.

Hans-Peter Messmer. *The indispensable PC hardware book*. Reading, MA: Addison-Wesley Longman Publishing Co., Inc., 2001.

D Michael Miller, Dmitri Maslov, and Gerhard W Dueck. A transformation based algorithm for reversible logic synthesis. In *Proceedings 2003. Design Automation Conference (IEEE Cat. No. 03CH37451)*, pages 318–323. IEEE, 2003.

Mehdi Mirzaee et al. DNA and quantum theory. In *2007 First International Conference on Quantum, Nano, and Micro Technologies (ICQNM'07)*, pages 4–4. IEEE, 2007.

Neeraj Kumar Misra, Subodh Wairya, and Vinod Kumar Singh. Approaches to design feasible error control scheme based on reversible series gates. *European Journal of Scientific Research*, 129(3):224–240, 2015.

Sajib Kumar Mitra and Ahsan Raja Chowdhury. Minimum cost fault tolerant adder circuits in reversible logic synthesis. In *2012 25th International Conference on VLSI Design*, pages 334–339. IEEE, 2012.

Sajib Kumar Mitra, Lafifa Jamal, Mineo Kaneko, and Hafiz Md Hasan Babu. An efficient approach for designing and minimizing reversible programmable logic arrays. In *Proceedings of the great lakes symposium on VLSI*, pages 215–220. ACM, 2012.

Payman Moallem, Maryam Ehsanpour, Ali Bolhasani, and Mehrdad Montazeri. Optimized reversible arithmetic logic units. *Journal of Electronics (China)*, 31(5):394–405, 2014.

Majid Mohammadi and Mohammad Eshghi. On figures of merit in reversible and quantum logic designs. *Quantum Information Processing*, 8(4):297–318, 2009.

Majid Mohammadi, Mohammad Eshghi, Majid Haghparast, and Abbas Bahrololoom. Design and optimization of reversible bcd adder/subtractor circuit for quantum and nanotechnology based systems. *World Applied Sciences Journal*, 4(6):787–792, 2008.

Gordon E Moore et al. Cramming more components onto integrated circuits, 1965.

Nobuhiko Morimoto, Masanori Arita, and Akira Suyama. Solid phase DNA solution to the hamiltonian path problem. In *DNA Based Computers*, pages 193–206, 1997.

Kenichi Morita. Reversible computing and cellular automata-a survey. *Theoretical Computer Science*, 395(1):101–131, 2008.

Matthew Morrison and Nagarajan Ranganathan. Design of a reversible ALU based on novel programmable reversible logic gate structures. In *2011 IEEE computer society annual symposium on VLSI*, pages 126–131. IEEE, 2011.

Matthew Morrison, Matthew Lewandowski, and Nagarajan Ranganathan. Design of a tree-based comparator and memory unit based on a novel reversible logic structure. In *2012 IEEE Computer Society Annual Symposium on VLSI*, pages 231–236. IEEE, 2012.

Man Lung Mui, Kaustav Banerjee, and Amit Mehrotra. A global interconnect optimization scheme for nanometer scale VLSI with implications for latency, bandwidth, and power dissipation. *IEEE Transactions on Electron Devices*, 51(2):195–203, 2004.

Jumana A Muwafi, Gerhard Fettweis, and Howard W Neff. Circuit for rotating, left shifting, or right shifting bits, November 2 1999. US Patent 5,978,822.

Michael Nachtigal and Nagarajan Ranganathan. Design and analysis of a novel reversible encoder/decoder. In *2011 11th IEEE International Conference on Nanotechnology*, pages 1543–1546. IEEE, 2011.

AN Nagamani, HV Jayashree, and HR Bhagyalakshmi. Novel low power comparator design using reversible logic gates. *Indian Journal of Computer Science and Engineering (IJCSE)*, 2(4):566–574, 2011.

V Nagarajan, K Bharath, M Sharraa, and M Mani. Digital model for interconnect analysis. In *2005 NORCHIP*, pages 106–109. IEEE, 2005.

NM Nayeem and JE Rice. A simple approach for designing online testable reversible circuits. In *Proceedings of 2011 IEEE Pacific Rim Conference on Communications, Computers and Signal Processing*, pages 85–90. IEEE, 2011a.

Noor M Nayeem and Jacqueline E Rice. Online fault detection in reversible logic. In *2011 IEEE International Symposium on Defect and Fault Tolerance in VLSI and Nanotechnology Systems*, pages 426–434. IEEE, 2011b.

Noor M Nayeem, Lafifa Jamal, and Hafiz MH Babu. Efficient reversible montgomery multiplier and its application to hardware cryptography. *Journal of Computer Science*, 5(1):49, 2009a.

Noor Muhammed Nayeem, Adnan Hossain, Mutasimul Haque, Lafifa Jamal, and Hafiz M Hasan Babu. Novel reversible division hardware. In *2009 52nd IEEE International Midwest Symposium on Circuits and Systems*, pages 1134–1138. IEEE, 2009b.

Michael A Nielsen and Isaac Chuang. Quantum computation and quantum information, 2002.

Hooman Nikmehr, Braden Phillips, and Cheng-Chew Lim. A novel implementation of radix-4 floating-point division/square-root using comparison multiples. *Computers & Electrical Engineering*, 36(5):850–863, 2010.

Stuart F Oberman. Floating point division and square root algorithms and implementation in the amd-k7/sup tm/microprocessor. In *Proceedings 14th IEEE Symposium on Computer Arithmetic (Cat. No. 99CB36336)*, pages 106–115. IEEE, 1999.

Stuart F Oberman and Michael J Flynn. Design issues in division and other floating-point operations. *IEEE Transactions on Computers*, 46(2):154–161, 1997.

Howard Ochman, Anne S Gerber, and Daniel L Hartl. Genetic applications of an inverse polymerase chain reaction. *Genetics*, 120(3):621–623, 1988.

Mitsunori Ogihara and Animesh Ray. Simulating boolean circuits on a DNA computer. *Algorithmica*, 25(2-3):239–250, 1999.

Behrooz Parhami. Fault-tolerant reversible circuits. In *2006 Fortieth Asilomar Conference on Signals, Systems and Computers*, pages 1726–1729. IEEE, 2006.

Raj B Patel, Joseph Ho, Franck Ferreyrol, Timothy C Ralph, and Geoff J Pryde. A quantum Fredkin gate. *Science Advances*, 2(3):e1501531, 2016.

Gheorghe Păun, Grzegorz Rozenberg, and Arto Salomaa. *DNA computing: new computing paradigms*. New York: Springer, 1998.

Asher Peres. Reversible logic and quantum computers. *Physical Review A*, 32 (6):3266, 1985.

Marek Perkowski. Reversible computation for beginners. *Lecture Series*, 2000.

Marek Perkowski, Lech Jozwiak, Pawel Kerntopf, Alan Mishchenko, Anas Al-Rabadi, Alan Coppola, Andrzej Buller, Xiaoyu Song, Svetlana Yanushkevich, Vlad P Shmerko, et al. A general decomposition for reversible logic. *Proceedings of the International Workshop on Applications of the Reed-Muller Expansion in Circuit Design (RM'2001)*, Starksville, MI, 2001.

Marek Perkowski, Martin Lukac, Pawel Kerntopf, Mikhail Pivtoraiko, Michele Folgheraiter, Yong Woo Choi, Jung-wook Kim, Dongsoo Lee, Woong Hwangbo, and Hyungock Kim. A hierarchical approach to computer-aided design of quantum circuits. 2003.

Imre Petkovics, Ármin Petkovics, and János Simon. A survey of ICT: evolution of architectures, models and layers. In *2016 IEEE 14th International Symposium on Intelligent Systems and Informatics (SISY)*, pages 215–220. IEEE, 2016.

P Picton. Multi-valued sequential logic design using Fredkin gates. *Multiple-Valued Logic Journal*, 1(4):241–251, 1996.

PD Picton. Fredkin gates as the basic for comparison of different logic designs. *Synthesis and Optimization of Logic Systems*, 1994.

Xuemei Qi, Fulong Chen, Kaizhong Zuo, Liangmin Guo, Yonglong Luo, and Min Hu. Design of fast fault tolerant reversible signed multiplier. *International Journal of Physical Sciences*, 7(17):2506–2514, 2012.

Xuemei Qi, Fulong Chen, Liangmin Guo, Yonglong Luo, and Min Hu. Efficient approaches for designing fault tolerant reversible bcd adders. *J. Comput. Inf. Syst*, 9(14):5869–5877, 2013.

Rubaia Rahman, Lafifa Jamal, and Hafiz Md Hasan Babu. Design of reversible fault tolerant programmable logic arrays with vector orientation. *International Journal of Information and Communication Technology Research*, 1(8), 2011.

V Rajmohan and V Ranganathan. Design of counters using reversible logic. In *2011 3rd International Conference on Electronics Computer Technology*, volume 5, pages 138–142. IEEE, 2011.

V Rajmohan, V Renganathan, and M Rajmohan. A novel reversible design of unified single digit BCD adder-subtractor. *International Journal of Computer Theory and Engineering*, 3(5):697, 2011.

TR Rakshith and Rakshith Saligram. Parity preserving logic based fault tolerant reversible ALU. In *2013 IEEE Conference on Information & Communication Technologies*, pages 485–490. IEEE, 2013.

Brian Randell, Pete Lee, and Philip C. Treleaven. Reliability issues in computing system design. *ACM Computing Surveys (CSUR)*, 10(2):123–165, 1978.

C Pandu Rangan, V Raman, and R Ramanujam. How to compute with DNA. *LNCS*, 1738:269–282, 1999.

HG Rangaraju, V Hegde, KB Raja, and KN Muralidhara. Design of low power reversible binary comparator. *Proc. Engineering (ScienceDirect)*, 2011.

HG Rangaraju, Vinayak Hegde, KB Raja, and KN Muralidhara. Design of efficient reversible binary comparator. *Procedia Engineering*, 30:897–904, 2012.

Zultakiyuddin Ahmad Rashid, Hamimah Adnan, and Kamaruzaman Jusoff. Legal framework on risk management for design works in malaysia. *J. Pol. & L.*, 1:26, 2008.

T Ravi, S Ranjith, and V Kannan. A novel design of D-flip flop using new rr fault tolerant reversible logic gate. *International Journal of Emerging Technology and Advanced Engineering Website*: www.ijetae.com (ISSN 2250-2459, ISO 9001: 2008 Certified Journal, 3(2), 2013.

Jacqueline E Rice. A new look at reversible memory elements. In *2006 IEEE International Symposium on Circuits and Systems*, pages 4. IEEE, 2006.

Jacqueline E Rice. An introduction to reversible latches. *The Computer Journal*, 51(6):700–709, 2008.

Jacqueline E Rice and V Suen. Using autocorrelation coefficient-based cost functions in esop-based toffoloi gate cascade generation. In *CCECE 2010*, pages 1–6. IEEE, 2010.

James E. Robertson. A new class of digital division methods. *IRE Trans. Electronic Computers*, 7(3):218–222, 1958. doi: 10.1109/TEC.1958.5222579. URL https://doi.org/10.1109/TEC.1958 .5222579.

Jonathan Rose and Stephen Brown. Flexibility of interconnection structures for field-programmable gate arrays. *IEEE Journal of Solid-State Circuits*, 26 (3):277–282, 1991.

Jonathan Rose, Abbas El Gamal, and Alberto Sangiovanni-Vincentelli. Architecture of field-programmable gate arrays. *Proceedings of the IEEE*, 81 (7):1013–1029, 1993.

Will Ryu. DNA computing: A primer. *Ars Technica*, 2000.

Mehdi Saeedi, Morteza Saheb Zamani, Mehdi Sedighi, and Zahra Sasanian. Reversible circuit synthesis using a cycle-based approach. *ACM Journal on Emerging Technologies in Computing Systems (JETC)*, 6(4):13, 2010.

Wolfram Saenger. *Principles of nucleic acid structure*. Springer Science & Business Media, 2013.

Jagannath Samanta, Bishnu Prasad De, Banibrata Bag, and Raj Kumar Maity. Comparative study for delay & power dissipation of cmos inverter in UDSM range. *International Journal of Soft Computing and Engineering (IJSCE)*, 1 (6):6, 2012.

Yasaman Sanaee and Gerhard W Dueck. ESOP-based Toffoli network generation with transformations. In *2010 40th IEEE International Symposium on Multiple-Valued Logic*, pages 276–281. IEEE, 2010.

Ankur Sarker, Tanvir Ahmed, SM Mahbubur Rashid, Shahed Anwar, Lafifa Jaman, Nazma Tara, Md Masbaul Alam, and Hafiz Md Hasan Babu. Realization of reversible logic in DNA computing. In *2011 IEEE 11th International Conference on Bioinformatics and Bioengineering*, pages 261–265. IEEE, 2011.

Ankur Sarker, Mohd Istiaq Sharif, SM Mahbubur Rashid, and Hafiz Md Hasan Babu. Implementation of reversible multiplier circuit using deoxyribonucleic acid. In *13th IEEE International Conference on BioInformatics and BioEngineering*, pages 1–4. IEEE, 2013.

Trailokya Nath Sasamal, Ashutosh Kumar Singh, and Anand Mohan. Design of two-rail checker using a new parity preserving reversible logic gate. *International Journal of Computer Theory and Engineering*, 7(4):311, 2015.

Tsutomu Sasao and Kozo Kinoshita. Conservative logic elements and their universality. *IEEE Trans. Computers*, 28(9):682–685, 1979. doi: 10.1109/TC.1979.1675437. doi.org/10.1109/TC .1979.1675437.

Abu Sadat Md Sayem and Masashi Ueda. Optimization of reversible sequential circuits. *arXiv preprint arXiv:1006.4570*, 2010.

Jeff Scott, Lea Hwang Lee, John Arends, and Bill Moyer. Designing the low-power m● core tm architecture. In *Power driven microarchitecture workshop*, pages 145–150. Citeseer, 1998.

Bibhash Sen, Manojit Dutta, Dipak K Singh, Divyam Saran, and Biplab K Sikdar. QCA multiplexer based design of reversible ALU. In *2012 IEEE International Conference on Circuits and Systems (ICCAS)*, pages 168–173. IEEE, 2012.

Bibhash Sen, Siddhant Ganeriwal, and Biplab K Sikdar. Reversible logic-based fault-tolerant nanocircuits in QCA. *ISRN Electronics*, 2013, 2013.

Bibhash Sen, Manojit Dutta, Mrinal Goswami, and Biplab K Sikdar. Modular design of testable reversible ALU by QCA multiplexer with increase in programmability. *Microelectronics Journal*, 45(11):1522–1532, 2014a.

Bibhash Sen, Manojit Dutta, and Biplab K Sikdar. Efficient design of parity preserving logic in quantum-dot cellular automata targeting enhanced scalability in testing. *Microelectronics Journal*, 45(2):239–248, 2014b.

Bibhash Sen, Manojit Dutta, Samik Some, and Biplab K Sikdar. Realizing reversible computing in QCA framework resulting in efficient design of testable ALU. *ACM Journal on Emerging Technologies in Computing Systems (JETC)*, 11(3):30, 2014c.

Md Shamsujjoha and Hafiz Md Hasan Babu. A low-power fault-tolerant reversible decoder using MOS transistors. In *2013 26th International Conference on VLSI Design and 2013 12th International Conference on Embedded Systems*, pages 368–373. IEEE, 2013.

Md Shamsujjoha, Hafiz Md Hasan Babu, and Lafifa Jamal. Design of a compact reversible fault tolerant field programmable gate array: A novel approach in reversible logic synthesis. *Microelectronics Journal*, 44(6):519–537, 2013.

Farah Sharmin, Md Masbaul Alam Polash, Md Shamsujjoha, Lafifa Jamal, and HM Hasan Babu. Design of a compact reversible random access memory. In *4th IEEE International Conference on Computer Science and Information Technology*, volume 10, pages 103–107, 2011.

Vivek V Shende, Aditya K Prasad, Igor L Markov, and John P Hayes. Reversible logic circuit synthesis. In *Proceedings of the 2002 IEEE/ACM international conference on Computer-aided design*, pages 353–360. ACM, 2002.

Vivek V Shende, Aditya K Prasad, Igor L Markov, and John P Hayes. Synthesis of reversible logic circuits. *IEEE Transactions on Computer-Aided Design of Integrated Circuits and Systems*, 22(6):710–722, 2003.

Peter W Shor. Polynomial-time algorithms for prime factorization and discrete logarithms on a quantum computer. *SIAM Review*, 41(2):303–332, 1999.

Love Singhal, Elaheh Bozorgzadeh, and David Eppstein. Interconnect criticality-driven delay relaxation. *IEEE Transactions on Computer-Aided Design of Integrated Circuits and Systems*, 26(10):1803–1817, 2007.

HP Sinha and Nidhi Syal. Design of fault tolerant reversible multiplier. *International Journal of Soft Computing and Engineering (IJSCE)*, 1(6): 120–124, 2012.

JL Smith and A Weinberger. Shortcut multiplication for binary digital computers. *NBS Circular*, 591:13–22, 1958.

John A Smolin and David P DiVincenzo. Five two-bit quantum gates are sufficient to implement the quantum Fredkin gate. *Physical Review A*, 53(4): 2855, 1996.

GL Snider, AO Orlov, I Amlani, GH Bernstein, CS Lent, JL Merz, and W Porod. Experimental demonstration of quantum-dot cellular automata. *Semiconductor Science and Technology*, 13(8A):A130, 1998.

Tao Song, Shudong Wang, and Xun Wang. The design of reversible gate and reversible sequential circuit based on DNA computing. In *2008 3rd International Conference on Intelligent System and Knowledge Engineering*, volume 1, pages 114–118. IEEE, 2008.

M Sudharshan, HR Bhagyalakshmi, and MK Venkatesha. Novel design of one digit high speed carry select BCD subtractor using reversible logic gates. 2012.

Leo Szilard. On the decrease of entropy in a thermodynamic system by the intervention of intelligent beings. *Behavioral Science*, 9(4):301–310, 1964.

Toshi Tajima. *Computational plasma physics: with applications to fusion and astrophysics.* Boca Raton, FL: CRC Press, 2018.

Fumiaki Tanaka, Takashi Tsuda, and Masami Hagiya. Towards DNA comparator: the machine that compares DNA concentrations. In *International Workshop on DNA-Based Computers*, pages 11–20. New York: Springer, 2008.

Himanshu Thapliyal and Nagarajan Ranganathan. Conservative QCA gate (cqca) for designing concurrently testable molecular QCA circuits. In *2009 22nd International Conference on VLSI Design*, pages 511–516. IEEE, 2009a.

Himanshu Thapliyal and Nagarajan Ranganathan. Design of efficient reversible binary subtractors based on a new reversible gate. In *2009 IEEE computer society Annual symposium on VLSI*, pages 229–234. IEEE, 2009b.

Himanshu Thapliyal and Nagarajan Ranganathan. Design of reversible latches optimized for quantum cost, delay and garbage outputs. In *2010 23rd International Conference on VLSI Design*, pages 235–240. IEEE, 2010.

Himanshu Thapliyal and MB Srinivas. The need of DNA computing: reversible design of adders and multipliers using Fredkin gate. In *Optomechatronic Micro/Nano Devices and Components*, volume 6050, page 605010. International Society for Optics and Photonics, 2005a.

Himanshu Thapliyal and MB Srinivas. A novel reversible TSG gate and its application for designing reversible carry look-ahead and other adder architectures. In *Asia-Pacific Conference on Advances in Computer Systems Architecture*, pages 805–817. Springer, 2005b.

Himanshu Thapliyal and MB Srinivas. An extension to DNA-based Fredkin gate circuits: design of reversible sequential circuits using Fredkin gates. *arXiv preprint cs/0603092*, 2006a.

Himanshu Thapliyal and MB Srinivas. A new reversible TSG gate and its application for designing efficient adder circuits. *CoRR*, abs/cs/0603091, 2006b. URL http://arxiv.org/abs/cs/0603091.

Himanshu Thapliyal and A Prasad Vinod. Design of reversible sequential elements with feasibility of transistor implementation. In *2007 IEEE International Symposium on Circuits and Systems*, pages 625–628. IEEE, 2007.

Himanshu Thapliyal and Mark Zwolinski. Reversible logic to cryptographic hardware: A new paradigm. In *2006 49th IEEE International Midwest Symposium on Circuits and Systems*, volume 1, pages 342–346. IEEE, 2006.

Himanshu Thapliyal, MB Srinivas, and Mark Zwolinski. A beginning in the reversible logic synthesis of sequential circuits. In *Military and Aerospace Applications of Programmable Devices and Technologies International Conference (MAPLD)*, 2005. URL https://eprints.soton.ac.uk/381081/.

Himanshu Thapliyal, Saurabh Kotiyal, and MB Srinivas. Novel BCD adders and their reversible logic implementation for IEEE 754r format. In *19th International Conference on VLSI Design held jointly with 5th International Conference on Embedded Systems Design (VLSID'06)*, pages 6–pp. IEEE, 2006.

Himanshu Thapliyal, Hamid R Arabnia, and MB Srinivas. Efficient reversible logic design of BCD subtractors. In *Transactions on Computational Science III*, pages 99–121. Springer, 2009.

Himanshu Thapliyal, Nagarajan Ranganathan, and Saurabh Kotiyal. Design of testable reversible sequential circuits. *IEEE transactions on very large scale integration (vlsi) systems*, 21(7):1201–1209, 2012.

Michael Kirkedal Thomsen and Robert Glück. Optimized reversible binary-coded decimal adders. *Journal of Systems Architecture*, 54(7):697–706, 2008.

Michael Kirkedal Thomsen, Robert Glück, and Holger Bock Axelsen. Reversible arithmetic logic unit for quantum arithmetic. *Journal of Physics A: Mathematical and Theoretical*, 43(38):382002, 2010.

Keith D Tocher. Techniques of multiplication and division for automatic binary computers. *The Quarterly Journal of Mechanics and Applied Mathematics*, 11(3):364–384, 1958.

E Fredkin T Toffoli et al. Conservative logic. *Int. J. Theor. Phys*, 21:219–253, 1982.

Tommaso Toffoli. Reversible computing. In *International colloquium on automata, languages, and programming*, pages 632–644. Springer, 1980.

Nirvan Tyagi, Jayson Lynch, and Erik D Demaine. Toward an energy efficient language and compiler for (partially) reversible algorithms. In *International Conference on Reversible Computation*, pages 121–136. Springer, 2016.

Yvan Van Rentergem and Alexis De Vos. Optimal design of a reversible full adder. *IJUC*, 1(4):339–355, 2005.

Dilip P Vasudevan, Parag K Lala, and James Patrick Parkerson. Online testable reversible logic circuit design using nand blocks. In *19th IEEE International Symposium on Defect and Fault Tolerance in VLSI Systems, 2004. DFT 2004. Proceedings*, pages 324–331. IEEE, 2004.

Dilip P Vasudevan, Parag K Lala, and James Patrick Parkerson. CMOS realization of online testable reversible logic gates. In *IEEE Computer Society Annual Symposium on VLSI: New Frontiers in VLSI Design (ISVLSI'05)*, pages 309–310. IEEE, 2005.

Dilip P Vasudevan, Parag K Lala, Jia Di, and James Patrick Parkerson. Reversible-logic design with online testability. *IEEE transactions on instrumentation and measurement*, 55(2):406–414, 2006.

Vlatko Vedral, Adriano Barenco, and Artur Ekert. Quantum networks for elementary arithmetic operations. *Physical Review A*, 54(1):147, 1996.

N Venkateswaran, RV Arjun, S Haswath Narayanan, and S Karthik. Towards predicting multi-million neuron interconnectivity involving dendrites-axon-soma synapse: A simulation model. In *7th International Conference on Computational Intelligence and Natural Computing in conjunction with 8th Joint Conference on Information Sciences (JCIS 2005) July*, pages 21–26, 2005a.

N Venkateswaran, S Balaji, and V Sridhar. Fault tolerant bus architecture for deep submicron based processors. *ACM SIGARCH Computer Architecture News*, 33(1):148–155, 2005b.

N Venkateswaran et al. Microprocessor design automation: A DNA-based evolutionary approach, 2006.

John Von Neumann. First draft of a report on the edvac. *IEEE Annals of the History of Computing*, 15(4):27–75, 1993.

Ioannis Voyiatzis, Dimitris Gizopoulos, and Antonis Paschalis. Accumulator-based test generation for robust sequential fault testing in DSP cores in near-optimal time. *IEEE Transactions on Very Large Scale Integration (VLSI) Systems*, 13(9):1079–1086, 2005.

Chetan Vudadha, P Sai Phaneendra, V Sreehari, Syed Ershad Ahmed, N Moorthy Muthukrishnan, and Mandalika B Srinivas. Design of prefix-based optimal reversible comparator. In *2012 IEEE Computer Society Annual Symposium on VLSI*, pages 201–206. IEEE, 2012.

John F Wakerly. *Digital design: principles and practices*. Englewood Cliffs, NJ: Prentice Hall, 2000.

James D Watson, Francis Crick, et al. A structure for deoxyribose nucleic acid. 1953a.

James D Watson, Francis HC Crick, et al. Molecular structure of nucleic acids. *Nature*, 171(4356):737–738, 1953b.

Katharina Wendler, Jens Thar, Stefan Zahn, and Barbara Kirchner. Estimating the hydrogen bond energy. *The Journal of Physical Chemistry A*, 114(35): 9529–9536, 2010.

Robert Wille and Rolf Drechsler. BDD-based synthesis of reversible logic for large functions. In *Proceedings of the 46th Annual Design Automation Conference*, pages 270–275. ACM, 2009.

Robert Wille, Daniel Große, Lisa Teuber, Gerhard W Dueck, and Rolf Drechsler. Revlib: An online resource for reversible functions and reversible circuits. In *38th International Symposium on Multiple Valued Logic (ismvl 2008)*, pages 220–225. IEEE, 2008.

Linda Wilson. International technology roadmap for semiconductors (itrs). *Semiconductor Industry Association*, 1, 2013.

D Harlan Wood and Junghuei Chen. Fredkin gate circuits via recombination enzymes. In *Proceedings of the 2004 Congress on Evolutionary Computation (IEEE Cat. No. 04TH8753)*, volume 2, pages 1896–1900. IEEE, 2004.

Frank Worrell. Microprocessor shifter using rotation and masking operations, March 17 1998. US Patent 5,729,482.

Guowu Yang, Xiaoyu Song, William NN Hung, and Marek A Perkowski. Bi-direction synthesis for reversible circuits. In *IEEE Computer Society Annual Symposium on VLSI: New Frontiers in VLSI Design (ISVLSI'05)*, pages 14–19. IEEE, 2005.

Guowu Yang, Xiaoyu Song, William NN Hung, and Marek A Perkowski. Bi-directional synthesis of 4-bit reversible circuits. *The Computer Journal*, 51 (2):207–215, 2008.

Yang Yingwei. DNA computing: DNA computers vs. conventional electronic computers. *University of Stuttgart*, 2002.

Nengkun Yu, Runyao Duan, and Mingsheng Ying. Five two-qubit gates are necessary for implementing the Toffoli gate. *Physical Review A*, 88(1):010304, 2013.

Rigui Zhou, Yang Shi, Jian Cao, et al. Transistor realization of reversible "zs" series gates and reversible array multiplier. *Microelectronics Journal*, 42(2): 305–315, 2011.

Rigui Zhou, Yancheng Li, Manqun Zhang, and BenQiong Hu. Novel design for reversible arithmetic logic unit. *International Journal of Theoretical Physics*, 54(2):630–644, 2015.

Zeljko Zilic, Katarzyna Radecka, and Ali Kazamiphur. Reversible circuit technology mapping from non-reversible specifications. In *Proceedings of the conference on Design, Automation and Test in Europe*, pages 558–563. EDA Consortium, 2007.

Christof Zwyssig, Simon D Round, and Johann W Kolar. An ultrahigh-speed, low power electrical drive system. *IEEE Transactions on Industrial Electronics*, 55(2):577–585, 2008.

Index

a

accelerating   194
addition   328
adenine   291
algorithms   231
archetype   300
architecture   249
area   52
arithmetic   249
artificial intelligence   365
asynchronous   101
autonomous   283, 329

b

barrel shifter   221
bases   291
Baugh-Wooley   43
bidirectional   113
binary   75
biochemical   292
biochemistry   353, 356
bit-indexing   221
bits   1
block   266
Boolean   275, 299
Boolean functions   231
Booth's   57
borrow   327
breadboard   129

c

carry   327
carry computation   19

carry look-ahead   177
carry-skip logic   177
cascade(ing)   158, 264
cell   50
circuit   6
clock frequency   300
CMOS   5
combination   211
combinational   15, 107, 216, 218
communication   1, 75
compact   255
comparator   76, 81, 331
complement   34, 109
complementary   137, 266, 317
complexity   224
components   161
composite   311, 321
computational   1, 187, 359
concentration   329
concurrent testing   269
conflicting   308
constituents   177
constructed   80, 266
consumption   187
control   318
control unit   161
correlation   308
correspondent   218
corresponding   135, 217
counter   99
cross-coupled   89
cross-points   137
cryptographic   366

Reversible and DNA Computing, First Edition. Hafiz Md. Hasan Babu.
© 2021 John Wiley & Sons Ltd. Published 2021 by John Wiley & Sons Ltd.

cryptosystems   365
current-flowing   359
cytosine   291

d
data processing   155
decimal   15
decoder   99, 149
demultiplexer   123
diagrams   211
digital   249
digital communications   5
dinucleotide   334
division   68
DNA-based Fredkin   341, 354
DNA-based Fredkin gate   347
DNA-based half-adder circuit   323
DNA-based multiplier   338
DNA-based Toffoli gate   347
DNA bases   318
DNA computing   362
DNA molecules   351
DNA squares   360
double helix   291, 359
double stranded DNA   347
double strands   311, 329, 334
duplication   249, 267

e
electrophoresis   328, 334, 356
electrophoretic   343
empty strand   311
enable signal   91
encoder   149
encoding   306
encryption   75
energy   369
environment   369
enzyme   296
error correction   319
ESOP circuit   271
eukaryotic   351

f
fabrication   323

fan-outs   267, 319
fault tolerant   235
faulty signal   221
Feynman, Richard   281
Feynman gate   10
field programmable   129
floating-point   15, 193
Fredkin   10, 351
friction-less   281
function generator   156
functions   152
fuzzy logic   364

g
garbage   5, 235, 319
gene pool   305
generalized   221
genes   302
genetic   302
genetic algorithm   363
genetic coding   360
greedy approach   139
guanine   291

h
half-subtractor   325
hardware   67, 193
hereditary   291
heuristic   236
high performance   249
hill-climbing   330
hybridization   326
hydrogen   291
hydrogen bonds   316

i
instruction   161
interfaces   323
inverters   126
irreversibility   1
irreversible   6, 261, 283

l
latches   87
left-shift   201

life-time   319
ligase   354
ligation   336
linear combination   305
linkage   302
living cells   288
logarithmic   221, 222
logic   231
logical AND   326
logic gates   211
low power   249

m
mapping   314
master-slave   100
meaningful shorter operand strand (MSOS)
      348
meaningless shortest input strand (MSIS)
      348
memory   129
memory cells   143
microarrays   364
microprocessors   75, 299, 359
migration   331
molecule   291
Moore's law   279
multiplexer   123, 341
multiplicand   56
multiplication   333, 336
multiprocessing   359

n
nanometers   11, 359, 367
nanotechnology   2, 5, 177
natural selection   309
negative   67
negative clocked   100
neural networks   363
non-concurrent testing   269
non-restoring   71
nucleotide   291

o
online testable   276
operation   190

operational speed   319
optimality   187
optimization   187
organism   291
overlapped   173

p
paradigm   124
parallel   118, 194, 287
parallel adder   68
parallelism   319
parity checking   221
parity preserving   177, 249
partial product generation   187
partial products   52
pattern   179
performance   130
performing   369
phosphate   291
phosphodiester   335
pipelining   300
potency   305
power   52
power consumption   319
powerful   359
prefetching   300
preservation   178
programmable   231
programmable gate array   129
Programmable read only memory   143
propagated   194

q
quantum   5
quantum costs   17, 54, 180
quantum theory   12
qubits   7, 370

r
random access memory   143
rearrangement   303
recoding   43
reconfigurable   129
recursive   214
register   2, 115

remainder   67, 210
replication   296
resembles   99
reversible   2, 370
reversible computing   261
reversible controlled unit   155
reversible full-adder   323
rotators   222

s
semiconductor   129
sequence counter   161
sequences   314
sequential   2, 67, 99
sequential circuit   316
set/reset   87
shifter   107, 333
shift register   107
signal integrity   300
significant   102, 193
silicon   359
silicon chip   359
single-bit   78
single strands   311, 329
straight   15

subtraction   328
subtractor   38
supercomputers   367
swarm intelligence   365
synchronous   102

t
temperature   261
testable   269
thymine   291
Toffoli   10
transcriptional   351
transistors   211, 299, 300, 359
traveling   360

u
unidirectional   221, 224

v
variables   325
vectors   6
von Neumann   346

w
wired together   129

Printed and bound by CPI Group (UK) Ltd, Croydon, CR0 4YY